QUANTITATIVE BIOSCIENCES

Dynamics across Cells, Organisms, and Populations

QUANTITATIVE BIOSCIENCES

Dynamics across Cells, Organisms, and Populations

JOSHUA S. WEITZ

Princeton University Press
Princeton and Oxford

Princeton University Press is committed to the protection of copyright and the intellectual property our authors entrust to us. Copyright promotes the progress and integrity of knowledge. Thank you for supporting free speech and the global exchange of ideas by purchasing an authorized edition of this book. If you wish to reproduce or distribute any part of it in any form, please obtain permission.

Requests for permission to reproduce material from this work
should be sent to permissions@press.princeton.edu

Published by Princeton University Press
41 William Street, Princeton, New Jersey 08540
99 Banbury Road, Oxford OX2 6JX

press.princeton.edu

ISBN 9780691181509
ISBN (pbk.) 9780691181516
ISBN (e-book) 9780691256481

Library of Congress Control Number: 2023946916

British Library Cataloging-in-Publication Data is available

Editorial: Sydney Carroll and Johannah Walkowicz
Production Editorial: Terri O'Prey
Text Design: Wanda España
Cover Design: Wanda España
Production: Jacqueline Poirier
Copyeditor: Jennifer McClain

Cover image: Simon Dack News / Alamy Stock Photo

MATLAB® is a registered trademarks of The MathWorks®, Inc.
For MATLAB® product information, please contact:
The MathWorks®, Inc.
3 Apple Hill Drive
Natick, MA, 01760-2098 USA
Tel: 508-647-7000 Fax: 508-647-7001
E-mail: info@mathworks.com
Web: https://www.mathworks.com
How to buy: https://www.mathworks.com/store
Find your local office: https://www.mathworks.com/company/worldwide

This book has been composed in MinionPro and Omnes

Printed on acid-free paper. ∞

Printed in the United States of America

10 9 8 7 6 5 4 3 2 1

Time will tell, and it is the measurements that tell us.
        —*Jim Peebles*

CONTENTS

    3.1   Living with randomness   57
    3.2   Stochasticity in gene regulation   60
    3.3   Characterizing dynamics of individual cells, given stochastic gene expression   64
    3.4   Is gene expression bursty?   68
    3.5   The geometry of bursts   72
    3.6   Take-home messages   77
    3.7   Homework problems   77
    3.8   Technical appendix   80

4   Evolutionary Dynamics: Mutations, Selection, and Diversity   87

    4.1   Evolution in action   87
    4.2   Selection and the disappearance of diversity   91
    4.3   Mechanisms that restore diversity   96
    4.4   Stochasticity in the evolution of populations—baseline expectations   99
    4.5   Evolutionary dynamics with stochasticity and selection   103
    4.6   Sweeps or hitchhiking or both?   107
    4.7   Take-home messages   110
    4.8   Homework problems   110
    4.9   Technical appendix   113

II   ORGANISMAL BEHAVIOR AND PHYSIOLOGY   119

5   Robust Sensing and Chemotaxis   121

    5.1   On taxis   121
    5.2   Why swim?   123
    5.3   The behavior of swimming E. coli   125
    5.4   Chemotaxis machinery   127
    5.5   Signaling cascades   129
    5.6   Fine-tuned adaptation   132
    5.7   Buffering and robust cellular adaptation   135
    5.8   Take-home messages   137
    5.9   Homework problems   138
    5.10   Technical appendix   142

QUANTITATIVE BIOSCIENCES AT ALL SCALES OF LIFE

In the late 1920s, Umberto D'Ancona had a problem. D'Ancona was an Italian biologist who had studied changes in the abundance of fishes in the Adriatic Sea during World War I. The war had closed major fisheries, but rather than increasing the number of all fish, the closures had seemingly paradoxical effects on fish populations. The closed fisheries seemed to benefit major marine predators—like shark and dogfish, whose numbers went up—while simultaneously hurting "food fish"—like sea bass and sea bream, whose numbers went down dramatically. When the fisheries reopened, the effects reversed. Puzzled, D'Ancona reached out to his father-in-law, the mathematician Vito Volterra. This correspondence—memorialized by Israel and Gasca (2002) in *The Biology of Numbers*—was the catalyst for a dramatic realization.

Volterra posited that the fluctuations were a result of nonlinear feedback between predators and prey. He envisioned a marine system in which prey could thrive in the absence of predators, predators consumed prey, and predators would have high mortality rates in the absence of prey. These three rules—growth, consumption and mortality—were enough to generate a broad set of conclusions (Volterra 1926). First, when prey were abundant, then predators should thrive; as predators grew in number, they would deplete prey; yet as prey became scarce, predators would start to die off, enabling prey to replenish again. This cycle implies that a system can oscillate without external oscillatory pressure. Instead, the oscillations emerge from endogenous and nonlinear feedback, specifically, the fact that the strength of predation increases as a product of the prey and predator populations. The model also seemed to reconcile his son-in-law's puzzle, in that it predicted that prey populations would go down while predator populations would go up as a function of the strength of external mortality.

This model, discovered and developed in parallel by the American physicist Alfred Lotka (1925), is now known as the Lotka-Volterra model of predator-prey dynamics. It does not just explain a potentially paradoxical facet of changes in the early twentieth-century Adriatic fisheries but is a harbinger of a far broader phenomenon. Yet the model also has its faults. The original version of the model introduced by both Lotka and Volterra had an unusual feature. The initial conditions mattered, a lot. That is, if one were to begin a system with a particular number of predators and prey, the model predicted oscillations in both that would return precisely to that initial point. Similarly, if the system had started with 10-fold as many predators, then the model would again predict oscillations that would return precisely to that new starting point. Formally speaking, these are termed "neutral" oscillations and point to conserved quantities in the system. Such conservation laws may be important in physics, but in this context they point to pathologies in the ecological model. Later refinements to the Lotka-Volterra model showed how predator-prey dynamics could give rise to robust limit cycles, in which a system would exhibit regular oscillations of the same nature, irrespective of where it began (Rosenzweig and MacArthur 1963).

The beauty of such models also poses a danger. Once successful, it would appear that a valid path forward is to study every possible feature, variant, and case of the mathematical equations. In doing so, there is a serious risk of memorializing the equations rather than letting them, like their subject matter, evolve. With deification comes hagiography and increasingly sterile versions of mathematical biology. Indeed, caution had been made against exactly such a course by Nicolas Rashevksy, the founder of the *Bulletin of Mathematical Biophysics* (Cull 2007):

> the theory had to be quantitative, and
> the theorist had to be in contact with experiments.

This caution was not always respected, not even by Rashevsky. Yet the need remains, even if there are disagreements on the value of modeling and theory in the biosciences.

In response to a perceived valuation gap, Ray Goldstein (2018) wrote a thought-provoking piece entitled "Are Theoretical Results 'Results'?" The article had a single word abstract: Yes. The need for such a piece in other domains of science would be dismissed out of hand. Yet the biological sciences remain the least "quantitative" of the natural sciences. As a consequence it is hard for some to imagine the value of theory. Perhaps theory is meant to provide a nice story afterward to satisfy the theorists. Or perhaps theory is meant to fit data to a line, the closer the better, which turns theory into a form of higher-order polynomial fitting. Yet theory should aspire for more. Instead, theory, modeling, and quantitative analysis should be a partner in the doing of science, and the essential feedback and dialogue between hypotheses, mechanisms, experiments, and observations—and of course between scientists across disciplines.

Twenty years or so after Lotka and Volterra had their epiphanies, Salvador Luria, an Italian biologist, faced a different sort of problem. This was early days in molecular and cellular biology, decades before the genetic basis for heredity was discovered. Luria was interested in examining the basis for mutations in bacteria, so he initiated a series of careful experiments, counted cells, and thought. The particular case that troubled him involved bacteria grown overnight and then exposed to bacteriophage. Bacteriophage (or phage for short) are viruses that can exclusively infect and kill bacteria. Phage can, in principle, kill every susceptible bacteria they encounter. In his experiments, Luria added far more bacteriophage than susecptible bacteria, so that, in principle, every bacteria would be infected by a phage, and then lysed soon afterward. Instead, when Luria took this mixed culture of bacteria and phage and plated it out, he found "colonies" of bacteria that had survived and proliferated. These colonies were, as he subsequently learned, founded by resistant bacteria that phage could not infect and kill. Yet the real paradox began when he counted the number of such colonies in replicate experiments. Sometimes there would be 3, then 300, other times none at all. The experiments were in a fundamental sense unrepeatable. Luria turned to his colleague, the physicist Max Delbrück. And, for reasons explained in the next chapter, this lack of repeatability was precisely the evidence needed to decide whether or not mutations (e.g., from phage susceptibility to phage resistance) were dependent on, or independent of, selection. And, critically, one way to work through this evidence was via quantitative models of the two competing hypotheses. This mechanistic model has proved profoundly influential (Luria and Delbrück 1943), helping to confirm Luria and Delbrück's hypothesis that mutations are independent of selection—a Nobel Prize–winning discovery relevant to a diverse range of ongoing studies from virus-microbe dynamics to the rise and spread of cancer.

These stories are but two of many. They come from the past but are not confined to it. Indeed, challenges remain at all scales. For example, how can we preserve our antibiotics commons in the face of widespread proliferation of antibiotic-resistant genes that enable bacterial pathogens to survive and thrive even in the face of our last-line antibiotic compounds? This question will not be resolved through innovations in biochemistry and molecular biology on their own. Instead, there is a growing recognition of the importance of utilizing evolutionary models to robustly eliminate infections and ensure that microbes do not evolve resistance to the very compounds (or even viruses and commensal strains) used to combat them (Allen et al. 2014; Burmeister et al. 2021). Similar challenges exist at organismal scales: How do the actions of individual neurons lead to emergent cognition? How do the actions of individual agents drive coherent swarm intelligence? And how (and why) do groups of organisms cooperate given access to limited resources? Finally, I have been writing and revising this book amidst a global pandemic in which mathematical and computational models have played a critical role, ranging from forecasting to the development of strategies to control and prevent future spread. These challenges span scales from cells to individuals to ecosystems, and quantitative models have a critical role to play in each—especially if the next generation of life scientists are prepared to develop and integrate them into their research.

THE GOAL

The central objective of the course that inspired this book is to help students learn *how to reason quantitatively in the biosciences given uncertainty*. This central objective is made up of number of elements—quantitative reasoning, biosciences, and uncertainty. Taken alone, each has a limited value in transforming how biosciences is taught and how it is learned. Together, these elements provide the right balance to enable students to understand principles of how living systems work; and to catalyze a path for early career scientists to discover new principles of their own. In addition, these elements govern the course organization, the selection of examples, and the style in which each module builds up—from motivating biological problem, to theory, to integration of models, mechanisms, and data. This central objective functions, in some sense, like a credo for this book (and the course upon which it is based). It is worth explaining why.

Analytical reasoning is an essential element to science pedagogy and practice; it is vital to connect assumptions to conclusions, to weigh evidence, to consider and potentially eliminate one of two (or many) hypotheses or alternative mechanisms (Platt 1964). Such reasoning is central to nearly all life sciences, chemistry, earth sciences, physics, and engineering. Yet *quantitative reasoning* is not a characteristic of the vast majority of life science courses, nor is it a characteristic of the vast majority of foundational textbooks in the field. For example, a biology class may provide the context for understanding how a particular gene is transcribed into messenger RNA, which itself is translated into a protein, i.e., the building block of cellular functioning. The same class may require that students reason with respect to the temporal order of the appearance of transcripts and proteins. But how much of the protein should one expect, what is the correlation between RNA and protein dynamics, is there evidence that transcription and translation are both random processes, or do new RNA and protein molecules appear in clusters of short and productive "bursts"? In the absence of some kind of governing mathematical or quantitative framework, the answers to these questions each become sui generis and, strictly speaking, empirical. Instead, to connect these questions, to unify their answers, to reduce the quantitative state space of possible outcomes requires a different kind of logic and reasoning: one that includes mechanistic models with quantitative, explanatory, and predictive power.

Quantitative reasoning alone is critical but not sufficient. If it were, then the paradigm of mathematics and physics classes would seem to be a viable path forward for a parallel pedagogical track in the life sciences. That is to say, take a set of established equations, analyze them, explore the logical consequences of the relationships, and use their solutions as a proxy for the behavior of the natural world. However, unlike the constitutive equations of physics, a mathematical set of equations that describes a living system is not necessarily a hallowed object—not yet, at least. Models of living systems should not be put on pedestals nor conflated as substitutes for measurement. Models, mechanisms, and their predictions must engage with evidence taken from living systems in an iterative fashion—with far greater frequency than in certain branches of physics.

More than 10 years ago at a Burroughs Wellcome Fund–sponsored science gathering, I had the opportunity to discuss what was then a burgeoning idea for a new PhD program in quantitative biosciences at Georgia Tech with accomplished senior colleagues, including Prof. Emery Brown—a neuroscientist and physician jointly based at MIT and Harvard Medical School. I was excited. There was lots to do. At some level I felt that this was the right path to take and the right moment to take it. Yet when Emery asked me about the central aim of my approach, I flubbed. After some back-and-forth, he suggested that my description omitted a crucial last word: uncertainty. Emery is a statistician who is also a practicing anesthesiologist. He weighs evidence and uncertainty. But, then again, so do any of us who work to understand how living systems work, whether at molecular, cellular, individual, population, or ecosystem scales. Hence, to the extent that all models are wrong but some are useful—sensu George Box (1976)—a model's utility depends in large part on whether or not it can help resolve uncertainty in characterizing the driving mechanisms of a particular system or systems of interest. This uncertainty cuts to the very core of quantitative biosciences, in which it is necessary to ask: have we yet identified the right constitutive equations? In doing so, we are really asking: do we understand the mechanisms, processes, and rates underlying how a living system functions?

The process of learning to reason quantitatively in the biosciences given uncertainty can be taught. It takes time. This is a longish book, but even this is just a start. Students and early career researchers—like you—may come from a variety of backgrounds: biology, computing, engineering, mathematics, or physics. Some may feel more "prepared" than others; the reality is that this interdisciplinary material will be challenging yet purposeful. Some exposure to the life sciences, prior experience in coding, and coursework in mathematics that includes differential equations and probability would certainly facilitate engaging deeply with the material. Yet there are no formal prerequisites. Students have taken the course upon which this book is based with more or less of the necessary prerequisites in each of those facets and nonetheless thrived. How they did—how you will do—depends to a large extent on how much they embraced an even simpler credo:

You can do it.

The "doing" here represents a form of active learning, whether reading papers, engaging in discussions, and critically, building mathematical and computational models of living systems. These simulation models are not exact replicas of the living world. They are a means to represent mechanistic interactions at different scales of life. The course also emphasizes computation as a means to iterate between mechanisms, models, and data. The rationale is simple: there are many models of living systems, very few of which lend themselves to tractable analytical derivations. And the process by which a model of living systems can be made analytically tractable can often come with trade-offs: the simplification can lead to mathematically elegant solutions that no longer have direct bearing on the real world. Is it really worth the investment in solving a model in certain asymptotic limits? The answer to that question may be yes, as long as one keeps in mind a caution; that such limits guide a scientific process of inquiry but should not be conflated as reality without real-world measurements. As a result, computational simulation is often required to engage meaningfully with data. Simulations can help translate heterogeneous biological mechanisms into models, to identify testable hypotheses from simulations, and to integrate the inevitable differences between model predictions and data into the model structure itself. Surely there is a space for beautiful theory, but what is more beautiful than a theory that can explain the living world?

THE STRUCTURE OF THIS BOOK

Context

Understanding scientific principles underlying the structure, function, and behavior of living systems requires a combination of manipulative experiments, high-throughput observations, mechanistic models, and data-intensive algorithms. Yet the majority of life science undergraduates take a sequence of core science curricula that includes little to no quantitative reasoning. Instead, students take precalculus, calculus, or, in the most innovative of programs, a calculus or mathematical methods for the life sciences class (in the spirit of Adler 2012; Neuhauser 2011; (Bodine et al. 2014). These classes are meant to teach mathematics in a way that enhances students' breadth of training in the life sciences; critical first steps, but not in and of themselves capstone curricula. Moreover, advanced undergraduate and early graduate education is often centered on practical issues related to computational toolkits or direct practice in a specialized group. All of this leaves a gap, one that is not necessarily fillable by ad hoc approaches or short courses alone.

Instead, this book is centered on a counter-to-trend view: the ability to analyze one's data is not primarily limited by computational power, package availability, or size of data. Instead, advances in understanding are often limited by the quantitative reasoning skills of those students and scientists asking the questions in the first place. Improving such quantitative reasoning skills requires practical guides that bridge core curricula and independent research and training. This book attempts to do just that: by fostering a quantitative mindset rather than valorizing a quantitative toolset. Hence, the structure of this textbook (and associated computational guides) is intended to open the door to the field of "quantitative biosciences," providing trainees from multiple disciplines, including biology, a means to integrate quantitative methods into their life science research.

Pedagogical Structure

The book and associated course is organized around three major parts, moving from molecules and cells to tissues and organisms to populations, communities, and ecosystems. Parts I–III also represent different groups and perhaps even different cultures—broadly speaking, molecular and cellular biosciences, physiology and neuroscience, and finally, ecology and ecosystems science. In moving from small to large, my intention is not to create an ever increasingly complex series of models, which by the end would imply that students should model, or simulate, an ecosystem one messenger RNA at a time. To do so would be to recapitulate the failed project of trying to understand how an entire organism, population, or ecosystem works beginning with atomic-level dynamics—or better yet, quarks! Instead, I adopt a different credo, as expressed succinctly by the Nobel laureate P. W. Anderson (1972): "more is different." That is, by looking at complex systems involving heterogeneous agents, we are compelled to build appropriate and often different models, rather than modeling ecosystems as the aggregate of an Avogadro number's worth of proteins.

In practice, the book is organized around the understanding of key advances in the biosciences, one organizing unit at a time, in which foundational and recent advances depend critically on quantitative methods and reasoning. Each chapter combines:

- *Motivating questions* designed to represent key biological problems across scales, from molecules to ecosystems
- *Analytical methods* for developing and analyzing quantitative models
- *Quantitative logic* for how to reason given uncertainty in the biosciences

The main text is accompanied by a corresponding laboratory guide in one of three languages—MATLAB®, Python, or R. The lab modules extend the questions, methods, and logic in this book to include a critical new tool:

- *Computational skills* to implement and support a thorough understanding of stochastic and dynamic modeling at the interface between mathematical formalism and biological data

This structure builds upon practices developed in the context of Georgia Tech's cornerstone class for the Interdisciplinary Graduate Program in Quantitative Biosciences that has been offered annually since fall 2016. The structure ensures that concepts can be put into practice through weekly computational labs and homework tied to the motivating biological problem. The rhythm of the book will be more apparent as it unfolds, but each chapter strives to

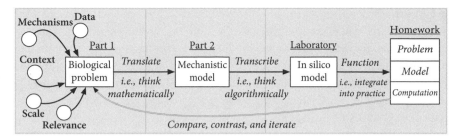

Figure P.1: Concept map of the modular structure of the course.

follow the concept map laid out in Figure P.1—particularly when used in concert with the accompanying computational guide. The methods also build upon each other, and the organization reflects transitions across scales and the accumulation of mathematical concepts and computational approaches.

Molecular and cellular biosciences

The series of modules comprising Part I of the book begins with Luria and Delbrück's question—are mutations dependent and independent of mutation?—and then grows in complexity. The chapters introduce how mathematical models can help resolve fundamental questions at the scale of molecules and cells: how do cells regulate which genes they express, how do they have memory, and how can they switch from one state to another? In doing so, the chapters emphasize that cells are not "perfect"; they grow, divide, and persist in a sea of molecules and are themselves made up of a froth of molecules, some abundant—like water—and some relatively rare—like ribosomes. The fact that the building blocks of cells are themselves discrete implies that cellular life has evolved both in spite of and because of the stochastic nature of the medium from which it is made. These concepts, both of stochasticity and of evolution, then build naturally to the concluding module: the neutral model of population genetics. These examples span more than 70 years of data, from Luria and Delbrück's seminal work to modern studies that utilize deep sequencing to examine evolutionary dynamics in the laboratory at unprecedented detail.

Tissues and organisms

Quantitative approaches are required at all scales of life. Part II focuses on organismal behavior and physiology, beginning with a module that bridges the gap between molecular, cellular, and organismal scales. The module focuses on the nanobrain of bacteria, i.e., the chemotaxis sensing machinery of *Escherichia coli*. In doing so, the module raises a question: how do organisms robustly deal with noise and uncertainty? The question of robustness is key as it provides an opportunity to address questions of how nonlinear feedback mechanisms can lead to new, emergent phenomena that are a feature of the system rather than any particular gene or protein. Robust adaptation is precisely the mechanism by which *E. coli* and other single-celled organisms react rapidly to changes in the chemical state of the environment and then return, robustly, to "business as usual" if there are no further changes to the background state. Once this concept is established, the text shifts to an entirely different cell type—animal neurons—as a means to address questions of how information and signals are processed, including how weak or unreliable signals can be ignored. Yet neurons do not act alone. Instead, individual neurons and cardiac cells can exhibit "excitability," which when coupled can then give rise to collective phenomena, including healthy beating as well as irregularities and dysregulation. Finally, the book moves even further outward to examine how information is processed into action. The focus in this last module is on a particular kind of behavior: movement. Movement may seem intuitive, but a full accounting of how organisms move, whether in water, in air, or over complex terrestrial terrains, remains a significant theoretical as well as engineering challenge. The reason, as is made clear, is that even if some of the constitutive equations are known from classic mechanics, knowing the parts list and component dynamics is not yet sufficient to understand the system as an integrated whole.

Populations and communities

Part III focuses on collectives, including many organisms of the same kind that behave similarly, many organisms of the same kind that nonetheless behave differently, or different organisms altogether (e.g., predator and prey, as well as hosts and parasites). The integration of mathematical methods in these areas is well developed. This is both a blessing and a curse. The benefit is that there is a significant body of mathematical work that can be drawn upon to understand collective dynamics, population dynamics, and community and ecosystem functioning, respectively. Yet this large body of prior mathematical work has, at times, veered into a self-referential state. Hence, this last section uses recent discoveries of exceptions to classic predictions as a means to reestablish a dialogue between models, mechanisms, and data. In doing so, the module turns to problems of collective behavior, recognizing that populations are not homogeneous and that individuals—like cells—differ in how they react and respond, often to the same stimulus. These examples, drawing upon evolutionary game theory and nonequilibrium statistical physics, are inspired by current discoveries, spanning scales of starling flocks to the dynamics of growing biofilms. These problems of collective behavior are magnified when the number of individuals (and not just their position or behavior) changes as a result of interactions. In this regard, the book focuses on predator-prey dynamics and its classic predictions of endogenous cycles contrasted with a large body of empirical work that shows that many observed population cycles have features that would seem impossible in classic Lotka-Volterra theories. Similarly, the chapter on epidemic dynamics reveals not only how much models have to offer in estimating otherwise hidden properties of epidemics but also how the assumption of models may limit efforts to predict (with certainty) how a disease may unfold in the face of stochasticity or changing behavior. Given the global impact of the SARS-CoV-2 epidemic, it is evident that mathematical models of living systems have the potential to help understand and even mitigate nonlinear dynamics occurring at large scales. It is for this reason that Part IV brings the book to a close with yet one more vista: drawing on theories of nonlinear dynamics to understand critical transitions, alternative stable states, and even catastrophes; precisely with the optimistic view that deeper understanding can help avert bad, even perilous, outcomes at the Earth systems scale.

YOU CAN DO IT

The book is about to begin. Yet the book (and accompanying computational guide) are not meant to impress you with how much others have done. Instead, the course is meant to help you on a journey that begins and ends in different locations, changing not only your perspective of how biological discoveries move forward but also how you can be part of this transformation. To do so requires that you come away with genuine skills, a certain degree of fearlessness, and hopefully a desire for even more integration of mathematical methods, computational models, and biological mechanisms. This holds whether or not you want to be a theorist, a modeler, or an experimentalist who strives to integrate models as part of your working approach.

To get there requires active immersion in the book and computational guide. For instance, once the formalities of the class logistics are out of the way, I begin one of my early lectures by asking a series of questions. No books, notes, computer, or calculator is required. These types of resources are wonderful, but better yet is the strength of trying out one's critical reasoning skills. Here are a few examples:

- How many *E. coli* are there in the room?
- How much time does it take for a single *E. coli* to divide and fill the room?
- How much time does it take for an *E. coli* to randomly diffuse across a room—if it were filled with liquid media?
- How much time does it take for an *E. coli* to swim across an environment the size of the room?
- How far could *E. coli* move in a division period?

These questions grow in difficulty. Surely it is not that hard to figure out how many *E. coli* are in a particular room. Doing so requires only that one reasons about the packing of very small objects into a very large space. Yet the point of even this simple question becomes apparent when, inevitably, a student asks: how big is *E. coli*? That is a good question. One might argue it is unfair to ask how many *E. coli* are in a room without first sharing the size of *E. coli*, especially to students from mathematics, physics, or engineering. Yet this is also part of the point.

Making progress in quantitative biosciences requires numbers as well as models. It is all well and good to know that something is possible in theory, but whether it happens in practice often depends on the magnitude of a size, speed, rate, or flux. It is precisely for this reason that the answers inevitably vary over *many orders of magnitude*. The variation reflects, in part, uncertainty in how big *E. coli* is. *E. coli* is a rodlike Gram-negative bacterium approximately 1–2 μm in length and 1 μm in diameter. Its volume is approximately 1 μm^3. Does this matter? Surely it does if one wants to account for its biomass, surface area, maximal rate of encounter with viruses, speed of replication, and metabolic flux. Hence, an implicit goal of this book is to inculcate a sense of bionumeracy, following on the precedent of the BioNumbers effort and associated database (Milo et al. 2010). The other reason why students often report highly disparate answers to even the first question—how many *E. coli* are in the room—has to do with units. In a field where numbers are rarely reported, units are often neglected as well. As a consequence, there is the tendency amongt life science students to invoke a process that resembles "Random Formula Recall" when faced with a question for which a formula is required. In essence, students have been taught that a certain formula is the right formula for that class of problem. But what happens when you are faced, in a laboratory or field context, with a problem for which a model is needed? Where will you turn—and to which page, of which chapter, of which book?

My hope is that by emphasizing a process of beginning each model with a biological problem, identifying underlying mechanisms, and translating those mechanisms into a model—including parameters and units—you will have a chance to integrate this model-building process into your core practice. And, with some effort, perhaps these quantitative models will become part of an integrative and interdisciplinary approach to discover new principles of how living systems work.

ACKNOWLEDGMENTS

This textbook and the associated computational laboratories represent an effort to establish an intellectual foundation for the study of living systems across scales. The material would not have been possible without the input of colleagues, students in the Quantitative Biosciences PhD program at Georgia Tech, and those who utilized material under development as part of other undergraduate and graduate programs and as part of short courses in the United States, France, Italy, and Brazil. Special thanks go to the QBioS and Georgia Tech students who provided critical feedback that has shaped the need for the book and the nature of the book itself: Qi An, Akash Arani, Emma Bingham, Pablo Bravo, Alfie Brownless, Rachel Calder, Alexandra Carruthers-Ferrero, Hyoann Choi, Ashley Coenen, Shlomi Cohen, Raymond Copeland, Sayantan Datta, Kelimar Diaz, Marian Domínguez-Mirazo, Robert Edmiston, Nolan J. English, Shuheng Gan, Namyi Ha, Hayley Hassler, Maryam Hejri Bidgoli, Lynn (Haitian) Jin, Elma Kajtaz, Cedric Kamalseon, Katalina Kimball-Linares, Tucker J. Lancaster, Daniel A. Lauer, Alexander (Bo) Lee, Zewei Lei, Guanlin Li, Hong Seo Lim, Ellen Liu, Lijiang Long, Katie MacGillivray, Jiyeon Maeng, Andreea Magalie, Pedro Márquez-Zacarías, Zachary Mobille, Daniel Muratore, Carlos Perez-Ruiz, Aaron R. Pfennig, Rozenn Pineau, Brandon Pratt, Joy Putney, Aradhya Rajanala, Athulya Ram, Elisa Rheaume, Rogelio Rodriguez-Gonzalez, Benjamin Seleb, Varun Sharma, Benjamin Shipley, Cassie Shriver, Michael Southard, Sarah Sundius, Disheng Tang, Stephen Thomas, Kai Tong, Akash Vardhan, Hector Augusto Velasco-Perez, Ethan Wold, Fiona Wood, Leo Wood, Siya Xie, Seyed-Alireza (Ali) Zamani-Dahaj, Christopher Zhang, Mengshi Zhang, Conan Y. Zhao, and Baxi Zhong.

Likewise, I am indebted to colleagues who shared their feedback, references, counsel, and advice on how to communicate principles of living systems rigorously and meaningfully. Thank you to Francois Blanquart, Hannah Choi, Michael Cortez, Antonio Costa, Silvia De Monte, Ido Golding, Jacopo Grilli, Jeremy Harris, Oleg Igoshin, Jacopo Marchi, Matteo Marsili, Henry Mattingly, Daniel Rothman, and Andrew Zangwill for their detailed examination of individual chapters and their willingness to read the text closely and with care. Thank you also to Van Savage, David Murrugarra, Rafael Peña Miller, and Carles Tardío Pi for their comprehensive review of the book. Van gets a double thanks for his willingness to try out this material in formative stages—thanks also to Tianyun Lin for facilitating feedback from UCLA students. Throughout, I have tried to follow the good counsel of my colleagues in communicating both the biological and theoretical concepts—any mistakes are mine alone.

The book has been shaped by many conversations developing the Foundations of QBioS course at Georgia Tech. Many thanks to Ned Wingreen for sharing his perspsective on what a course like this should cover, and for sharing materials that were invaluable at early stages of development. Thank you to colleagues at Georgia Tech, especially those in the Physics of

Living Systems program, for their input on the material over many years, especially Flavio Fenton, J. C. Gumbart, Simon Sponberg, and Daniel Goldman, as well as to colleagues in Biological Sciences, especially Liang Han, Will Ratcliff, and Soojin Yi, as well as multiple group members who helped teach part of the course and whose input was critical to improve the material: Stephen Beckett, David Demory, Jeremy Harris, Joey Leung, Adriana Lucia Sanz, and Jacopo Marchi. Because this book is meant to be part of a combination package— each chapter is accompanied by a computational lab in MATLAB, Python, or R—I want to extend my deepest gratitude to the multiple computational laboratory book authors who brought the code to life, including Bradford Taylor (MATLAB), Nolan English (Python), Alexander Lee (Python), Ali Zamani (Python), and Marian Domínguez-Mirazo (R).

The time and resources to develop this book have been made possible, in part, by grants and support from the National Science Foundation (through programs in Physics of Living Systems, Biological Oceanography, Dimensions of Biodiversity, and Bridging Ecology and Evolution), National Institutes of Health (through programs in General Medical Sciences as well as Allergy and Infectious Diseases), Army Research Office, Charities in Aid Foundation, Marier Cunningham Foundation, Chaire Blaise Pascal Program of the Île-de-France, Mathworks Corporation, Simons Foundation, Centers for Disease Control and Prevention, A. James & Alice B. Clark Foundation, and the Burroughs Wellcome Fund. Special thanks to the staff, scientists, and advisory board members at the Burroughs Wellcome Fund who supported my interface research at a critical moment in my career. That opportunity has shaped many of my choices since, including the decision to move ahead with a new graduate program in quantitative biosciences, develop its cornerstone class, and eventually write this book.

Thank you to the staff at Princeton University Press for your collective patience in seeing this project take off and in supporting me as I made it to the finish line. Particular thanks to Alison Kalett for believing in the book at an early stage and to Sydney Carroll for continuing to believe in this book while nudging me in the right direction as I wrote and revised amidst a global pandemic. Finally, many thanks to the capable team of copyeditors, illustrators, indexers, and production specialists who have elevated this material into an integrated whole.

This book has taken up many days and nights. I am deeply grateful for Maira's support and patience and for the encouragement of Ilan and Noa, who wanted to know for quite some time if I was done yet. Yes, I am done.

Part I

Molecular and Cellular Biosciences

Chapter One

Fluctuations and the Nature of Mutations

1.1 CHANCE FAVORS THE INDEPENDENT MUTATION

Evolution via natural selection denotes any nonrandom change in the genetic makeup of a population due to the differential reproduction and/or survival of individuals. As such, evolution via natural selection requires standing variation to facilitate dynamic change in populations, again and again, over generation and generation. Mutations in the genome of replicating organisms are the grist for this long wheel of evolutionary change. Yet, in the early part of the twentieth century, scientists had not yet identified the molecular basis for heredity. Big questions remained in the field. Big questions that, for us, have become matters to be read and memorized in textbooks. But to start in the process of integrating quantitative methods into our study of living systems requires that we try, however difficult, to displace ourselves from the present time and put ourselves in the mindset of others.

Early molecular biologists faced a profound challenge: what was the basis for the generation of individual variation? The existence of diversity was never in question, but how such diversity came into being was. The two major theories differed radically with respect to the nature of the link between the introduction of variation and its differential selection. Are mutations dependent on selection or independent of selection? The idea that mutations depend on selection seems heretical to modern practitioners of quantitative biology. Yet it was not certainly not always the case. Charles Darwin's theory of evolution via natural selection presumed that variation was introduced in some kind of heritable material. The differential success in survival and reproduction became the mechanism to "select" for a subset of variants. Then those more fit variants would produce new offspring, different again from them, and so on. In essence, mutations are independent of selection.

The contrasting idea is often attributed to Jean-Baptiste Lamarck, a French biologist active in the late eighteenth and early nineteenth centuries. To understand Lamarckian evolution, it is worth sharing a few examples. First, consider a parent who decides to join a gym. She (or he) gets strong. Will the child be more likely to have bigger muscles than if the parent had skipped the gym and stuck to a steady diet of barbecue and ice cream? It seems unlikely, but according to Lamarckian evolution the answer would, in fact, be yes. Another example. The classic one. Consider a female giraffe grazing in the Serengeti. Food is sparse, so the female giraffe must stretch and stretch to reach its preferred acacia leaves. One day the giraffe has a calf. Would the calf have a shorter neck had the mother not had to stretch as far? This is the essence of Lamarckian evolution: it posits that experiences that change the phenotypic state of a parent will be passed on heritably to its offspring. In other words, "mutations depend on selection." The examples of the gym aficionado and the long-necked giraffe seem improbable. But as anyone who follows the field of epigenetics knows, present experiences can shape the phenotypes of offspring, often in profound ways.

But Luria and Delbrück did not work with humans or giraffes. Instead, working with microbes and their viruses afforded them a quantitative framework to directly address these two hypotheses. It was already known from the work of Frederick Twort (1915) and Felix d'Herelle (1917) that viruses could infect, lyse, and propagate on bacteria (for a historical perspective, see Summers 1999). These bacteriophage were relatively specific in their activity. That is, some phage could spread on certain bacteria but not others. The difference between a phage-resistant and phage-susceptible strain could be identified through a simple colony assay where the number of resistant bacteria were measured on agar plates. Hence, in the case of microbes and viruses, the two hypotheses can be summarized as follows:

Spontaneous mutation The change from virus sensitivity to virus resistance happens spontaneously to cells, irrespective of their interaction with viruses. This spontaneous change is rare.

Acquired heritable immunity A small fraction of infected cells survive and acquire an immune state, which can be passed on heritably to daughter cells. This acquired change is rare.

The hypotheses of spontaneous mutation and acquired heritable immunity map roughly to Darwinian and Lamarckian evolution, respectively (Figure 1.1). Yet these two hypotheses should have profoundly different consequences on the variability expected in colony counts of cells that can no longer be killed by viruses, even if their recent ancestor could.

The Luria and Delbrück paper is a seminal event in the history of biological sciences. It proved transformative to understanding the nature of evolution—showing that heritable changes in cellular state were independent of, rather than dependent on, selective forces. The finding is particularly striking given that the work was completed 10 years before the discovery of the double-helix structure of DNA was made possible by Francis Crick, Rosalind Franklin, James Watson, and others (Judson 1979). For Luria

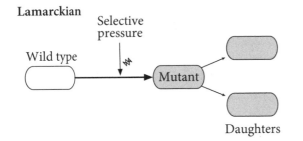

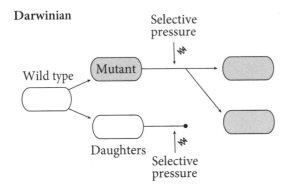

Figure 1.1: Schematic of Lamarckian and Darwinian views of selection and mutation. (Top) Lamarckian selection changes the cell, and daughter cells heritably retain the characteristic of the mother. (Bottom) Darwinian selection changes the functional traits/phenotypes of the cell, but differences in daughter cells (when they arise via mutation; see gray versus white) are independent of the selective experience of the mother.

and Delbrück, the selective force was the killing power of bacteriophage. Bacteriophage ("phage") are viruses that exclusively infect and kill bacteria. Yet the first image of a virus was only seen under a microscope 4 years before! Nonetheless, phage and bacteria were already becoming the workhorses driving discoveries into cellular function.

It was in this context that Salvador Luria—a biologist from the University of Indiana—and Max Delbrück—a physicist at the University of Vanderbilt—initiated what was to become a long-term and deservedly famous collaboration. (The rest of this chapter refers to their collaborative work with the initials LD.) Yet the success of LD's ideas came slowly. Their 1943 paper, "Mutations of Bacteria from Virus Sensitivity to Virus Resistance," is difficult to read (Luria and Delbrück 1943). The difficulty is not ours alone, separated as we are in time by 70 years and missing context. Perhaps the authors simply wrote in different ways and the hints at their underlying methods in the text would have been well understood by their peers. This seems unlikely.

Recall that the work of Alfred Lotka and Vito Volterra on predator-prey dynamics was not yet 20 years in the past. The integration of mathematics and biology was hardly commonplace. Moreover, unlike the bulk of models of biological systems, the work of Luria and Delbrück combines elements of both continuous and discrete mathematics. Perhaps it was only Delbrück who truly understood the mathematical nature of his arguments. Indeed, 10 years later, Esther Lederberg and Joshua Lederberg leveraged their ingenious idea of replica plating to show the clonal nature of virus resistance in bacteria (Lederberg and Lederberg 1952). It was then that LD's ideas began to gain acceptance not only because of authority but through the adage "seeing is believing."

The following sections lay out the core arguments to decide whether mutations are dependent on or independent of selection. In doing so, it is critical to review the nature of the heritable state as well as the experimental details and mechanistic hypotheses at stake. This chapter reviews multiple lines of evidence in support of the competing hypotheses, including the quantitative predictions for both the mean and variance of mutant colonies. As we will see, the history of the LD experiments and their outsized influence on the foundations of molecular biology lie in an "irreproducibility opportunity" (Figure 1.2). This schematic provides a visual recapitulation of the kind of data that LD observed—in which some of their experimental replicas included zero (or very few) resistant colonies and others included hundreds. As it turned out, the large-scale disagreement among replicate experiments was precisely the evidence needed to distinguish between the Darwinian and Lamarckian hypotheses. And, by the end of this chapter, you will have a sense of how important this variability was (and is) to understanding something fundamental about how life works.

Experimental replicates

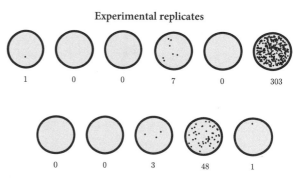

Figure 1.2: Schematic of colony assays illustrating the apparent lack of reproducibility in the Luria-Delbrück experiment. The number of colonies is listed below each plate; these numbers in each experimental replica correspond to experiment 17 (Luria and Delbrück 1943). What would you have done with such large variation between experiments? Is this failure? Or, instead, something more profound? How and why this lack of reproducibility explains the very nature of mutations forms the centerpiece of this chapter.

1.2 CELLULAR PHENOTYPES

Bacteriophage (or phage) are viruses that exclusively infect and lyse bacteria. Infection is initiated via encounter between the virus particle and the surface of the bacterial cell. After encounter and successful adsorption, the genetic material of the phage is injected into the cytoplasm of the bacteria where phage genes redirect the bacterial machinery, including transcriptional enzymes and ribosomes, to copy the viral DNA and produce viral proteins. These viral proteins self-assemble into capsids, which are then packed with viral DNA; through a timed process, viral encoded enzymes—including holins and lysins—make small holes in the inner membrane and cell wall of the bacterial cell. As a consequence, the cell explodes and dozens, if not hundreds or more, virus particles are released. The infection and lysis of bacteria, like *E. coli B*, by phage, like phage α, is depicted in Figure 1.3. This process can be scaled up, millions and billions of times over. Indeed, that is precisely what Luria did.

He did so, as the story goes, on a Sunday. The Saturday night before, in January 1943, Luria had been at a faculty dance and social (those were different times (Luria 1984)). The social included slot machines, which generally yield nothing but occasionally pay off in large jackpots. Luria had observed similarly large, rare events in his experiments to probe the change from virus sensitivity to resistance among bacteria (see an example in Figure 1.2). He reasoned: what if such events were not a mistake in his experimental design but rather a feature of the resistance process itself? These slot machines and their jackpots were the catalyst Luria needed to revisit his own thinking on the nature of mutations (Judson 1979). To test the idea, Luria returned to his laboratory and conducted the prototype of what became the experimental observations at the core of the 1943 LD paper. It is worth explaining precisely what those experiments entail.

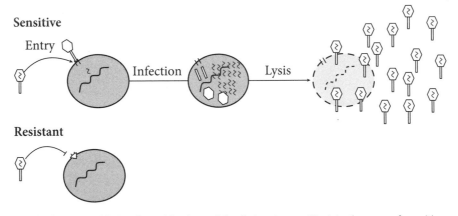

Figure 1.3: Infection and lysis of sensitive bacterial cells by viruses. (Top) In the case of sensitive cells, viruses inject their genetic material into hosts, the virus genome replicates, virus capsids self-assemble and are packed with virus genomes, and then virus particles are released back into the environment following the lysis of cells. (Bottom) In the case of resistant cells, viruses are unable to adsorb, infect, and lyse the cell. Note that, more generally, resistance to infection can be due to extracellular and/or intracellular mechanisms (Labrie et al. 2010).

Luria's experimental design culminates in the interaction of phage T1 with bacteria (now known as *E. coli B*) on agar plates. Despite having many more viruses than bacteria, Luria had observed that bacteria can and do survive the interaction. To reach this point requires the following steps (Figure 1.4). First, a culture of bacteria is grown up overnight. Such cultures typically include bacteria at densities on the order of 10^8 per ml. If grown in a 100 ml flask, this represents over 10 billion bacteria. In parallel, viruses are added to a culture of bacteria. The replication of viruses inside sensitive bacteria leads to the release of large numbers of viruses, which reinfect new cells and release more viruses, such that total virus densities can rapidly exceed 10^9 per ml. Ensuring the culture exclusively contains viruses requires another step. Chloroform is often added to eliminate any remaining bac-

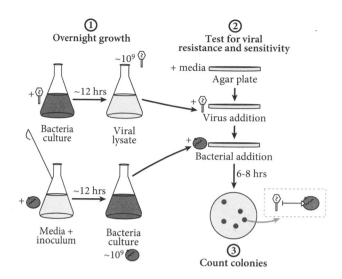

Figure 1.4: The Luria-Delbrück experiment, including overnight growth of bacteria and phage, mixing on agar plates, and colony counting.

teria, the culture spun down, and the supernatant removed to extract a *viral lysate*, i.e., a culture of virus particles. This is how the experiment starts. Next, the viral lysate is poured atop agar plates and bacteria are added. The vast majority of bacteria should be infected by viruses and lyse. Yet a few, sometimes hundreds, are not killed. These bacteria replicate, beginning with just a single cell until they form a clustered group of thousands to tens of thousands of bacterial cells on the plate. These dense assemblages of bacteria that arise from a single bacterium are termed a *colony*. How many colonies appear, how often no colonies appear, and the variation in colony counts between replicate plates forms the heart of the Luria and Delbrück experiment.

The results from different experiments are shown in Table 1.1, where the columns denote distinct experiments and the rows denote distinct counts of the number of colonies in a series of biological replicates as measured in distinct agar plates. There are many striking features of these results. First, there are many replicas with zero resistant colonies. Yet there are also many replicas with dozens if not hundreds of resistant colonies. Imagine yourself staring at this very data, not knowing what Luria and Delbrück discovered. What would you have done? If you are a PhD student, ask yourself: would you show these results to your adviser? Or, instead, would you have thought: there's a mistake in the experiment. It's not repeatable. Yet that lack of repeatability, i.e., the zeros and the jackpots together, is the critical clue to understanding the nature of mutations.

1.3 MUTATIONS THAT DEPEND ON SELECTION

What if Lamarck was right and mutations depend on selection? For a moment, disregard the potential mechanism by which bacteria acquire resistance and/or immunity. Instead, consider what would happen in the event that N bacteria on the agar plate were each exposed

Table 1.1: Number of resistant colonies observed in Luria and Delbrück's 1943 experiment.

Experiment	Replica																			
	1	2	3	4	5	6	7	8	9	10	11	12	13	14	15	16	17	18	19	20
1	10	18	125	10	14	27	3	17	17											
10	29	41	17	20	31	30	7	17												
11	30	10	40	45	183	12	173	23	57	51										
15	6	5	10	8	24	13	165	15	6	10										
16	1	0	3	0	0	5	0	5	0	6	107	0	0	0	1	0	0	64	0	35
17	1	0	0	7	0	303	0	0	3	48	1	4								
21a	0	0	0	0	8	1	0	1	0	15	0	0	19	0	0	17	11	0	0	
21b	38	28	35	107	13															

Note: The table here includes a subset of the original data published in (Luria and Delbrück 1943). Each row is a different experiment and each column is a different replicate within that experiment. The number of replicates for each experiment was not fixed. Each replicate denotes a biological replicate referring to distinct numbers of cultures examined in each experiment.

to one or more viruses, and that such interactions can trigger a heritable immunity mechanism. To do so, assume that the probability of an acquired mutation is μ_a. These mutant bacteria are not killed by viruses and pass on this resistance trait to daughter cells. What then is the probability of observing m resistant bacteria, i.e., the "mutants"? This probability is equivalent to flipping a biased coin N times, with the successful outcome occurring quite rarely; e.g., if $\mu_a = 10^{-8}$, then the mutation occurs one in every one hundred million trials. Formally, the probability can be written as follows:

$$p(m|N, \mu_a) = \overbrace{\frac{N!}{(N-m)!m!}}^{\text{permutations}} \overbrace{\mu_a^m}^{\text{mutation}} \overbrace{(1-\mu_a)^{(N-m)}}^{\text{sensitive}}. \tag{1.1}$$

The first term denotes the number of ways to choose exactly m of N individuals. For example, when $m = 1$, then this combinatorial prefactor is N, i.e., the mutation that does occur could have occurred to any one of the N bacteria in the population. Similarly, when $m = 2$, then this combinatorial prefactor is $N(N-1)/2$, i.e., the number of ways to choose two unique members of a population of size N, and so on. The remaining factors correspond to the probability that a mutation with probability μ_a occurs precisely m times and, by virtue of the size of the population, that a mutation does not occur—with probability $1 - \mu_a$—precisely $N - m$ times (explaining why these counts appear as exponents).

This formula denotes the binomial distribution, but saying that does not seem particularly helpful. If you were to calculate this formula on the computer, you might find that calculating massive factorials is not altogether helpful. Instead, we should consider the fact that the experiment was done in a particular regime, that is, when N is a very, very large number, on the order of 10^8 or greater. Similarly, the mutation rate, although unknown, was almost certainly a very small number, on the order of 10^{-8}. In other words, the probability for observing m mutants can be readily calculated in certain limits, e.g., when $N \gg 1$

and $0 < \mu_a \ll 1$. In that regime, the binomial distribution reduces to

$$p(m|N, \mu_a) = \frac{(N\mu_a)^m e^{-N\mu_a}}{m!}. \tag{1.2}$$

This formula denotes the Poisson distribution and is the limit of a binomial distribution given many trials and small probability of success (see the technical appendix for a detailed derivation).

The Poisson distribution has a number of interesting properties. For one, the mean is equal to the value of the argument in both the exponential and the polynomial term: $\bar{m} = N\mu_a$. Moreover, the variance of a Poisson distribution is equal to the mean. Hence, the standard deviation—the square root of the variance—should be $\sigma = \bar{m}^{1/2}$. This scaling implies that replica experiments should yield small fluctuations, such that the standard deviation σ increases more slowly than the mean $\bar{m}$. One way to measure the smallness of fluctuations is to consider the ratio of the standard deviation to the mean, termed the *relative error*. For example, if there are 10 mutants on average, then the standard deviation should be 3 and the relative error, or $\sigma/\bar{m}$, should scale like $\bar{m}^{-1/2}$, or 1/3. Similarly, if there are 100 mutants on average, then the standard deviation should be 10 and the standard error should be 1/10. In essence, there should be a relatively consistent number of mutants between trials. As a result, the acquired immunity hypothesis predicts that repeated experiments should tend to have similar levels of colonies despite the randomness associated with the mutational process.

With this model in hand, let us now adopt the perspective of an experimentalist and try to infer the most likely mutation rate given measurements. In essence, rather than asking how many mutants we expect to see if we know the true mutation rate, we would like to ask: what is the most likely mutation rate compatible with the observations we make? To find the answer, we must turn to the data.

One set of data is reproduced in Table 1.1. The table includes the numbers of resistant colonies in a series of replicate experiments. The number of resistant colonies differ. Some are small, in some cases there are no resistant colonies whatsoever, and some are large, quite large compared to others, e.g., hundreds versus a handful. There are at least two ways to use this data. First is to note that if the process of mutation depends on selection, then we should expect sometimes not to see any mutants at all. This probability is

$$p(0|N, \mu_a) = \frac{(N\mu_a)^0 e^{-N\mu_a}}{0!} = e^{-N\mu_a}. \tag{1.3}$$

Hence, given an observation, f_0, of the fraction of replicates with zero colonies, then the best estimate of the acquired mutation rate should be

$$\hat{\mu}_a = -\frac{\log f_0}{N}. \tag{1.4}$$

This method has advantages but also drawbacks. First, in the event that $f_0 = 0$, then the mutation rate is undefined. In the event that occurs, it is still possible to use a bound, e.g., the frequency of zero events should be $f_0 < 1/s$ where s is the number of replicates. There are other approaches. Note that the average number of mutant colonies, $\bar{m}_{obs}$, is another feature of the Poisson distribution. It is predicted to be $N\mu_a$. Hence, it may be reasonable to assume that the best estimate of the acquired mutation rate should be $\hat{\mu}_a = \frac{\bar{m}_{obs}}{N}$. This is in fact sensible. Using the mean as the basis for estimating the mutation rate is equivalent to the *maximum likelihood* estimate of the unknown rate.

Formally, we would like to estimate the value of the "acquired" mutation rate that is the most likely value given observations. The choice of the adverb "most" implies there is a range of potential values to choose from. To begin, denote the joint probability of mutation rate and observed number of mutants as $P(\mu_a, m)$. This joint probability can be written as

$$P(\mu_a|m)p(m) = L(m|\mu_a)q(\mu_a). \tag{1.5}$$

This expression leverages the law of total probability such that P denotes the posterior probability of the parameter given the data, p denotes the probability of the data, L denotes the likelihood of the data given a parameter, and q denotes the prior probability of the parameter. Here P is the posterior probability and and L is the likelihood function of observing a certain number of mutant colonies given a known mutation rate. This equation can be rewritten as follows:

$$P(\mu_a|m) = \frac{L(m|\mu_a)q(\mu_a)}{p(m)} \tag{1.6}$$

where $p(m)$ and $q(\mu_a)$ are probability distributions of the data and of the prior of the parameter to be estimated, respectively. Now consider two values of the mutation rate—μ_a and μ'_a—and ask: which is more compatible with observations? To answer this question requires comparing the ratio of the posterior probabilities,

$$\frac{P(\mu_a|m)}{P(\mu'_a|m)} = \frac{L(m|\mu_a)q(\mu_a)p(m)}{L(m|\mu'_a)q(\mu'_a)p(m)}. \tag{1.7}$$

In the event there is no a priori reason to favor one mutation rate over another, then $q(\mu_a) = q(\mu'_a)$. This is what statisticians mean by *uninformed priors*. Using such uninformed priors yields

$$\frac{P(\mu_a|m)}{P(\mu'_a|m)} = \frac{L(m|\mu_a)}{L(m|\mu'_a)}. \tag{1.8}$$

In other words, to find the μ_a that is most likely, in a posterior sense, one should find the value of μ_a that maximizes the likelihood. As shown in Figure 1.5, the likelihood function has a zero derivative in μ_a at its maximum (more generally, this is true of both local minima and maxima for functions of one variable). Hence, rather than simulating the likelihood for every possible observation, it is possible to identify a general formula for the maximum likelihood estimate, $\hat{\mu}_a$. The technical appendix explains how to take a first derivative of this likelihood, yielding the maximum likelihood estimate

$$\hat{\mu}_a = m/N. \tag{1.9}$$

In this case, the value as inferred by the mean is the right choice. Caution: This equivalence between the mean and maximum likelihood solution need not always be the case.

But herein lies the problem. The variance of the Poisson distribution is equal to the mean. So we should expect that estimated variances are similar to estimated means. Moreover, we should expect that the standard deviation, which is the square root of the variance, should be smaller than the mean. For example, if there are 10 colonies on average per plate, then the acquired heritable immunity hypothesis predicts that plates will have nearby values, e.g., 5, 12, 7, 9, and so on. Moreover, large deviations should be very rare. This is not the case in the experiments of LD (as seen in Table 1.1). Hence, although it may be possible to estimate an acquired mutation rate through the zero-colony or average-colony methods, the data already suggests that these rates correspond to a quantitative feature of the incorrect mechanism. To consider another mechanism requires that we evaluate the number of mutations that would arise in different replicates if mutations were independent of selection.

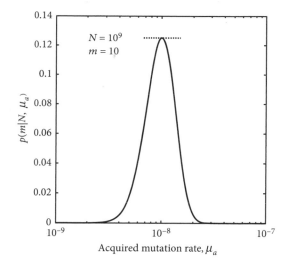

Figure 1.5: Likelihood, $L(m|\mu_a)$ from Eq. (1.2), given variation in μ_a from 10^{-9} to 10^{-7} and the observation that $m = 10$ when $N = 10^9$. As expected, the maximum likelihood corresponds to $\mu_a = m/N$ or $\hat{\mu}_a = 10^{-8}$. Note that the dashed lines provide a visualization of the zero first derivative corresponding to the value of $\hat{\mu}_a$.

1.4 INDEPENDENT MUTATIONS: A CONTINUOUS MODEL

1.4.1 Spontaneous mutations–dynamics

LD proposed a different approach to the origin of resistant mutants in their experiment. Perhaps resistance mutations in the bacteria did not arise due to interactions with the virus. Instead, what happens if the mutant bacteria were already there, waiting, as it were, to be revealed through the process of interacting with viruses that would otherwise kill them? This is the core idea of mutations being independent of selection—and of the Darwinian concept of evolution via natural selection. But how many mutants should there be? This is

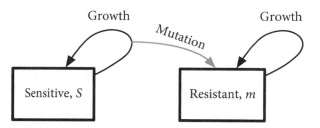

Figure 1.6: Population model of the growth of susceptible bacteria, S, and resistant bacterial mutants, m. Here the S population divides, sometimes yielding mutant bacteria that are resistant to viral infection. The m population also divides and back-mutations leading to virus sensitivity are ignored.

where theory becomes essential. If mutants arise independent of selection, then in principle they could have arisen very early in the experiment or perhaps near the end, in the very last generation of bacteria to divide before viruses were added to the agar plate. If they arose early, then a single resistant cell could divide many times before being plated on a lawn covered in viruses. Such an experiment would yield a *very large* number of resistant colonies. This possibility is worth exploring in detail.

To address this possibility, LD proposed a continuous model of bacterial population dynamics including two populations: sensitive cells and resistant mutants (Figure 1.6). (In practice, it was Delbrück who proposed the mathematical model.) In this continuous model, susceptible bacteria grow at a rate r, but a small fraction of the offspring, μ, mutate to become resistant. These mutant bacteria also grow, and in the absence of other evidence that there is a link between resistance and growth rates, then LD assume that mutants also grow at a rate r. This model defines a linear dynamical system involving two population types S and m, the number of susceptible and mutant individuals in the population:

$$\frac{dS}{dt} = \overbrace{rS(1-\mu)}^{\text{growth of sensitives}} \tag{1.10}$$

$$\frac{dm}{dt} = \overbrace{\mu rS}^{\text{new mutations}} + \overbrace{rm}^{\text{growth of mutants}} \tag{1.11}$$

This model seems simple in many ways. However, it contains subtleties both in terms of the biology and in terms of the dynamical system itself. First, the model assumes that mutations occur during reproduction. Similar results could hold if mutations occur at any moment. Second, for biologists, writing equations in this way is not necessarily intuitive. In my experience, the instinct of most biologists when asked to translate a mechanism into a model is to think in terms of update rules, i.e., the value of the population at the next time, $x(t+1) = \ldots$, rather than in the changes in the population at the current time, dx/dt. Hence, if you share that instinct, consider taking a diversion to the technical appendix for how to move between the update perspective and the dynamical systems perspective. Finally, the text sometimes uses the notation dx/dt to denote the derivative of a population/variable with time, and sometimes uses the notation $\dot{x}$ to denote the same thing.

The next challenge is to solve this dynamical system to quantify both the sensitive and mutant cells as a function of time given the mutation rate μ. Try to solve by stepping away from the text with a blank piece of paper and keeping in mind only these two rules: (i) Susceptibles divide and sometimes generate mutants. (ii) Mutants also divide.

Now, if you tried and got stuck, keep in mind that there is a helpful trick: add the two derivatives together to find that $\dot{S} + \dot{m} = r(S + m)$. In other words, the entire population $N = S + m$ is growing at a rate r, while the balance of individuals shifts between S and m. The solution to this exponential growth problem for total cells is $N(t) = N_0 e^{rt}$. There is another observation: the mutants—despite being far more rare—are actually growing faster in a per capita sense than the residents. This is true even before selection was applied! Returning to the equations and solving the $\dot{S}$ equation yields the following (where $m(t) = N(t) - S(t)$):

$$S(t) = N_0 e^{r(1-\mu)t}, \tag{1.12}$$

$$m(t) = N_0 e^{rt}\left(1 - e^{-r\mu t}\right). \tag{1.13}$$

The length of the experiment is on the order of 10–20 generations, i.e., the product rt is a dimensionless number of that magnitude. Hence, we can approximate $e^{-r\mu t} \approx 1 - r\mu t$ given that $\mu \ll 1$, such that the number of mutants is predicted to grow faster than exponentially:

$$m(t) = N_0 e^{rt} \mu r t. \tag{1.14}$$

Note at this point that it is important to reconcile this finding with the objective of LD: to determine whether large fluctuations are consistent with mutations being independent of or dependent on selection. The notion of consistency implies a particular experimental design in which LD performed a series of replicates—many, but certainly not infinite, a point that we will revisit later in the chapter. Hence, to begin to compare theory and experiment, it is worth generalizing this model to apply not only to *E. coli* and phage but to a larger class of problems. Thus far we have retained the growth rate *r*. A generalization is enabled by noting that the rate *r* and time over which the experiment is conducted *t* also appear together— suggesting that the growth rate is not a particularly important feature of the phenomena. The key is that *r*, as measured in inverse time, and *t*, as measured in time, must share the same units—hours, minutes, etc. If they do, then their product will remain the same even if we change units. That gives another clue. The value of *r* doesn't matter that much; it is the product of *r* and *t* that matters. If *r* is the effective inverse of the division period, then *rt* is simply a measure of the effective number of divisions in this growing population. Hence, it would seem that we would be better off developing a general theory, rather than one tuned for a particular value of *r*. The fact that LD could manipulate bacteria over dozens of generations reinforces their prudence in working with bacteria and not giraffes for this particular class of problems.

To formally work in this direction, denote a rescaled time, $\tau = rt$, such that $d\tau = rdt$. Hence, $\tau = 1$ is effectively one division, $\tau = 10$ is effectively 10 divisions, and so on. The dynamical equations can be written initially as

$$\frac{dS}{dt} = rS(1 - \mu) \tag{1.15}$$

$$\frac{dm}{dt} = \mu rS + rm \tag{1.16}$$

while dividing both sides by *r* yields

$$\frac{dS}{rdt} = S(1 - \mu) \tag{1.17}$$

$$\frac{dm}{rdt} = \mu S + m. \tag{1.18}$$

Next, replace $d\tau = rdt$, yielding

$$\frac{dS}{d\tau} = S(1 - \mu) \tag{1.19}$$

$$\frac{dm}{d\tau} = \mu S + m \tag{1.20}$$

This last set of equations implies that, irrespective of the growth rate, the sensitive population will grow more slowly than that of the mutation population of resistant bacteria. Using the same logic as before, albeit forgoing the explicit inclusion of the growth rate, yields a prediction for the number of resistant mutants expected after a dimensionless time τ:

$$m(\tau) = N_0 e^\tau \mu \tau. \tag{1.21}$$

Therefore, at the final time, the number of mutants is expected to be

$$m(\tau_f) = N_f \mu \tau_f \tag{1.22}$$

where $N_f = N_0 e^{\tau_f}$ is the total number of bacteria exposed to viruses. This is, in modern terms, equivalent to Eq. (6) of LD's paper. This equation can be put into practice. Given an observation of the average number of mutants in replicate experiments, then it is possible to estimate the mutation rate:

$$\hat{\mu} = \frac{m_{obs}}{N_f \tau_f}. \tag{1.23}$$

There are two key caveats here. The first caveat is that this estimate of a mutation rate simply becomes an alternative estimate to that obtained assuming the acquired immunity hypothesis. It may be right, but the fact that we can make such an estimate does not provide the necessary evidence in favor of the hypothesis. The second caveat is that the approach to solving this problem is somewhat nonintuitive, i.e., involving mathematical tricks that tend to obscure the key biological drivers of the variation. Let's try another way, hopefully one that helps build intuition.

1.4.2 Spontaneous mutations—a cohort perspective

According to the spontaneous mutation hypothesis, mutants emerge in the growing bacterial culture before viruses are added. Hence, if a single resistant mutant appeared five generations before bacteria were mixed with viruses, then that single mutant would have given rise to $2^5 = 32$ new mutants, each of which corresponds to an observed, resistant colony on the agar plate. Likewise, if a single resistant mutant appeared seven generations before bacteria were mixed with viruses, then that single mutant would have given rise to $2^7 = 128$ new mutants. Hence, the older a mutant is, the more daughter cells appear in that lineage. Yet there is also a counterbalancing force. Given that the population is growing, it is far more likely that mutants will appear near the end of the experiment, even if those mutants have less time to reproduce. It is possible to formalize this by moving from non-overlapping generations to continuous dynamics and by estimating the number of mutants in terms of cohorts, grouped by their age of first appearance.

To do so, it is essential to recognize that the rate of appearance of mutants is $\mu S(\tau)$. Hence, in a small interval of time $d\tau$, a total of $\mu S(\tau) d\tau$ mutants will emerge (at least on average). Each of these cohorts of new mutants will grow exponentially, reaching a final size $e^{\tau_f - \tau}$ greater by the end of the experiment. Hence, given that mutants can appear at any time, we can write

$$
\begin{aligned}
m &= \int_0^{\tau_f} d\tau \;\; \overbrace{(\mu S(\tau))}^{\text{new mutant cohort}} \cdot \overbrace{e^{\tau_f - \tau}}^{\text{growth of mutant cohort}} \\
&= \int_0^{\tau_f} d\tau \, \mu N_0 e^{(1-\mu)\tau} e^{\tau_f - \tau} \\
&= \int_0^{\tau_f} d\tau \, \mu N_f e^{-\tau_f} e^{(1-\mu)\tau} e^{\tau_f - \tau} \\
&= \int_0^{\tau_f} d\tau \, \mu N_f e^{-\mu \tau} \\
&= \left[-N_f e^{-\mu \tau} \right]_0^{\tau_f} \\
&= \left[-N_f \left(e^{-\mu \tau_f} - 1 \right) \right]
\end{aligned}
\tag{1.24}
$$

and using the approximation $e^x \approx 1 + x$ for $|x| \ll 1$ yields

$$m = \mu \tau_f N_f, \tag{1.25}$$

precisely what was derived in the dynamical systems approach in the prior section! This is the same answer, but with less mathematical trickery and more intuition.

Now, of course, there is a problem. Working with continuous dynamics also risks unwittingly creating a "continuous fallacy." The continuous fallacy assumes implicitly that fractions of organisms can grow. But fractional organisms do not grow; they don't even exist! Yet the continuous model described above suggests they do (Figure 1.7). For example, what does it mean if at some point $m = 0.0001$? There is not one-ten-thousandth of a mutant proliferating in the flask before viruses are added. Hence, a deterministic model that assumes the growth of fractional organisms may pose problems when trying to compare results to experiments in which rare events matter. One way to address this issue would be to transform the model from a continuous framework to an entirely stochastic framework (in fact, later work did that (Lea and Coulson 1949)). That approach is one that is amenable to the use of a fully stochastic treatment as well as to computation— as realized through the homework problems recommended for this chapter. However, such analysis is mathematically far more difficult (Lea and Coulson 1949; Kessler and Levine 2013). Instead, another way to address this issue is to use a continuous model, albeit to incorporate the stochastic nature of the appearance of mutants by applying a deterministic growth model only to periods in which at least one mutant is likely to be present. That is the tack taken in the next section.

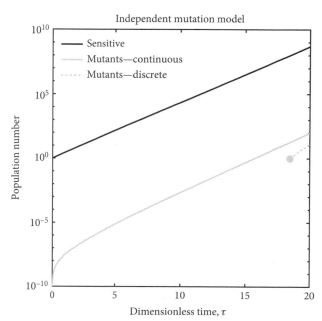

Figure 1.7: Contrasting dynamics of mutants given continuous dynamics and "discrete" approximations. The continuous model assumes that mutants are continuously generated, even at fractional levels. The discrete model assumes that mutants grow continuously only after a single first mutant appears. The variation in timing of the appearance of the first mutant underlies the "jackpot" effect described earlier. Dynamics are simulated assuming $N_0 = 1$ and $\mu = 10^{-8}$.

1.5 MODELING THE GROWTH OF (DISCRETE) MUTANTS

Understanding the implications of the independent mutation hypothesis requires building upon the continuous model while recognizing that mutants are discrete. For example, consider beginning an experiment with approximately 1000 bacteria founded by a single susceptible cell. If the mutation rate is of the order 10^{-8}, it would seem highly unlikely that one of those cells is resistant, indeed with odds on the order of 1/100,000. Yet, in the continuous model, the mutant population is immediately generated, albeit fractionally, and allowed to grow. This may lead to an overestimate of the expected size of the mutant population, and, by extension, biases in estimating the actual mutation rate (see Figure 1.7).

Instead, to overcome the continuous fallacy, it is imperative to estimate the time τ_0 where the first mutant is likely to appear. In this model, rather than assuming that fractional mutants grow, we expect there should not be mutants, i.e., $m(\tau) = 0$, for $\tau < \tau_0$ and otherwise:

$$S(\tau) = N_{\tau_0} e^{(1-\mu)(\tau-\tau_0)} \tag{1.26}$$

$$m(\tau) = N_{\tau_0} e^{(\tau-\tau_0)} \mu (\tau - \tau_0) \tag{1.27}$$

Connecting theory and experiments requires estimating the realized number of mutants at the end of the experiment, i.e., when $\tau = \tau_f$ and given a final number of cells $N_f = N_{\tau_0} e^{(\tau_f-\tau_0)}$. Hence, we can write

$$m(\tau_f) = N_f \mu (\tau_f - \tau_0) \tag{1.28}$$

Now recall also that $\tau_f - \tau_0 = \log(N_f/N_{\tau_0})$, which is a feature of the exponential growth of cells. It would seem that we are nearly there in terms of incorporating the discrete nature of mutations in the estimation procedure for μ. Altogether, the experiment yields an observed number of mutants m, a total number of bacteria N_f, as well as the duration of growth τ_f. If we only knew the approximate time at which resistant mutants appear, we would be able to also estimate μ. That time is related to N_{τ_0}. This is where the final puzzle is solved.

If one in a million offspring yielded a resistant bacteria, then one would expect to wait until there were on the order of one million bacteria before finding a mutant. In other words, the time of the first mutant appearance should satisfy $N_{\tau_0} \mu \approx 1$ or alternatively that $N_{\tau_0} \approx 1/\mu$—this is the circular gray point noted in the demonstration example in Figure 1.7. The time (on the x axis), τ, of this point corresponds to the moment when it is likely that a mutant first appears. The number of mutants (on the y axis) is set to 1. The smaller μ is, the larger the population must get before the first mutant appears, and therefore there is less time for this clonal population of mutants to grow exponentially. Substituting this time yields a new estimate of the expected number of mutants at the end of the experiment:

$$m(\tau_f) = N_f \mu \log(N_f \mu). \tag{1.29}$$

Note that if there are C multiple replicates, then the first time a mutant would appear in one of the replicates would be of the order $1/(C\mu)$ such that

$$m(\tau_f) = N_f \mu \log(C N_f \mu). \tag{1.30}$$

Eq. (1.30) can be put into practice. Given an observed average number of mutants m as well as the number of replicates C and population size N_f, it can be used to identify a unique value of μ. This equation is implicit. Nonetheless, it can be "inverted" so as to solve the problem numerically. But even if we have an estimate, this doesn't answer the deeper question: is there sufficient evidence to accept the independent mutation hypothesis and reject the hypothesis that mutations are dependent on selection?

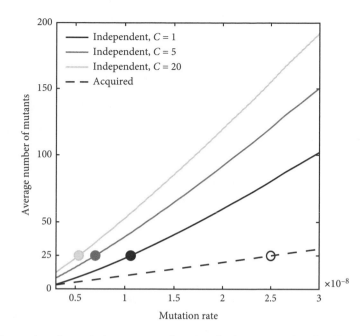

Figure 1.8: The number of expected mutants as a function of unknown mutation size. Here, Eq. (1.30) and Eq. (1.31) are used for the independent and acquired hypotheses, respectively. For the independent hypothesis, the average number of mutants increases logarithmically with C, the number of replicate cultures. Hence, given an observation, it is possible to "invert" the curves and find an estimate of μ_f or μ_a given an observation of m, and the values N_f and C. Here $N_f = 10^9$. The circles denote estimated mutation rates given an observation of 25 for the average number of mutants, solved using a nonlinear zero finding method for the independent mutation case.

Favoring one hypothesis over another requires not just alternative estimates but evidence of the incompatibility of one hypothesis to explain the observed data. Thus far, the theory presented here only utilizes the mean number of colonies to provide two alternative estimates of the mutation rate. Eq. (1.30) links data to an estimate of the mutation rate when mutations are independent of selection. Yet we already derived an alternative equation for the estimated mutation rate when mutations are dependent on selection:

$$m(\tau_f) = N_f \mu_a. \tag{1.31}$$

Figure 1.8 shows the expected number of mutants as a function of μ for three cases of C for the independent mutation hypothesis in contrast to the expected number of mutants in the acquired mutation hypothesis. The same figure also shows how the estimate of μ varies with C and the underlying mutational mechanism given the same observation. That is, if 25 resistant colonies were observed on average, then using the mean information alone would simply lead to distinct estimates of the mutation rate but would not be sufficient to distinguish between the two classes of hypotheses. One measurement, multiple estimates. Distinguishing them requires going beyond means, all the way to the variation.

1.6 VARIANCE OF MUTANTS WHEN MUTATIONS ARE INDEPENDENT OF SELECTION

How much variation is expected among resistant colonies if mutations arise spontaneously, independent of selection? We have already shown that mutants arise in different cohorts, e.g,. from time τ_0 to τ_f. Earlier cohorts may be less likely to arise, but when they do, they lead to a larger number of mutants. Later cohorts are more likely to arise, and when they do, they lead to a smaller number of mutants. Altogether, these cohorts contribute to the expected variation in outcomes across replicate experiments. For example, if there were only mutants at τ_0 and at some other point τ_1, then the total variance in the number of mutants would be

$$\text{Var}(m) = \text{Var}(m|\tau_0) + \text{Var}(m|\tau_1), \tag{1.32}$$

which is to say that expected *variances add!* There are two contributions to the variance of a cohort. First, the number of new mutants generated in a given generation is itself a Poisson random number whose expected value is $N(\tau)\mu$. However, this Poisson random number is multiplied by an exponential factor, corresponding to the proliferation of the cohort. If $x \sim \text{Poisson}(N, \mu)$, then the variance of that random variable multiplied by a constant factor is $\text{Var}(\alpha x) = \alpha^2 \text{Var}(x)$ where α is a constant. In other words, if the cohort grows by a factor of 16, its mean increases that much, but the variance (involving squared values) goes up by a factor of 256! This hints at the possibility that variation in the emergence of early mutants in a growing population before exposure to viruses could underlie the large variation in observed outcomes.

It is possible to assess the variance expected in outcomes by focusing on the case where there are only two potential times when mutants arise:

$$\text{Var}(m) = \left(e^{\tau_f - \tau_0}\right)^2 \mu N(\tau_0) + \left(e^{\tau_f - \tau_1}\right)^2 \mu N(\tau_1). \tag{1.33}$$

However, recall that the numbers of cells are themselves growing exponentially, such that $N(\tau_i) = N_f e^{-(\tau_f - \tau_0)}$, and so

$$\text{Var}(m) = \mu N_f \left[e^{\tau_f - \tau_0} + e^{\tau_f - \tau_1}\right]. \tag{1.34}$$

We can generalize this idea to any value of τ_i between τ_0 and τ_f, i.e., moving to the continuum limit, such that

$$\text{Var}(m) = \mu N_f \int_{\tau_0}^{\tau_f} d\tau \, e^{\tau_f - \tau}$$
$$= \mu N_f \left[e^{\tau_f - \tau_0} - 1\right] \tag{1.35}$$

for which we should recall that $\tau_f - \tau_0 \sim \log C\mu N_f$. Finally, we can write

$$\text{Var}(m) = \mu N_f \left(C\mu N_f - 1\right) \approx C\left(\mu N_f\right)^2. \tag{1.36}$$

This equation implies that the variance grows faster than the mean, unlike in the case of mutations that are dependent on selection.

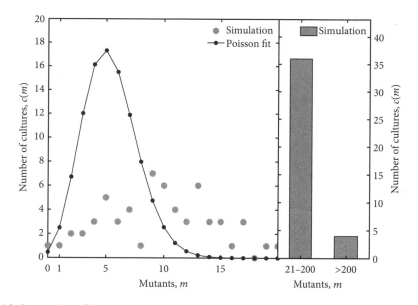

Figure 1.9: Comparison of the expected distribution of mutants assuming the acquired mutation hypothesis to the realized number of mutations given an independent mutation hypothesis. The black solid line denotes the Poisson fit assuming mutation depends on selection. The gray circles denote the results of an LD simulation, with final population size of $\sim 5.4 \times 10^8$ given $\mu = 10^{-8}$. The right panel denotes the large number of jackpots, including four cases where there are far more than 200 mutants in a single experiment out of 100 experiments. As is apparent, such jackpots are wholly unexpected given the Poisson assumption that would arise if mutations were dependent on selection.

Table 1.2: Hallmark features of the acquired mutation hypothesis and spontaneous resistance hypotheses

	Acquired	**Spontaneous continuous**	**Spontaneous discrete**
Mean	$\mu_a N_f$	$\mu N_f \log N_f$	$\mu N_f \log C\mu N_f$
Variance	$\mu_a N_f$	μN_f^2	$C\mu^2 N_f^2$
$\frac{\text{Variance}}{\text{Mean}}$	1	$\frac{N_f}{\log N_f}$	$\frac{C\mu N_f}{\log C\mu N_f}$

Note: The terms *continuous* and *discrete* refer to whether mutant cohorts are assumed to begin at $\tau = 0$ or $\tau = \tau_0$; see text for details. As is apparent, the independent case leads to variance:mean ratios far above 1.

Summarizing these findings requires a focus on the qualitative differences implied by the link between the expected variance and the mean number of resistant colonies. Table 1.2 compares and contrasts the mean, variance, and ratio of variance to mean for both hypotheses. The variance as estimated for the case of mutations independent of selection is $C\left(\mu N_f\right)^2$, whereas the mean is $\mu N_f \log C\mu N_f$ such that the ratio is

$$\frac{\text{Var}}{\text{Mean}} = \frac{C\mu N_f}{\log C\mu N_f}. \tag{1.37}$$

This ratio includes a relatively large number over its log, which should yield a ratio much larger than 1. In the case of mutations that are independent of selection, the bulk of the variation stems from the very earliest of mutant cohorts, because when they do occur, they grow exponentially, leading to jackpots and large variation. These large jackpots are incompatible with the acquired immunity hypothesis (Figure 1.9). The finding and interpretation of large variances in repeated experiments of phage lysis of bacteria has remained a salient example of the integration of quantitative reasoning of a living system given uncertainty—and such work continues to inspire. As but one example, the interested reader may want to explore recent work showing how delays between the onset of a mutation and change in phenotype impact the population-level mutant distribution (Sun et al. 2018).

It is now up to you to work computationally to help build your intuition as to whether the fluctuations observed are large enough to reject the mechanism that mutations are dependent on selection in favor of the mechanism that mutations are independent of selection. In doing so, a full stochastic framework is used in the homework problems to test the limits of both the simple models and the theoretical predictions—which come with a caveat. The scaling presented in the last column of Table 1.2 differs from the scaling found in a fully stochastic treatment (Lea and Coulson 1949; Zheng 1999). The reason is that correcting for the appearance of the first mutant applies to the sample statistics with a finite number of replicates and not necessarily to the expected mean and variance in the limit of infinite replicates. This claim is equivalent to considering the limit that $CN_0\mu \ll 1$ or equivalently that $CN_0 \ll \frac{1}{\mu}$, i.e., a mutant is unlikely to have already been present at the start of the experiment. As C gets very large then it is more and more likely that a mutant will be present at the very start of one of the C replicates—and the continuous fallacy stops being a fallacy. Problem 6 in this chapter explores how the mean and variance increase with mutation rate given sufficiently small C. Although the quantitative details differ, the fluctuations remain large.

1.7   ON (IN)DIRECT INFERENCE

This chapter has explored a key concept in modern biology. The work of LD is particularly notable for its integration of mathematical theory, physical intuition, and model-data integration as a means to understand the nature of mutation. The conceptual notion of spontaneous mutation versus acquired hereditary immunity can be seen in the schematic in Figure 1.10. As is apparent, the possibility for jackpots is enhanced when lineages (i.e., a bacterium and its descendants) all have the property of resistance. This possibility of jackpots is to be expected when mutations arise spontaneously during the growth process and are unrelated to the selection pressure. Hence, irreproducibility is a hallmark of a particular biological mechanism. The work of LD showed that mutations arose independent of selection and were not acquired as a result of interaction with a selective pressure. Their 1943 paper and its findings were cited when Luria and Delbrück received their Nobel Prize in Physiology or Medicine along with Alfred Hershey in 1969.

We now accept this paper as having established the independence of mutation from selection. It informs not just foundational work but interpretation of the emergence of the frequency and variation in cancer cells (Fidler and Kripke 1977). Yet it took a decade for the "biometric" approach of Luria and Delbrück (sensu Esther Lederberg and Joshua Lederberg) to be accepted. The acceptance was not just because of a gradual increase in

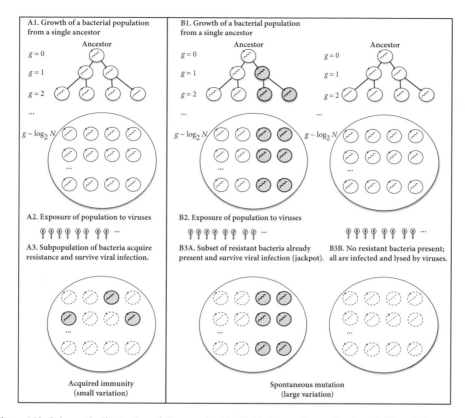

Figure 1.10: Schematic illustration of the acquired heritable immunity mechanism (left) and the spontaneous mutation mechanism (right), including differences in the number of resistant colonies—adapted from J. S. Weitz, *Quantitative Viral Ecology* (with permission) (Weitz 2015).

quantitative rigor in cellular and molecular biology. Indeed, the subject of understanding the basis for the "Luria-Delbrück" distribution continues even now (Kessler and Levine 2013). Instead, fellow researchers were eventually convinced by the dissemination of the elegant replica plating method of Esther Lederberg and Joshua Lederberg (1952), likely even more so than by beautiful mathematics (Figure 1.11; note that Esther Lederberg has been underappreciated for the scope of her contributions (see Schaechter (2014)). The idea of the replica plating method is that a bacterial lawn, likely with preexisting resistant mutants, can be transferred to multiple plates. The transfer is meant to preserve the existing spatial structure of bacteria, including bacterial mutants. These multiple replicate plates are then exposed to a phage lysate. Hence, if the position of the resistant mutants in each replicate plate were similar, that would show—visually—that the property of phage resistance was already present before the interaction with the virus. The numbered colonies in Figure 1.11 demonstrate this very point—many appear in exactly the same position in at least two plates—and are an example of how much a beautiful experiment design can offer.

Despite this chapter's singular focus on evidence building toward a conclusion that mutations are independent of selection, there is a caveat to this seminal discovery.

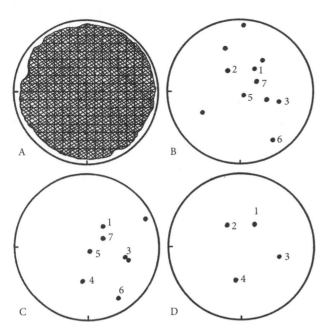

Figure 1.11: Replica plating method to demonstrate viral-resistant mutants are independent of selection. The original plate is shown in A. Colonies' cells resistant to phage T1 are numbered in replica plates B, C, and D. As noted in the original caption, the colocation of resistant colonies in the same location implies they are "derived from small clones of resistant mutants already present at corresponding sites on the plain agar plate, A." Reproduced from Figure 2 of Lederberg and Lederberg (1952).

This caveat is shaped by new research into the origins of genetic variation in microbes. To understand the caveat, it is worth considering the Gedankën experiment: what would have happened to the history of molecular biology if Luria and Delbrück had used *Streptococcus thermophilus* and its phage rather than *E. coli B* and phage T1? The *S. thermophilus* strain utilizes an acquired immune defense system known as CRISPR (clustered regularly interspaced palindromic repeats) or CRISPR-Cas (Barrangou et al. 2007; Makarova et al. 2011). Although CRISPR-Cas is known as the basis for a revolution in genome engineering and biotechnology, at its heart the CRISPR system is a de facto adaptive immune system in bacteria and archaea that enables microbes that survive an infection to become heritably resistant. These microorganisms seem, in some sense, akin to giraffes reaching for acacia leaves and passing on their longer necks to their offspring (Koonin and Wolf 2009). A strange world, but it is the one we live in.

1.8 TAKE-HOME MESSAGES

- Mutations are the generative driver of variation in the evolutionary process.
- Prior to the work of Luria and Delbrück, there was a major unanswered question: are mutations independent of or dependent on selection? Afterward, the consensus shifted: mutations are independent of selection.
- Experimental evidence using phage and bacteria showed that the number of mutational events varied significantly between experiments.
- This large variation, i.e., a lack of reproducibility, was a key hallmark of the independent mutation hypothesis, and counter to predictions of the acquired mutation hypothesis.
- Solving for the Luria and Delbrück distribution is non-trivial; nonetheless, the central concepts of proliferation and mutations among clones is readily analyzed, simulated, and compared to data.
- Although exceptions abound (including CRISPR-Cas immunity), the concept that mutations are independent of selection remains the paradigm in biology.

1.9 HOMEWORK PROBLEMS

A central goal of this book is to help readers develop practical skills to quantitatively reason about living systems given uncertainty. However, each chapter is only part of this process (just like listening to lectures, paper discussions, and in-class work help solidify understanding). Moreover, for many readers, the mathematical and biological insights provide a partial guide. If seeing is believing, then coding and simulation are a central path to build intuition and insight on the themes developed in this and subsequent chapters. The following homework operates in that spirit and is best approached after working through the exercises in the accompanying computational lab guide. The laboratory guides—in MATLAB, Python, and R—provide insights into how to:

- Sample from random distributions
- Utilize the properties of uniform random distributions to generate random distributions that are nonuniform, e.g., exponential distribution
- Compare and contrast the Poisson with the binomial distribution
- Develop stochastic simulations of growing populations

The homework helps leverage this "toolkit" in order to build intuition on the core ideas of Luria and Delbrück's seminal paper.

The overall objective of these problems is to reproduce the "irreproducibility" of the number of resistant mutants, as observed by LD, and to begin to reach tentative conclusions regarding the confidence on estimated mutation rates and mechanisms in inferring the basis of mutation from resistant colony data. The problems utilize a common set of assumptions initially. That is, in these problems, consider an experiment with C cultures, each of which has N sensitive cells. Every time a cell divides there is a probability μ that one, and only one, of the daughter cells mutates to a resistant form. We assume that the offspring of resistant cells are also resistant, i.e., there are no "back-mutations." Good luck. And remember, you can do it!

PROBLEM 1. Simulating the Luria-Delbrück Experiment over One Generation

Write a program to simulate just one generation of the LD experiment—stochastically. Simulate $C = 500$ cultures, each of which has $N = 1000$ cells and $\mu = 10^{-3}$, i.e., a very high mutation rate. What is the distribution of resistant mutants that you observe across all the cultures? Are they similar or dissimilar to each other? Specify your measurement of $c(m)$, i.e., the number of cultures with m resistant mutants. Is this distribution well fit by a Poisson distribution? If so, what is the best fit shape parameter of the Poisson density function and how does that relate to the microscopic value of the mutation you used to generate the output? Finally, to what extent are the fluctuations "large" or "small"?

PROBLEM 2. Simulating the Luria-Delbrück Experiment Forward One Generation at a Time

Extend the program in Problem 1 by setting $C = 1000$, $N = 400$, and $\mu = 10^{-7}$, while having the population grow over $g = 15$ generations. What choice did you make with respect to modeling the population? If you decided to model each individual cell in each individual culture, explain your rationale. Next, develop a model that represents the emergence of new resistant cells in each generation in each culture en masse (that is, all at once). (Hint: Think about how prudent use of the Poisson random generation function could help.) The objective here is to develop a working simulation that is both accurate and efficient—in doing so, compare the speed when you use Poisson versus binomial random number generating functions.

PROBLEM 3. Characterizing the Mutant Distribution

Using the simulation in Problem 2, report and describe the shape of $c(m)$—the number of cultures with m resistant mutants. Describe and characterize the shape of this distribution and contrast it to that expected under the acquired hypothesis. What is the mean and variance? Are fluctuations large or small? Finally, to what extent are there jackpot cultures? Can you define a principled way to identify those jackpots? To what extent do fluctuations become small when jackpots are excluded?

PROBLEM 4. Why Are There Jackpots?

Characterize the "age" of each mutant using your LD simulation. That is, if you have not already done so, keep track of the first appearance of each mutant and characterize the relative importance of mutants of different "ages" to the total number at the end. Do early/late mutants contribute disproportionately to the total number? Do early/late mutants more strongly influence variation?

PROBLEM 5. Moving Backward like Luria and Delbrück from Observations to Estimates

Take the results from the first 100 of your experiments in Problem 2, that is, the number of resistant mutants, $m_1, m_2, \ldots, m_{100}$. Treat these 100 numbers as data. Now write a program that leverages results of new simulations to infer the most likely value of μ, the mutation rate, and (if possible) a 95% confidence interval for it—in doing so, pretend that you do not know what μ is in advance. Consider using two pieces of evidence: (i) the number of cultures with no resistant mutants; (ii) the mean number of resistant mutants. Finally, ask: is your data also consistent with the acquired resistance hypothesis? Why or why not?

PROBLEM 6. Scaling of the Mean and Variance of Mutants with μ

Using a stochastic simulation of the LD problem (ideally, using a Poisson approximation to the mutant generation at each step), simulate the expansion of a bacterial population with zero mutants at $t = 0$ and $N_0 = 1000$ over 18 generations so that the final population has more than two billion cells. Upon division of a wild-type cell, there is a μ probability that one of the daughter cells will be a mutant—both mutant and wild-type cells divide each generation. Modulate the mutation rate from $\mu = 10^{-8}$ to $\mu = 10^{-5}$ and use a value of $C = 20$. If your model works, it should look like the following:

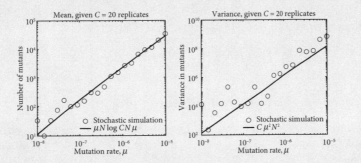

Using your model, explore lower and higher values of C, e.g., $C = 10$ and $C = 40$. Compare and contrast your findings with the scaling in Table 1.2, accounting for the emergence of mutants after the start of the experiment. Using these same settings, compare the LD estimates, assuming mutants grow deterministically from the start. In discussing your results, provide a rationale for gaps between theory and simulation.

1.10 TECHNICAL APPENDIX

One of the key features of a compelling theory and model is that it is often easy to understand in retrospect but hard to complete in the absence of the solution. Such theories are like puzzles and any good solver has a repertoire of techniques. This appendix reviews basic techniques and provides additional information to help fill in the gaps in the main text. These reviews may be of particular value to readers with strong biological training who may not have seen these mathematical methods used in concert with the analysis of biological problems. Indeed, why remember how to do a Taylor expansion or calculate the binomial coefficient if it is never called upon in your research?

Factorials The term $N!$ denotes a factorial, or $N \times (N-1) \times (N-2) \times \cdots \times 2 \times 1$.

Binomial coefficients The binomial coefficient counts the number of ways to permute the location of k events of N trials. Formally, the coefficient can be written as $\binom{N}{k}$ such that

$$\binom{N}{k} \equiv \frac{N!}{(N-k)!k!}. \tag{1.38}$$

For example, consider the case where $N = 5$ and $k = 2$, for which there are $5! / (3! \times 2!) = 10$ distinct combinations of arranging two positive events out of 5 trials. The binomial coefficient can be understood as follows. First, consider the way in which k events can be selected out of N trials in a particular sequence. The first event has N options, the second $N-1$, the third $N-2$, and so on, so that the total number of events is $N(N-1)(N-2)\cdots(N-(k-1))$. However, the resulting placement overcounts the number of unique configurations; e.g., for $N = 5$ and $k = 2$, it is possible to first select trial 1 and then trial 2 or to first select trial 2 and then trial 1. The degree of overcounting can be calculated by counting the combinations of the identity of the positive events; i.e., the first event has k options, the second has $k - 1$, and so on. That is to say, there are $k!$ ways to permute the identity of the positive events. Hence, the total number of ways to permute the location of k positive events in N trials is

$$\frac{N(N-1)(N-2)\ldots(N-(k-1))}{k!} = \frac{N!}{(N-k)!k!}, \tag{1.39}$$

precisely the binomial coefficient listed above.

From the binomial to Poisson Consider the probability that k events take place, each with probability μ, out of N trials, i.e,

$$p(k|N,\mu) = \frac{N!}{(N-k)!k!}\mu_a^k(1-\mu_a)^{(N-k)}. \tag{1.40}$$

In the Luria experiments, there are many bacteria ($N \gg 1$) and the probability of mutation is rare, $\mu \ll 1$. Denote the average number of events as $\hat{m} = N\mu$. In the limit of large bacterial populations and rare acquisition of virus resistance, then

$$\begin{aligned}
p(k|N,\mu) &= \frac{N(N-1)(N-2)\cdots(N-(k-1))}{k!}\mu^k(1-\mu)^{(N-k)} \\
&= \frac{N^k\mu^k(1-1/N)(1-2/N)\cdots(1-(k-1)/N)}{k!}(1-\mu)^{(N-k)} \\
&\approx \frac{\hat{m}^k}{k!}\left(1-\left(\sum_{i=0}^{k-1}i\right)/N\right)e^{-\hat{m}}(1+\mu k) \\
&\approx \frac{\hat{m}^k e^{-\hat{m}}}{k!}
\end{aligned} \tag{1.41}$$

Note that this approximation utilizes a Taylor expansion of exponentials, i.e., $e^{-x} \approx 1 - x$ and $e^x \approx 1 + x$ for small values of x, and also ignores small terms, e.g., those in which a number is divided by N or multiplied by a factor of μ. The resulting Eq. (1.41) corresponds to the Poisson distribution with an expected value of $\hat{m}$ for the number of successful events given N trials.

Taylor expansion A Taylor expansion represents a formal mechanism to approximate the values of arbitrary functions near a reference point using a sequence of polynomials. It is easiest to illustrate the concept of Taylor expansions for functions of one variable, like $f(x)$, but the concepts can be extended to multiple dimensions, e.g., $g(x,y)$. In the one-dimensional case, the value of $f(x)$ near the reference point x_0 can be approximated as

$$f(x) \approx f(x_0) + \frac{df}{dx}\bigg|_{x=x_0}(x-x_0) + \frac{d^2f}{dx^2}\bigg|_{x=x_0}(x-x_0)^2 + \cdots. \tag{1.42}$$

In essence, the Taylor expansion assumes that function values near a reference point can be approximated by starting at the function value of the reference and then fitting a tangent to it. The slope of the tangent defines the "rise" of the function, i.e., df/dx evaluated at $x = x_0$, and the "run" is the difference between the point of interest and the reference, i.e., $(x - x_0)$. The linear approximation can be extended using the curvature—the second derivative—multiplied by the difference squared, i.e., $(x - x_0)^2$. As a result, the Taylor expansion approximates arbitrary functions as a combination of polynomials. In practice, we will rarely utilize Taylor expansions other than to first (i.e., linear) or second (i.e., quadratic) order. For example, e^x for values of x near 0 can be represented as $1 + x + x^2/2 + \cdots$. Similarly, $\log(1 + x)$ for values of x near 0 can be represented as $x - x^2/2 + \cdots$. Note that retaining higher-order terms can sometimes prove essential (as will be shown in later chapters).

Solving the exponential growth equation Consider the dynamical system

$$\frac{dx}{dt} = rx. \tag{1.43}$$

This exponential growth equation can be solved, first by dividing both sides by x and multiplying both sides by dt, and then by integrating:

$$\int \frac{dx}{x} = \int r\, dt$$

such that

$$\log x = rt + C \tag{1.44}$$

where C is a constant of integration. Exponentiating both sides leads to the relationship $x(t) = \tilde{C}e^{rt}$ where $\tilde{C} = e^C$. Given the initial conditions, $x(t = 0) = x_0$, then one can write the complete solutions $x(t) = x_0 e^{rt}$.

Maximum likelihood of the Poisson distribution Consider the Poisson distribution, with expected mean $\lambda = \mu N$, such that

$$p(m|N, \lambda) = \frac{\lambda^m e^{-\lambda}}{m!}. \tag{1.45}$$

The most likely value of λ given an observation of m corresponds to a *maximum* in p. To find such a maximum, first take a derivative with respect to λ, set the derivative to zero, and then double-check that the second derivative at this point is negative (which it will be). The derivative of Eq. (1.45) yields

$$\frac{m\lambda^{m-1}e^{-\lambda}}{m!} = \frac{\lambda^m e^{-\lambda}}{m!}. \tag{1.46}$$

Canceling factors on both sides yields

$$\frac{m}{\lambda} = 1. \tag{1.47}$$

Recall that $\lambda = \mu N$, so that if $m = m_{obs}$, then

$$\hat{\mu} = \frac{m_{obs}}{N} \tag{1.48}$$

where $\hat{\mu}$ is the most likely estimate of the mutation rate. This maximum likelihood estimate of the mutation rate is the ratio of the observed number of mutants to the total number of cells. Technically, one should double-check that this value is the maximum likelihood—it is.

Law of total probability Consider the joint probability of two variables, $p(x, y)$. This joint probability can be written in two ways: $p(x|y)p(y)$ or $p(y|x)p(x)$. The equivalence of these two forms is the basis for Bayes' rule.

From updates to derivatives Consider a population x, which experiences changes due to interactions with other populations and environmental factors. Denote the rate of change as $f(x)$ such that the rate of change depends on parameters but also on the population level x. How can we build a model that describes the trajectory of the population over time? To do so, consider changes that occur over some interval Δt. In that case,

$$\overbrace{x(t + \Delta t)}^{\text{next value}} = \overbrace{x(t)}^{\text{current value}} + \overbrace{f(x) \times \Delta t}^{\text{increment}}. \tag{1.49}$$

Yet the value $x(t + \Delta t)$ can be thought of as a function, the value of x changing over time. A Taylor expansion can be used to approximate the future value of this function given the current value, i.e.,

$$x(t + \Delta t) \approx x(t) + \left(\frac{dx}{dt}\right)\Delta t. \tag{1.50}$$

Substituting in this expansion to the update rule yields

$$x(t) + \left(\frac{dx}{dt}\right)\Delta t = x(t) + f(x)\Delta t, \tag{1.51}$$

which after canceling terms yields

$$\frac{dx}{dt} = f(x). \tag{1.52}$$

Hence, the continuous rate of change of a population can be expressed in terms of a series of small increments. The usual way to obtain the dynamical systems representation is to note that the derivative is the limit of the finite differences of x, i.e.,

$$\frac{dx}{dt} = \lim_{\Delta t \to 0} \frac{x(t + \Delta t) - x(t)}{\Delta t}. \tag{1.53}$$

Nonetheless, it has been my experience in teaching that biologists prefer to begin with the concept of an update rule and reduce to the continuous limit from there, rather than formally beginning the other way around.

Chapter Two

Bistability of Genetic Circuits

2.1 MORE IS DIFFERENT

The inner workings of cells represent the central occupation of the biological sciences. This is not meant to valorize nor indict that state of affairs. It is what it is. Moreover, cellular life is arguably the building block of all life, or as some might call it "the basic unit" of life. But in exploring cells, there is a temptation: to continue to reduce the cell into constituent parts until one reaches molecules. There are many molecules inside a cell: nucleotides, proteins, metabolites, lipids, sugars, and water. And why stop with molecules? Why not go all the way to atoms or even quarks? Although the laws of chemistry and biology must be consistent with those of the fundamental laws of physics, this does not mean that we will glean insights from an endless reduction ad absurdum. Instead, we should heed the caution in P. W. Anderson's framing of an alternative to reductionism in science: "more is different" (Anderson 1972). In the present context, asking how cells work may be limited to understanding chemical reactions inside cells. Instead, our goal is broader, more ambitious: what enables cells to have so many different forms and functions and self-organize into tissue as part of complex, multicellular organisms? To begin to answer this question requires that we start at the appropriate scale and with a focus on the appropriate process—that of gene regulation.

Indeed, molecular and cellular biology does have its own organizing "law," what is called the central dogma, i.e., that information encoded in DNA is transcribed into a message, mRNA, which is then translated into proteins that carry out the basic functions of cellular life. There are many such functions (Figure 2.1). Cells can divide, grow, search for food, deal with stresses and shocks, enter periods of stasis, emerge from quiescence, incorporate and release small molecules, and survive (and even "learn" from) interactions with viruses and other mobile genetic elements. Indeed, taking just one of this long list, or an even longer list, of functions in the context of a particular microbe or cell type in a multicellular organism often constitutes the basis for a hopefully interesting career. But the longer the list becomes, the more evident it should be that these functions represent "potential" functions of a cell. A cell does not do each of these functions all at once. The strange cases of teratomas provide the exception that prove this rule. Teratomas are typically benign; they are tumor cells with multiple kinds of differentiated tissues organized together, e.g., a tooth

Motility

Switching between states

Sensing

Differentiation

Division

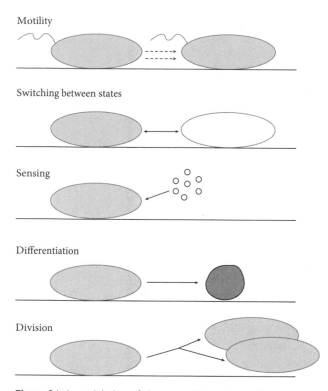

Figure 2.1: A partial view of the many things that cells can do, but not all at the same time—including motility, switching between states, sensing extracellular signals, differentiation, and division from one to two cells.

mixed with brain matter located in an unrelated organ. This sounds somewhat terrifying, and also provides context for the incredible coordination that must take place to ensure that differentiation unfolds during development, both in time and space. Yet the problems of development are, for now, a few steps ahead.

Here we approach the problem of how cells do some, but not all, things at once through the lens of bacterial gene regulation. In other words, the central question in this chapter is, "How can a cell have different, stable gene expression patterns?" The chapter follows the path laid down by many before to explain ways in which certain genes can turn on given the right signals or circumstances. Yet the other, perhaps more important, point to be made is that the regulation of individual components can lead to entirely new emergent phenomena not possessed by the components themselves. A key emergent phenomenon is that of bistability, the notion that cells, given the same environmental cues, can maintain distinct types of cellular states. Having such a property could be quite handy, not only to avoid teratomas but also to develop from one cell into many, and even for populations of cells to subdivide tasks or perform a form of evolutionary bet hedging.

Hence, this chapter provides context to understand core concepts of gene regulation, while moving quickly through many important aspects of molecular and cellular biology. Hopefully, the intuition you build will inspire you to go back and fill in the gaps. In doing so, you will be motivated by a relatively simple question: how can a cell exhibit bistability, with distinct phenotypes that have a memory of prior stimuli? These phenotypes include dormancy or growth, susceptibility or tolerance of antibiotics, and differential expression states (e.g., of toxins). In each case, the term *bistability* refers to the fact that individual cells possess a single, characteristic phenotype even if they have the same genotype. These phenotypes—the *states*—are durable over time, but cells can switch between states, sometimes at random and sometimes due to sensing and response from cues in the environment. Indeed, bistability extends even to viruses. For example, bacteriophage lambda can infect bacteria of the species *E. coli* and, in some instances, the viral genome is integrated into that of the host. This integration is relatively stable, in part, because of a gene regulatory circuit that represses the genes in the phage genome that would lead to excision from the host and initiation of lysis. Yet such systems—however interesting they might be (Ptashne 2004)—do not necessarily contain a minimal set of ingredients to understand the dynamical origins of bistability.

For that, we turn to the breakthrough work on a genetic toggle switch by Timothy Gardner and colleagues in the lab of Jim Collins (Gardner et al. 2000). As this chapter explains, this genetic toggle switch enables the model-predicted control of a bacterium's phenotype resulting in alternative stable states. This example is one of a number of demonstrations of controllable bistability driven by regulatory feedback mechanisms—whether in bacteria (Tiwari et al. 2010; Chen and Arkin 2012) or yeast (Nevozhay et al. 2012). We also adopt many of the approaches to connecting mechanisms, models, and observations pioneered by Uri Alon and colleagues (Alon 2006).

The idea that one could rationally design a modular component to be inserted into populations of cells seems obvious in retrospect. Many students have either participated in or at least heard about iGEM, the international genetically engineered machines competition. iGEM has its roots in a 2003 course at MIT (Shetty et al. 2008; Canton et al. 2008; Smolke 2009). The basic premise is that it should be possible for undergraduates—initially those at MIT, but soon worldwide—to design, build, and implement "devices" made of modular building blocks to enable new or altered cellular functions. So easy anyone could do it. . . even a high school student. This vision is now a reality, at least in part. In practice, there are certain labs and companies whose expertise relies on combining a rational approach to developing parts that work and then connecting those parts to achieve a target outcome. As we will see, rationally designing a bistable genetic circuit requires more than just understanding the parts list; it requires understanding feedback of a nonlinear dynamical system. How then does one design a genetic toggle switch based on the principle that two genes can mutually inhibit each other? And even if such ideas are possible in theory, will they lead to the phenomenon that a stimulus that turns one of the genes on (and the other off) can also lead to cellular memory, i.e., the "On" gene stays on even when the stimulus is removed? Explaining the dynamics shown in Figure 2.2 represents the motivating challenge in this chapter. And, in focusing on a synthetic switch, this chapter also enables us to move from the lessons of P. W. Anderson to those of Richard Feynman, who famously cautioned: "What I cannot create, I do not understand."

2.2 MOLECULAR CAST AND SCENE

Cells are complex. They involve many components, and an assessment of their dynamics requires that we begin with a parts list. But a cell cannot be understood simply by counting these parts. Interactions matter, and feedback between components can give rise to new properties of the system that are not embedded in the "local" interactions. This concept of emergence, or "more is different," permeates much of this book. Moreover, given the realized complexity of cells, we will focus on a few components of the cellular landscape that form the essential components of a genetic toggle switch. Now to the play . . . beginning with the cast (Figure 2.3):

- *Gene*: Sequence of nucleotides that encodes a protein
- *Promoter*: Sequence of nucleotides upstream of a gene that allows gene transcription initiation and defines, in part, regulatory relationships
- *Operator*: Short sequence in the genome, typically adjacent to a promoter, where regulatory molecules can bind, thereby modulating gene expression

- *Protein*: Macromolecule that enables the structure, function, and regulation of cells
- *RNA polymerase*: Enzyme that transcribes mRNA
- *Transcription factor* (TF): Proteins that bind to promoters and/or operators and affect transcription initiation of target genes
- *Activators*: TFs that increase transcription of a target gene
- *Repressors*: TFs that decrease transcription of a target gene
- *Inducers* (signals): Small molecules that modify binding ability of regulatory proteins

This abbreviated summary of cellular roles helps us think about variations, exceptions, and ways in which the intrinsic dynamics govern evolved behavior or can be influenced or controlled for some engineered outcome. To do so, we focus on bacterial gene expression. For those familiar with the topic, the following sections should function like a Shakespearean play—sure, you know what happens in the end, but it is good enough to sit, watch,

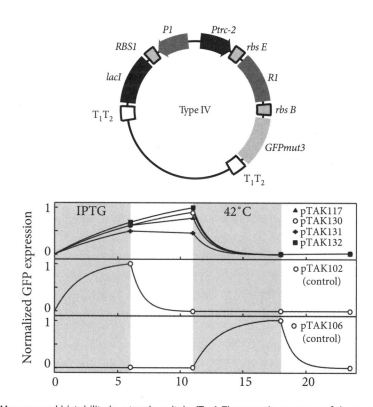

Figure 2.2: Memory and bistability in a toggle switch. (Top) The genetic structure of the toggle switch on a plasmid, including mutually repressing genes *lacI* and *R1*, which bind and inhibit expression at the promoters *Ptrc-2* and *P1*, respectively—full details of the switch are described later in this chapter. (Bottom) Demonstration of memory and bistability. The key experimental observation is that an inducer (isopropyl βD-1-thiogalactopyranoside, or IPTG) was added and then removed, yet the behavior of the genetic circuit continued unabated (as seen in the increased in the normalized GFP expression). The expression decreased when a new temperature perturbation was made, denoted by the shaded region (42°). This holds true for each of the strains that contains the synthetic toggle switch design (top row) but either of the strains that contains only one of the two toggle switch components (middle and bottom rows). This chapter explains how and why these simple experiments reveal both memory and bistability as characteristics of genetic circuits and not just the genes themselves. Figures reproduced from Gardner et al. (2000).

and listen. For those relatively new to the life sciences, this brief refresher should help acclimate you to new surroundings. Either way, take a moment to admire a few billion years of evolution's handiwork.

For bacteria, genes are (largely) continuous sequences of DNA, often about 1000 base pairs long. These sequences have a start and stop. This information, stored in DNA, is transcribed into a messenger RNA by an enzyme, RNA polymerase. The process is termed *transcription*. For physicists, mathe-

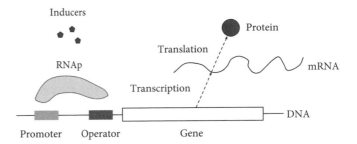

Figure 2.3: The intracellular cast: a preamble to understanding the basics of gene expression dynamics and transcriptional networks (layout based on the design of Uri Alon (2006)).

maticians, and others not versed in the life sciences, it may sometimes seem that terms in biology are invoked out a sense of whim, whimsy, or perhaps even negligence. (And, before casting stones, it may be worthwhile to consider the logic underlying the fanciful description of quarks—though doing so seemed to have been prudent for selling the public on the concepts.) Yet the term *transcription* is apt. To transcribe is to copy, faithfully. Here DNA, and its molecular alphabet of As, Ts, Cs, and Gs (i.e., adenines, thymines, cytosines and guanines, respectively), is copied into RNA, and its molecular alphabet of As, Us, Cs, and Gs. The swap of T for U reflects the chemical change of the molecule thymine (T) for uracil (U). In essence, DNA and RNA are written in the same language. The message encoded in RNA is then "translated" into protein at the ribosome. Again, the term is apt. Proteins comprise a linear chain of amino acids. There are 20 amino acids. Information is translated from the 4-letter alphabet of DNA/RNA into the 20-letter alphabet of amino acids. Three DNA or RNA letters encode one amino acid. Given the $64 = 4^3$ possibilities of combining three DNA/RNA letters, it is evident that multiple "codons" encode for the same amino acid. In addition, the start and stop of such amino acids are signified by distinct "start" and "stop" codons—a different combination of three DNA/RNA letters. This constitutes the central dogma, that information stored in DNA is transmitted via an RNA message into a protein, which enables cellular function.

But it turns out that sequence is not enough. Instead, for 70 or more years, the science of gene regulation has laid out principles and mechanisms by which what happens at one gene affects the behavior of another gene, and possibly many others. These principles are dynamic, and they are what we turn to next.

2.3   THE FIRST INGREDIENT: REGULATION OF A TARGET GENE

Consider a gene X that regulates a gene Y. This phrase—X regulates Y—means that a protein produced via the translation of the mRNA transcribed from gene X is a transcription factor, and that this transcription factor binds to a site (typically) upstream of gene Y. This binding modulates the RNA polymerase activity of gene Y, modulating the transcription initiation of the gene. That's a mouthful. Instead, let's stick with the phrase "gene X regulates gene Y." The gene X is called an *activator* if its binding increases the gene expression of Y and is called a *repressor* if its binding decreases the gene expression of Y (see schematics in

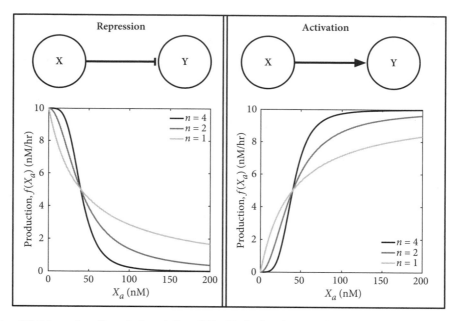

Figure 2.4: Schematics of targeted regulation of Y by X, whether via repression or activation. In both examples, X is a transcription factor that modulates the expression of another gene. The functional form can vary, as seen in the graphs, in which $\beta = 10$ nM/hr, $K = 40$ nM and $n = 1, 2,$ and 4. The functions for the activator and repressor are found in Eq. (2.2) and Eq. (2.3), respectively.

Figure 2.4). For now, we will focus on the concentration of proteins and ignore many of the other processes. These processes range from very fast (like binding of signals to transcription factors that can transform their activity and the binding of active transcription factors to DNA) to very slow (like the global evolution of transcriptional modification). For reference, Box 2.1 provides a sense of scale typical of bacterial gene regulation when division rates are on the order of 30 minutes–1 hour (for more, see Alon 2006).

BOX 2.1. Time Scales in the Regulation and Evolution of Bacteria

Bacteria are diverse. Hence, any discussion of time scales must be treated with some caution; what is true for *E. coli* growing in a rich medium in the laboratory need not apply precisely to the life of a marine cyanobacterium growing in the surface ocean with scarce nutrients or even to a soil microbe that faces severe variation in environmental conditions conducive to growth versus dormancy. Nonetheless, the following provide some estimates of time scales of fast, slow, and very slow processes for bacteria about the size of *E. coli* growing in a rich medium.

Fast
- Binding of a signal to a transcription factor (1 msec)
- Binding of active transcription factor to DNA (1 sec–1 min)

Slow
- Transcription and translation of a gene (5 min)
- Turnover (50% change) of a protein (1 hr)

Slower still
- Evolutionary modification of promoter (w/fixation) (days ???)
- Evolutionary modification of TFs and genes (w/fixation) (days ???)
- Global evolution of transcriptional regulation (Myr ???)

Of note, the binding of an individual protein to DNA will be slower, but time between binding events decreases as TF concentration increases.

The basic processes by which a gene X regulates a target gene Y can be viewed as a dynamical system. A dynamical system is one in which the state of the system, e.g., the intracellular concentration of proteins, changes with time as a result of feedback and interactions with other components. A simplified form of such a dynamical system can be written as

$$\frac{dy}{dt} = \overbrace{f(x)}^{\text{production}} - \overbrace{\alpha y}^{\text{degradation}} \tag{2.1}$$

where x and y denote the concentration of regulatory and target proteins in the cell. In doing so, the sequential processes of transcription and translation of gene Y are simplified into a net "production" rate. The production term depends on binding of X proteins to operator sites adjacent to gene Y. It also depends on the degree of cooperativity of binding; e.g., if X is an activator, then

$$f(x) = \frac{\beta x^n}{K^n + x^n}, \tag{2.2}$$

whereas if X is a repressor, then

$$f(x) = \frac{\beta}{1 + x^n/K^n}, \tag{2.3}$$

with examples shown in Figure 2.4. These "Hill functions" (discovered by Archibald V. Hill) are often invoked in models of living systems—but where do they come from?

The choice of saturating and declining response functions is biophysically motivated. Sigmoidal activation and repression functions emerge naturally from "cooperative" binding of transcription factors to promoters, where the exponent n denotes the effective degree of cooperativity in the system. As n increases, the system responds ever more sharply to changes in x below and above the transition threshold K (a molecular concentration). In biological terms, one can think of n as indicating the effective polymer state of the transcription factor, e.g., $n = 1$ corresponds to monomers, $n = 2$ to dimers, $n = 4$ to tetramers, and so on. Hence, as n increases, the change from a gene that is on to one that is off (and vice versa) becomes ever more stark, and starts to resemble a Boolean logic gate—either on or off. The detailed derivation of how to go from the molecular cast of players to these sigmoidal functions is described in the technical appendix. There is one other note required before approaching the question of how to analyze a dynamical system comprising one gene regulating another gene. Typically, the activity of proteins is controlled by signals, e.g., IPTG and other inducers. These small molecules can bind to the protein changing their activity, enabling them or disabling them from binding to a promoter region and regulating a target gene. The subsequent analysis presumes that the signal is present, but changes in signal presence can also be denoted by modulating the effective level of x. How do we analyze such a mathematical model of a dynamical system of the kind shown in Eq. (2.1)? For those with experience in the study of dynamical systems, this may appear simple; nonetheless, the models build in complexity in this chapter. As such, it's important to have a good plan. Here is a good plan:

First, find the steady states. In this chapter, we focus on steady states that denote the combination of concentration (here y^*) in which there is no further change in concentrations (here $dy/dt = 0$).

Then determine the stability of steady states. The stability of a steady state can be analyzed by evaluating the response of the system given small changes in its concentrations. This is technically a form of "local" stability analysis.

Finally, analyze the global dynamics of the system. In some cases, the response of the system to small perturbations is sufficient to understand the dynamics of the system given large perturbations. In other cases, such local stability analysis is the first step in a systematic examination of system behavior.

The following sections unpack each of these concepts, and the interested reader is encouraged to refer to the technical appendix and even full-length treatments of dynamical systems (Strogatz 1994) or introductions to models of population biology (Hastings 1997) for more details.

2.3.1   Finding steady states

Consider a cell with a certain amount of a transcription factor but with no inducer. Hence, the active concentration begins at $x = 0$. Adding an inducer rapidly changes the feedback properties of the transcription factor, so that the protein can turn on gene Y. This will then change the concentration of the proteins of Y. Eventually, in theory at least, the concentration of y will reach a "steady state." Alternatively, one can imagine preparing an ensemble of cells, each with different fixed levels of y, and then adding an inducer and checking which, if any, remain the same after activation, i.e., time invariant. In the present context, the term *steady state* denotes the fact that there is a particular value $y = y^*$ such that the concentration does not change. In other words, $\frac{dy}{dt} = 0$. In general, there can be more complex kinds of steady states, including oscillatory steady states. But the present kind, what one may call an equilibrium, will suffice for now. (The use of the term *equilibrium* reflects a convention from the study of dynamical systems such that when $y(t) = y^*$, then $y(t)$ will remain invariant for all times (Strogatz 1994).) The status of activation will become important later on in the discussion of the toggle switch.

For the activator, $\frac{dy}{dt} = 0$ of Eq. (2.1) occurs when

$$y^* = \frac{\beta x^n}{\alpha \left(K^n + x^n \right)}. \tag{2.4}$$

For example, if $\beta = 10$ nM/hr, $\alpha = 1$ hr^{-1}, $K = 40$ nM, and $n = 4$, then $y^* = 0.58$ and 8.4 when $x = 20$ nM and 60 nM, respectively. Hence, there is an approximately 14-fold increase in output given a 3-fold change in input (Figure 2.5). Similarly, for the repressor, $\frac{dy}{dt} = 0$ when

$$y^* = \frac{\beta}{\alpha\left(1 + x^n/K^n\right)}. \qquad (2.5)$$

In this case, if $\beta = 10$ nM/hr, $\alpha = 1$ hr^{-1}, $K = 40$ nM, and $n = 4$, then $y^* = 9.4$ and 1.6 when $x = 20$ nM and 60 nM, respectively. Hence, there is an approximately 6-fold decrease in output given a 3-fold change in input. These nonlinear input-output relationships are an example of one modest consequence of cooperative binding, e.g., by dimers or tetramers rather than by monomers.

2.3.2 Stability of steady states

The identification of steady states is only part of the puzzle. Imagine in the case of the activation of Y by X that the system is close to but not quite at y^*. If so, will the deviation, $|\epsilon_y| \equiv |y(t) - y^*|$ increase or decrease with time? This "local" change in behavior of the dynamical system defines its local stability. For most applications, equilibria can be classified as either unstable or stable. Occasionally, they will be marginally (or neutrally) stable. An unstable fixed point is one in which the deviation grows in time. A stable fixed point is one in which the deviation diminishes in time. A marginal/neutrally stable fixed point is one in which the deviation neither grows nor diminishes.

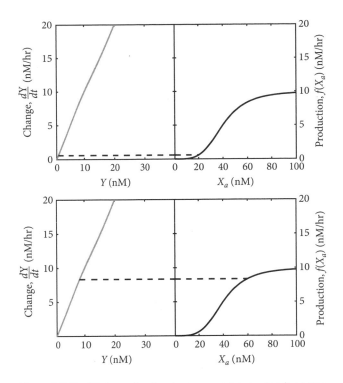

Figure 2.5: Equilibrium of a simple gene regulatory circuit occurs when degradation balances production. In both panels, degradation (left) is compared to production (right). The degradation depends only on y, whereas production depends on x. The two situations contrast different values of input transcription factor, x, from 20 nM (top) to 60 nM (bottom). The resulting output in y^* increases more than 14-fold (from 0.58 nM to 8.4 nM), a strongly *nonlinear* response given the 3-fold change in input transcription factor levels.

There are multiple ways to characterize local stability, particularly for a one-dimensional system. In this case, given that X is constant, then $f(X)$ is also a constant. Hence, irrespective of whether X is an activator or a repressor, there will be more production than degradation when $y < y^*$ and more degradation than production when $y > y^*$. As a consequence, y will converge to y^* over time, i.e., the equilibrium is stable.

Another way to assess the stability of a dynamical system near its fixed point is to "linearize" the system in terms of its dynamic variables, in this case y. By linearizing a system, what was previously an intractable problem can be reduced to a linear differential equation that can be solved completely (albeit insofar as the dynamics remain close to the fixed point). This method is powerful, despite its apparent limitations in scope. The technical appendix to this chapter reviews the procedure in detail for systems with one or two dynamic variables. Similar concepts underlie systems with an arbitrary number of variables. The present system is already "linear" in y, which means that such an approach is better reserved for a different type of model. At present, given the monotonicity of either the activation or inhibition case, there can only be one equilibrium when X regulates Y and this equilibrium is stable. To see why, consider that x is fixed, i.e., it is controlled separately. In that case,

$$\frac{dy}{dt} = C(x) - \alpha y \tag{2.6}$$

such that $y^* = C(x)/\alpha$. Denote $u = y - y^*$ as the deviation of the concentration of Y away from its equilibrium value such that $y(t) = y^* + u(t)$. As a result, we can rewrite the dynamics of the deviations, i.e.,

$$\frac{du}{dt} = C(x) - \alpha y^* - \alpha u. \tag{2.7}$$

However, because $C(x) - \alpha y^* = 0$ at steady state, this implies that the dynamics of the deviations are simply

$$\frac{du}{dt} = -\alpha u. \tag{2.8}$$

This equation implies that small deviations exponentially decay back toward equilibrium with a time scale of $1/\alpha$. Analyzing the full dynamics of the system involves a bit more work.

2.3.3 Dynamics of a simple gene regulatory system

Thus far we have focused on the limited dynamics near a fixed point, but for a single gene regulatory system it is possible to explore further. The prior section provided evidence that a system initialized near $y = y^*$ will approach y^*. In fact, the claim is more general. There is only a single value y^* where the system is at equilibrium, and so the dynamics will always approach the same cellular state irrespective of the initial prepared state, $y = y_0$. But even if the system eventually reaches this equilibrium, there is still a new question to ask: what is the characteristic time scale at which it will approach this equilibrium? In this case, the question can be answered "by hand," e.g., by rewriting this equation as

$$\frac{dy}{C(x) - \alpha y} = dt \tag{2.9}$$

and then integrating by parts (details of which are found in the technical appendix). If the cell is initially prepared without protein (i.e., the signal is off prior to being induced), then the solution is

$$y(t) = \frac{C(x)}{\alpha} \left(1 - e^{-\alpha t}\right) \tag{2.10}$$

where $C(x)$ is a nonlinear function of the transcription factor concentration. The convergence of proteins to the equilibrium value for different values of x can be seen in Figure 2.6. The takeaway from this analysis is that, for directional

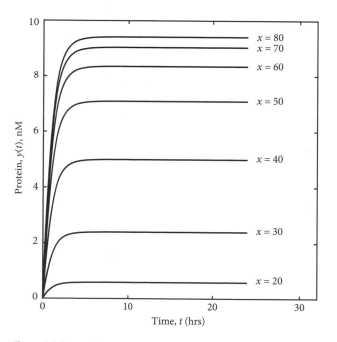

Figure 2.6: Dynamics given positive gene regulation for $x = 20$ to $x = 80$ via Eq. (2.1). The dynamics of $y(t)$ converge to y^*, which is a nonlinear function of the input x. In these series of plots, the input value of x varies from 20 nM to 80 nM even as the output varies 15-fold.

gene regulation (X regulates Y), convergence to an equilibrium is typical and such convergence can be nonlinearly related to input. Such conclusions must be viewed as tentative, however, given that dynamics can get far more interesting when there is feedback between output and input.

2.4   FEEDBACK AND BISTABILITY—AUTOREGULATION

How can a cell exhibit bistable gene expression? There are multiple paths to choose from—using either positive or negative feedback. Yet what works in theory may not always work in practice (Cherry and Adler 2000). Gardner et al. (2000) provide one clue for their choice of using a pair of negative feedback loops as part of their article:

> Although bistability is theoretically possible with a single, autocatalytic promoter,
> it would be less robust and more difficult to tune experimentally.

Yet, as other follow-up work has shown, sometimes it can be more robust and perhaps even easier to tune a bistable switch experimentally using positive, autocatalytic promoters (Nevozhay et al. 2012). In theory (and sometimes in practice), a "single, autocatalytic promoter" provides a more direct route to understand core concepts that will be preserved even when shifting toward related designs—like the toggle switch.

Consider a case in which a gene encodes for a transcription factor that positively regulates its own expression. If a cell has relatively low levels of gene expression, then it will not express; if it has relatively high levels of gene expression, then it will express even more. The consequence is that low states get lower and high states get higher, potentially diverging into alternative stable states. This section explains and explores these mechanisms in more detail by contrasting cases in which a gene can self-regulate positively ($X \rightarrow X$) or negatively ($X \dashv X$) (Figure 2.7). The former is termed an *autocatalytic loop* and the latter an *autoinhibitory loop*—and as it turns out, only the former can potentially exhibit bistability. Motivated by the prior section's plan for analyzing nonlinear dynamical systems, we will define the model, identify steady states, and classify stability. This analysis will form the basis for considering behavior of the toggle switch in the next section.

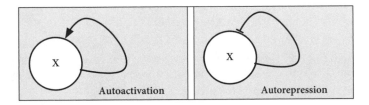

Figure 2.7: Schematic of an autocatalytic and an autoinhibitory loop, in which a gene positively (left) or negatively (right) regulates its own expression. In the autocatalytic loop, the arrow head denotes that increasing levels of x increase its own expression. In the autoinhibitory loop, the perpendicular line head denotes that increasing levels of x decrease its own expression.

2.4.1   Negative autoregulation

Consider the dynamics of a self-inhibitory loop, $X \dashv X$:

$$\frac{dx}{dt} = \frac{\beta}{1 + (x/K)^n} - \alpha x \tag{2.11}$$

where, as before, β denotes the maximum production rate, K is a half-saturation constant, α is a decay rate, and n is the degree of cooperativity. In this case, the steady state of the system must satisfy

$$\frac{\beta}{1 + (x^*/K)^n} = \alpha x^*. \tag{2.12}$$

The term on the left-hand side is a monotonically decreasing function of x. In contrast, the term on the right-hand side is a monotonically increasing function of x. When $x = 0$, then production is β and degradation is 0. Hence, production will steadily decrease from β and degradation will increase from 0. There can only be one crossing of these two curves. This crossing denotes the equilibrium (Figure 2.8, left). For values of $x < x^*$ concentration will increase, and for values of $x > x^*$ degradation will exceed production and concentration will decrease. We expect that the equilibrium is both locally stable and, in fact, globally stable. That means that the system will converge to x^* no matter what its initial conditions. The technical appendix provides extended details on how to assess the stability of 1D and 2D dynamical systems.

2.4.2   Positive autoregulation

Consider the dynamics of a self-catalytic loop, $X \to X$:

$$\frac{dx}{dt} = \frac{\beta x^n}{K^n + x^n} - \alpha x \tag{2.13}$$

where, as before, β denotes the maximum production rate, K is a half-saturation constant, α is a decay rate, and n is the degree of cooperativity. In this case, the steady state of the system, $x = x^*$, must satisfy

$$\frac{\beta x^n}{K^n + x^n} = \alpha x. \tag{2.14}$$

There are multiple possibilities for the number of times that the function on the left-hand side can intersect the function on the right-hand side. First, consider the case when $n = 1$. In that case, when $x \ll K$, then an equilibrium must satisfy $\beta x/K = \alpha x$. This is only true in the special case when $\beta/K = \alpha$—implying that there should only be a single fixed point, either at $x^* > 0$ when $\beta/K > \alpha$ or $x^* = 0$ when $\beta/K < \alpha$. However, for larger values of n, then the degradation rate will exceed that of the production rate for small values of x. The question then becomes, is there a value at which production exceeds degradation? If so, then because of the saturating production levels, there is the possibility of two stable states: one low and one high. The dynamics of this system are shown in Figure 2.8, in which a change in the strength of the promoter when turned on leads to a qualitative switch in outcome, i.e., from a

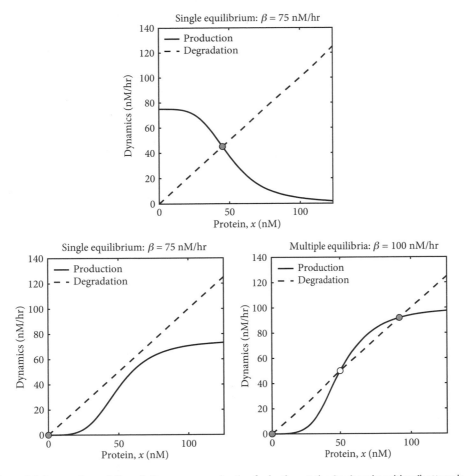

Figure 2.8: Comparison of degradation versus production for both negative (top) and positive (bottom left and bottom right) feedback loops. For negative autoregulation, there is only a single stable equilibrium. In contrast, in this example, there is a single stable equilibrium for positive autoregulation when $\beta = 75$ nM/hr and two stable equilibria along with one unstable equilibrium when $\beta = 100$ nM/hr. Here $K = 50$ nM, $n = 4$, $\alpha = 1$ hr^{-1}. In each case, the solid line denotes the corresponding autoregulatory feedback function (whether autocatalytic or autoinhibitory) and the dashed line denotes the net degradation, i.e., αx. Stable and unstable equilibria are denoted with solid circles and open circles, respectively.

system with a single stable equilibrium in the Off state ($x^* = 0$) to one in which there are three equilibria—off (x_{off}), on (x_{on}), and an intermediate unstable state (x_{mid}). The basis for the bistability is now apparent. When the strength of autocatalytic feedback is sufficiently high, then an On state continues to generate more protein, which ensures that the system stays on. In contrast, when the density of proteins drops below x_{mid}, then protein production drops (nonlinearly), causing x to drop even lower, leading to less production, and so on until the gene is in the Off state again. Hence, the outcome depends on initial conditions in a way that the system can retain a memory of its prepared, initial state (Figure 2.9). These results also suggest that changes in parameters of this autocatalytic loop can lead to qualitative changes in system dynamics.

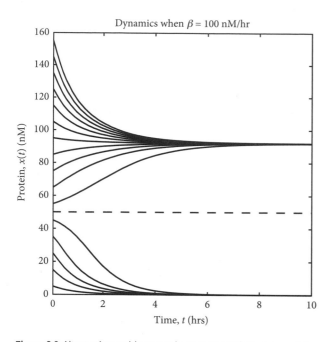

Dynamics when $\beta = 100$ nM/hr

Figure 2.9: Alternative stable states in an autocatalytic gene regulatory loop. In this example, the dynamics correspond to simulations of Eq. (2.13) given positive feedback when $\beta = 100$ nM/hr, $K = 50$ nM, $n = 4$, and $\alpha = 1$. The initial conditions vary from near 0 to nearly 160 nM. The system converges to either $x^* = 0$ (the Off state) or $x^* > 0$ (the On state) depending on whether the initial conditions are less than or greater than K, respectively.

2.4.3   Qualitative changes in dynamics

The preceding section suggests that a population of cells can undergo a qualitative change of state as a function of some parameter or set of parameters. For an autocatalytic loop, that may be α, β, K, or even n. A particularly interesting feature is the existence of "critical points" at which a small change in a parameter leads to a large change in outcome. To explore this concept, let us make the simplifying assumption that $n \gg 1$, such that the autocatalytic production function looks like a "step function," with a sharp transition of production, from 0 when $x < K$ to β when $x \geq K$. Formally, we write this as

$$\frac{dx}{dt} = \beta \Theta(x - K) - \alpha x \qquad (2.15)$$

where Θ is the Heaviside step function, which is 1 when its argument is non-negative and 0 when its argument is negative. Given this definition, it is apparent that there is always at least one fixed point, when $x = 0$. In that case, there is no production nor degradation because there is no transcription factor present to degrade. Because of the autocatalytic nature of the gene circuit, then no new production will take place. When $x = K$, then the production is β. Hence, production will exceed degradation when $\beta > \alpha K$. If this condition is met, then instead of just one steady state, there will be three:

$$x^* = 0 \text{ Off state, stable} \qquad (2.16)$$

$$x^* = K \text{ Borderline state, unstable} \qquad (2.17)$$

$$x^* = \beta/\alpha \text{ On state, stable} \qquad (2.18)$$

Of note, these results hold even if there is some basal level of production. The stability, denoted here, is shown graphically in Figure 2.9. The important point is that bistability is not inevitable given positive feedback. Instead, bistability also depends on the quantitative rates, i.e., $\beta/\alpha > K$; that is, the concentration at the On state must exceed the half-saturation constant. This intuition generalizes to realistic cases with finite levels of cooperativity, i.e., when $n > 1$ (Box 2.2). This intuition will also be useful as we move on to the circuit designed, implemented, and tested by Gardner and colleagues: the toggle switch.

BOX 2.2. Bifurcation Diagram in Depth

A bifurcation diagram denotes the relationship between output (e.g., steady state) and input (e.g., a driving parameter) in which there is a sudden change in the output as a function of the input. To begin, consider a positive gene regulatory feedback loop as described in the text, characterized by a maximum production rate β, half-saturation constant K, degradation rate α, and Hill coefficient n. To consider how output gene expression relates to a governing parameter, hold all other parameters fixed while changing β incrementally. For each value of β, simulate the dynamics given a range of initial conditions (or, if possible, identify the fixed points numerically). Then collect the final states of the simulation and plot those final states as a function of the parameter. In this particular case, it is also possible to solve for the intersection of the production and degradation functions, e.g., via a zero-finding algorithm, and use these solutions as equivalents to the expected steady states. The final element is to denote which of the fixed points are stable or unstable. In this case, the low and high steady states are both stable and the intermediate value is unstable (denoted in the diagram below by solid and open circles, respectively). This approach is not a comprehensive solution to bifurcation diagrams, but it provides the essential ideas and helps identify the existence of a critical point. In this particular example, when $K = 50$ nM, $\alpha = 1$ nM/hr, and $n = 4$, then $\beta_c \approx 88$ nM/hr (left). When there is no nonlinear feedback in the system (i.e., $n = 1$), then bistability is no longer possible (right).

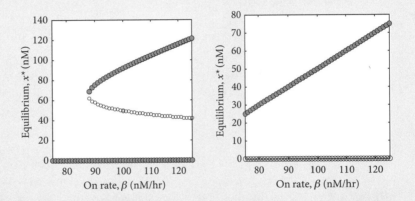

The nonlinear positive feedback loop leads to a saddle-node bifurcation in which one unstable equilibrium and one stable equilibrium "collide," leaving only a single fixed point for values of $\beta < \beta_c$. Hence, even though the structure of the system is the same throughout, the system can exhibit the feature of alternative stable states only when the strength of the promoter exceeds the critical value. For more details on bifurcation theory, consult the excellent introductory text to *Nonlinear Dynamics and Chaos* by Steve Strogatz (1994).

2.5   THE DYNAMICS OF A GENETIC TOGGLE SWITCH

The toggle switch: A network in which two genes mutually repress one another, that is, $X \dashv Y$ and $Y \dashv X$. The toggle switch was synthesized using existing components, i.e., promoter-repressor pairs and green fluorescent reporters (as reproduced in Figure 2.10). How did the team know that such a switch could possibly work, and what are the relevant

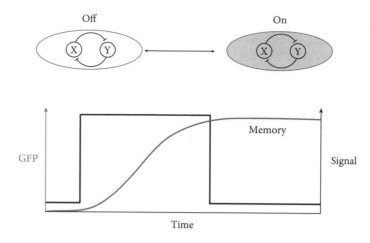

Figure 2.10: A genetic toggle switch in which the genes X and Y mutually inhibit each other can lead to bistability (top) and cellular memory, even when stimuli are removed (bottom). The top schematics indicate the state of the cell as a whole, in which there is not (left) or there is (right) green fluorescent protein (GFP) in the system. In a system with alternative stable states, the system can remain in a new state (e.g., on) even after the signal is removed; this is a form of cellular memory. Given this book's color scheme, the G in GFP is operationally represented as gray.

governing parameters that could ensure that the toggle switch really does exhibit bistability and memory in practice, not just in theory?

To begin, consider the mathematical representation of an idealized toggle switch, in which x and y denote the concentrations of proteins associated with the two genes:

$$\frac{dx}{dt} = \frac{\beta_1}{1 + (y/K)^\epsilon} - \alpha x \tag{2.19}$$

$$\frac{dy}{dt} = \frac{\beta_2}{1 + (x/K)^\gamma} - \alpha y \tag{2.20}$$

This simplified model assumes potentially different levels of cooperativity in the mutual repression. Note that the final engineered form of the switch was represented by a slightly more complex version (Gardner et al. 2000). To get there, rescale time such that $\tau = \alpha t$, just as was shown in Chapter 1's analysis of Luria and Delbrück's experiments (Luria and Delbrück 1943). In this case, $d\tau = \alpha dt$, and by dividing the above equations by, α we find

$$\frac{dx}{d\tau} = \frac{\beta_1/\alpha}{1 + (y/K)^\epsilon} - x \tag{2.21}$$

$$\frac{dy}{d\tau} = \frac{\beta_2/\alpha}{1 + (x/K)^\gamma} - y \tag{2.22}$$

Next, rescale the concentrations relative to a common half-saturation constant—though in general it is important to keep in mind that such affinities could differ. That is, define $u = x/K$ and $v = y/K$ such that $dx = Kdu$ and $dy = Kdv$. The dynamics of the toggle switch can now be written as

$$K\frac{du}{d\tau} = \frac{\beta_1/\alpha}{1 + v^\epsilon} - Ku \tag{2.23}$$

$$K\frac{dv}{d\tau} = \frac{\beta_2/\alpha}{1+u^\gamma} - Kv \tag{2.24}$$

and by dividing both sides by K, we find

$$\frac{du}{d\tau} = \frac{\beta_1/(\alpha K)}{1+(v)^\epsilon} - u \tag{2.25}$$

$$\frac{dv}{d\tau} = \frac{\beta_2/(\alpha K)}{1+(u)^\gamma} - v \tag{2.26}$$

Finally, denote a dimensional production rate $\tilde{\beta}_1 = \beta_1/(\alpha K)$ and $\tilde{\beta}_2 = \beta_2/(\alpha K)$ such that

$$\frac{du}{d\tau} = \frac{\tilde{\beta}_1}{1+(v)^\epsilon} - u \tag{2.27}$$

$$\frac{dv}{d\tau} = \frac{\tilde{\beta}_2}{1+(u)^\gamma} - v \tag{2.28}$$

This is precisely the model proposed for the genetic toggle switch, albeit using different constant names. At this point, to stay consistent, we will use $\alpha_1 \equiv \tilde{\beta}_1$ and $\alpha_2 \equiv \tilde{\beta}_2$. With this definition in place, it is possible to identify steady states and learn about the overall system dynamics.

2.5.1 Theory of robust, toggle switch design

The identification of steady states requires that all variables of the system remain fixed. Another way to say this is that the nullclines must intersect. A *nullcline* defines a set of points (u, v) such that one of the variables does not change. Hence, the nullcline of u must satisfy $du/dt = 0$ and the nullcline of v must satisfy $dv/dt = 0$. Both variables remain fixed at the intersection of the nullclines, implying that there must be a combination of values (u^*, v^*) such that

$$u^* = \frac{\alpha_1}{1+(v^*)^\epsilon} \tag{2.29}$$

$$v^* = \frac{\alpha_2}{1+(u^*)^\gamma} \tag{2.30}$$

These two functions define the $\dot{u} = 0$ and $\dot{v} = 0$ nullclines, respectively. The nullclines can, potentially, intersect at three points—two of these fixed points denote stable states and one is an intermediate unstable state. To see this, substitute the first nullcline relationship to get a single relationship for v^*:

$$v^*\left[1 + \left(\frac{\alpha_1}{1+v^\epsilon}\right)^\gamma\right] = \alpha_2. \tag{2.31}$$

The left-hand side is a function of v^*. When $v^* \to 0$, then the function is approximately $v(1 + \alpha_1^\gamma)$. When $v^* \to \infty$, then the function is approximately v. In between, the function has a local maximum and minimum. Hence, depending on the value of α_2, there can be a region in which there are three possible steady states (Figure 2.11). However, if α_2 is either too small or too large, then there will not be multiple equilibrium points and the toggle switch

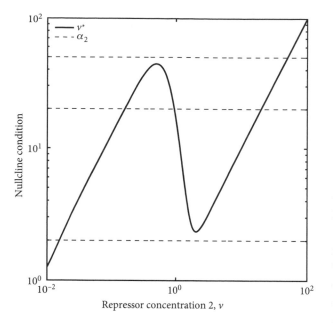

Figure 2.11: A potential bifurcation given variation α_2 for the toggle switch. Here $\epsilon = 3$, $\gamma = 3$, and $\alpha_1 = 5$. The solid line denotes the left-hand side of Eq. (2.31), and the dashed lines denote the value $\alpha_2 = 2$, 20, and 50, moving from bottom to top.

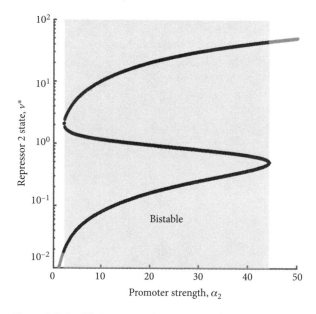

Figure 2.12: Equilibrium state of repressor 2, v^*, as a function of the input promoter strength, α_2. For intermediate values of α_2, there are multiple solutions, corresponding to a bistable region where v^* is either low or high (note the log axes). In contrast, when $\alpha_2 \to 0$, then there is only a single stable state corresponding to low v^*, and when $\alpha_2 \to \infty$, then there is only a single stable state corresponding to high v^*. The formulation for this condition is found in Eq. (2.31).

will only have one potential state. Notice that the single states corresponding to monostability have v^* either quite high or quite low. Only one repressor wins. As intuition suggests, for lower values of α_2 (corresponding to the maximum strength of v), the repressor concentration v is forced into an Off state, and for sufficiently high values of α_2, the repressor concentration v is forced into an On state. On the other hand, in the bistable region both repressors can win—in exact analogy to the Off and On states of the autocatalytic loop. Figure 2.11 illustrates that point, given $\epsilon = 3 = \gamma$. This relationship can be evaluated in a different order and, in so doing, we can ask instead, what values of v^* are possible given a fixed value of α_2? There are many ways to do this, but one can think of this as turning Figure 2.11 on its side and then finding the intersections of v^* as α_2 increases. Figure 2.12 shows the result of such a procedure, where the shaded region is bistable and the unshaded region is monostable. Hence, in theory, this toggle switch has a robust region of multistability for a broad range of promoter strengths—the precise feature that the genetic toggle switch was designed to replicate.

2.5.2 Synthetic biology—nonlinear dynamics in the real world

What did Gardner and colleague do? They built a genetic construct on a plasmid that had an antibiotic resistance cassette and then grew cells with antibiotics to ensure that the plasmid would be retained by the cells. The construct included two promoters, each of which turned on a different repressor. That is, X_1 represses P_2 and X_2 represses P_1. One would expect that either X_1 or X_2 could be turned on, but not both. One such construct included the lac repressor *lacI* and the temperature-sensitive phage λ cI repressor *cIts*. In *E. coli*, the lac repressor controls the levels of enzyme expression to process sugars. But the lac repressor can bind to galactose or a galactose mimic, like the molecular inducer IPTG. Similarly, in phage λ, the repressor CI can bind to a core operator site at the center of the phage λ genetic circuit, enhancing the likelihood that phage will integrate into the bacterial host rather than killing it. But *cIts* is a mutant repressor that loses its ability to bind at high temperatures. This setup, as depicted in Figure 2.2, underlies the tension of the toggle switch.

When experimentalists add IPTG, LacI proteins bind to the IPTG molecules and do not repress expression of *cIts*. In this case, the GFP protein associated with the P2 direction of the construct is expressed and cells turn green. Similarly, the CIts protein binds to the P1 promoter, keeping the system on, even if IPTG is removed (see Figure 2.2). Next, when the system temperature increases, the temperature-sensitive repressor will no longer function, and then expression via *lacI* will subsequently repress expression at the P2 promoter. The green GFP protein will dilute away, and the cell will remain off, even after temperature is returned to normal.

This construct and dynamic results illustrate the core findings: that the system does have bistability and that the bistability has a memory-like feature, retaining an imprint of past environmental changes due to the mutual inhibition and feedback embedded into the design of the plasmid and of the nonlinear dynamical system itself. Indeed, the genetic toggle switch is remarkable in many ways, not least of which because it leverages concepts of nonlinear dynamics to design and implement a controllable biological system. The implementation resides inside *E. coli* cells. The components of the toggle switch are themselves derived from organisms. The components are well-characterized parts of the *E. coli* lactose utilization system and of the gene circuit of the bacteriophage λ. The advantages of using these parts is that their behavior is relatively well understood in isolation and in their original context—but the combination yields a reproducible genetic circuit that performs an altogether new function in a different context.

That the genetic toggle switch can exhibit bistability mutually reinforces the utility of quantitative approaches in characterizing and engineering living systems while also creating a pathway for a field of study: engineered cellular systems biology. The finding that adding a stimulus can induce a state and that removal of the stimulus does not change the state implies that it should also be possible, in theory and in practice, to program cellular states. Notably, the process is reversible, in that applying a different stimulus (e.g., increasing temperature, which deactivates a temperature-dependent repressor molecule) can lead to a switch back to the other state. This example of robust switching between alternative stable states reveals that individual cells with the same genetic architecture need not have the same phenotype. Instead, phenotypes may be dependent on past stimulis, precisely because of intrinsic feedback in regulatory circuits. In the next chapter, we will see that such variation extends even further given that stochastic expression of genes involved in regulation, whether for the toggle switch or other circuits, can lead to sustained differences in cells as a result of intrinsic noise.

2.6   TAKE-HOME MESSAGES

- Regulation of gene expression enables cells to respond to internal and external stimuli.
- Gene regulation is fundamentally nonlinear, i.e., small differences in input can lead to large differences in output.
- Nonlinear gene regulation is the basis for cellular bistability, i.e., well-separated gene expression states can remain durable over time, even given the same genotype.

- Bistability can be recapitulated and engineered in synthetic switches, in which a stimulus can induce a change in gene expression that is retained over multiple cell generations, even after the stimulus is removed.
- Although not covered in this chapter, memory and bistability in cells provide a basis for mechanisms of differentiation in single- and multicellular organisms.

2.7   HOMEWORK PROBLEMS

These problems are intended to deepen your understanding of bistability both in theory and in practice, i.e., as part of in silico experiments. The overall objective of this problem set is to explore the principles by which a system may exhibit bistability, with an emphasis on synthetically designed bistable gene regulatory networks. The problem sets builds upon a set of methods developed in the corresponding computational laboratory associated with this chapter, including

- Simulating coupled systems of differential equations
- Phase plane visualization
- Qualitative analysis of nonlinear dynamic systems (including nullcline analysis)
- Linear stability analysis, in silico
- Bistability and bifurcation diagrams

PROBLEM 1. Characterizing a "Bifurcation" Given Bistability

Write a program to simulate an autoregulatory, positive feedback loop $X \rightarrow X$. Assume that the protein dilution rate is α, the max production rate is β_+, the basal level is β_-, and the half-saturation concentration is K. Here you will explore the consequences of adding in proteases that may specifically target and degrade the protein.

- First, explore the theoretical limits to bistability given Boolean logic, i.e., is there a range of values for which the system should be bistable?
- Assume that $\beta_- = 5$ nM/hr, $\beta_+ = 50$ nM/hr, and $K = 25$ nM. Find a range of α values in which the long-term dynamics exhibit monostability and bistability—keep in mind that you cannot go below 0.5 /hr if cells divide at this rate. Explain and contrast your results to theory.
- Relax the "Boolean" assumption and explore whether or not your conclusions guide similar results when the cooperativity of the feedback is of order 1, 2, 3, and 4.

PROBLEM 2. Simulating the Idealized "Toggle Switch"

Write a program to simulate the toggle switch—as described in an idealized fashion in Box 1 of Gardner et al. (2000) (call this Model A):

$$\frac{du}{dt} = \frac{\alpha_1}{1+v^\epsilon} - u \tag{2.32}$$

$$\frac{dv}{dt} = \frac{\alpha_2}{1+u^\gamma} - v \tag{2.33}$$

Here you will explore the consequences of changes in activation, degradation, and cooperativity in modulating bistability.

- Using Model A, and using the values $\gamma = 3$, $\epsilon = 3$, and $\alpha_1 = 5$, $\alpha_2 = 5$, (i) numerically identify nullclines and visualize them; (ii) numerically identify fixed points; (iii) confirm the stability of each of the identified fixed points.
- Again, using Model A, fix the value of $\alpha_1 = 5$ and modulate α_2 across a range of values relative to that of α_1, simulating each and identifying the relevant steady state(s). Visualize and describe your findings.
- Given the results of the prior analysis, either confirm or refute the claims of Gardner et al. with respect to the role of degradation time scales in driving bistability.

PROBLEM 3. Toggle Switch in Practice

In this problem, write a program to simulate the toggle switch as described in the caption to Figure 5 of Gardner et al. (2000) (call this Model B):

$$\frac{du}{dt} = \frac{\alpha_1}{1+v^\epsilon} - u \tag{2.34}$$

$$\frac{dv}{dt} = \frac{\alpha_2}{1+\left(\frac{u}{1+IPTG/K)^\eta}\right)^\gamma} - v \tag{2.35}$$

Using Model B and parameters from Gardner's Figure 5 ($\alpha_1 = 156.25$, $\alpha_2 = 15.6$, $\epsilon = 2.5$, $\gamma = 1$, $\eta = 2.0015$, and $K = 2.9618 \times 10^{-5}$), try to recapitulate the following predictions, providing visual evidence and supporting analysis (when feasible):

- Cells can be put into the high GFP and low GFP states by the use of inducers.
- Cells remain in such states even when the inducers are removed.
- There exist up to three steady states of the cell's GFP expression as a function of changing IPTG concentrations.

Once you have identified the steady states, describe each and identify the region of bistability.

Additional Problems

- What features are required for bistability in networks with only one or two components?
- How robust is the Model B switch to variations in cooperativity?
- How long is a typical residence time in an On or Off state for the reporter in Model B?
- For growing cells, how long would it take for the daughters of an On cell to lose the memory of their parental state?

2.8   TECHNICAL APPENDIX

Diffusive contact between TFs and DNA   A randomly diffusing protein travels a characteristic squared distance x^2 in a period of time $t \approx x^2/D$ where D is the diffusion constant. For globular proteins of approximately 5 nm in effective size, the diffusion rate is approximately $10 - 40 \ \mu m^2/sec$. Hence, given a characteristic distance of $x = 1$ μm, it should take $t \approx 0.025$–0.1 seconds for a TF to traverse a cell (though longer to find a particular target location). Note that the scaling of D was derived by Einstein, in his theory of Brownian motion (Einstein 1956), i.e.,

$$D = \frac{kT}{6\pi\eta a} \approx \frac{4 \times 10^{-14} \mathrm{g} \cdot \mathrm{cm}^2/\mathrm{sec}}{20 \times 10^{-2} \mathrm{g/cm} \cdot \mathrm{sec} \times 5 \ \mathrm{nm}} \approx 40 \mu m^2/\mathrm{sec} \tag{2.36}$$

where k is Boltzmann's constant, η is the viscosity of water, and a is the effective hydrodynamic radius. Cytoplasm is more viscous than water such that larger molecules may be "caged" and unable to move freely.

Analytical solution of constant on and off dynamics   Consider the dynamics in which a protein is turned on at a rate β and turned off at a per capita rate α:

$$\frac{dy}{dt} = \beta - \alpha y. \tag{2.37}$$

The dynamics can be solved explicitly via integration by parts, i.e.,

$$\frac{dy}{\beta - \alpha y} = dt \tag{2.38}$$

$$\frac{\log \beta - \alpha y}{-\alpha} = t + \mathrm{const} \tag{2.39}$$

$$\beta - \alpha y = \mathrm{const} \times e^{-\alpha t} \tag{2.40}$$

or

$$y(t) = \frac{\beta}{\alpha} - Ce^{-\alpha t} \tag{2.41}$$

where the value of C must be determined from initial conditions. In the event that $y(t=0) = 0$, then $C = \beta/\alpha$ such that

$$y(t) = \frac{\beta}{\alpha}\left(1 - e^{-\alpha t}\right). \tag{2.42}$$

Cooperativity in gene regulation The standard nonlinear functions of activation by a gene X whose concentration is x can be written as

$$f(x) = \frac{\beta x^n}{K^n + x^n}, \tag{2.43}$$

whereas if X is a repressor, then

$$f(x) = \frac{\beta}{1 + x^n/K^n}. \tag{2.44}$$

Where do these functions come from? At some level, we can think of them as phenomenological. The activator starts at $f(0) = 0$ and increases toward saturation at $f(x \gg K) \to \beta$. Likewise, the repressor starts at $f(0) = \beta$ and decreases as $f(x \to 0) \to 0$. The transitions happen close to the value $x = K$. The sharpness of the transition from low to high, or high to low, depends on n. This exponent characterizes the degree of nonlinearity in the system, but it also has a biophysical interpretation.

To understand how, consider the case of the activator, in which the expected expression is β when the promoter is occupied and 0 when it is not. In that event, the average expression is

$$f(x) = p_0 \times 0 + p_1 \times \beta \tag{2.45}$$

where p_0 and p_1 denote the probability the promoter is unoccupied and occupied, respectively. Yet promoter occupancy probabilities depend on interactions with transcription factors, e.g., X. Hence, consider the concentrations of unoccupied and occupied promoters as c_0 and c_1, respectively. In that case, the system can be represented in terms of a series of forward and backward chemical reactions:

$$X + C_0 \underset{k_-}{\overset{k_+}{\rightleftharpoons}} C_1. \tag{2.46}$$

This set of chemical reactions representing binding and unbinding can be written as a dynamical equation for the concentration c_0, i.e.,

$$\frac{dc_0}{dt} = -k_+ c_0 x + k_- c_1. \tag{2.47}$$

The solution is identified by noting that $c_0 + c_1 = c_T$, i.e., a fixed total concentration representing the density of promoter binding sites in the cell. Hence, reducing this one equation–two unknown problem into that of one equation and one unknown renders it solvable. The solution can be derived as follows:

$$k_+ c_0 x = k_-\left(c_t - c_0\right)$$

$$(k_+ x + k_-)c_0 x = k_- c_t$$

$$c_0 = c_T \left(\frac{k_-}{k_- + k_+ x} \right) \tag{2.48}$$

Given that $p_0 = c_0/c_T$, then

$$p_0 = \frac{1}{1 + x/K_D} \tag{2.49}$$

and

$$p_1 = \frac{x}{K_D + x} \tag{2.50}$$

where $K_D = k_+/k_-$ has units of a concentration. Even at this stage, this relatively minimal model of gene regulation leads to a nonlinear expression:

$$f(x) = \frac{\beta x}{K_D + x}. \tag{2.51}$$

Similar logic leads to the conclusion that, for repressors,

$$f(x) = \frac{\beta}{1 + x/K_D}. \tag{2.52}$$

Yet why are the sigmoidal functions so sharp? That is to say, why is $n > 1$?

One answer lies in cooperativity. Consider, for example, a transcription factor that binds to promoters not in the form of a monomer but instead in the form of a dimer. In that event, the concentration of dimers, x_2, should be in equilibrium with the concentration of monomers, x, i.e.,

$$X + X \underset{k_-}{\overset{k_+}{\rightleftharpoons}} X_2, \tag{2.53}$$

which can be written in terms of the dynamical system

$$\frac{dx}{dt} = -k_+ x^2 + 2k_- x_2 \tag{2.54}$$

such that at equilibrium

$$x_2 = k_+ x^2/(2k_-). \tag{2.55}$$

The rules for activation and repression apply as before, except that dimers (with concentration x_2) bind to promoters rather than monomers binding to promoters (with concentration x). We can rewrite the activation function as

$$f(x) = \frac{\beta x_2}{K_D + x_2} \tag{2.56}$$

or

$$f(x) = \frac{\beta k_+ x^2/(2k_-)}{K_D + k_+ x^2/(2k_-)}. \tag{2.57}$$

This shows that dimerization imposes sharper thresholds for activation. In practice, the value of n represents the effective degree of cooperativity and can be mapped explicitly to the number of monomers required for binding and regulation of expression downstream of a target promoter.

Linear stability analysis of a nonlinear dynamical system Most of the dynamic models in this book are *nonlinear*. A full analysis of nonlinear dynamical systems is outside the scope of this text and can be found in introductory texts such as *Nonlinear Dynamics and Chaos* (Strogatz 1994) or *Population Biology: Concepts and Models* (Hastings 1997). In practice, there is a standard approach to diagnose potential dynamics of a nonlinear system. This approach includes identifying fixed points, approximating the dynamics near the fixed point via a process termed *linearization*, and then solving for the expected qualitative dynamics near the fixed point by taking advantage of standard approaches to solve linear systems. The following sections demonstrate these in practice in the case of one-dimensional and two-dimensional systems. (Note: Portions of the following methods are taken directly from Weitz (2015).)

Linear stability analysis in 1D Consider an autocatalytic system as described in the text such that

$$\frac{dx}{dt} = \frac{\beta x}{x + K} - \alpha x \tag{2.58}$$

where β is the maximal production rate, K is the half-saturation constant, and α is the degradation/dilution rate. The steady state of this system can be found by setting the right-hand side to 0. In this case, there are two such states, $x^* = 0$ and $x^* = \frac{\beta}{\alpha} - K$. The stability of these states can be evaluated by considering the dynamics given small perturbations in x, i.e., $u = x(t) - x^*$.

First, in the event that the system is initiated near $x = 0$, then the concentrations can be assumed to be small, i.e., $x \ll K$. In that event, it is possible to approximate the dynamics, leveraging the fact that the deviation is small, i.e.,

$$\frac{dx}{dt} \approx \frac{\beta x}{K}(1 - x/K) - \alpha x, \tag{2.59}$$

using the approximation that $1/(1 + \epsilon) \approx 1 - \epsilon$ for $\epsilon \ll 1$. Eliminating higher-order terms of x/K leads to

$$\frac{dx}{dt} = \frac{\beta x}{K} - \alpha x. \tag{2.60}$$

This is a linearization of the original, nonlinear system, albeit near the fixed point $x^* = 0$. In this case, the stability of the system is determined by the value $\frac{\beta}{K} - \alpha$. When $\beta/\alpha > K$, then production is stronger than degradation and the protein concentration increases exponentially—this is termed an *unstable steady state*. When $\beta/\alpha < K$, then production is weaker than degradation and the protein concentration decreases exponentially back toward the steady state—this is termed a *stable steady state*.

Second, in the event that the system is initialized near $x^* = \frac{\beta}{\alpha} - K$, then we should consider changes in the deviations from the steady state, i.e., $u = x(t) - x^*$. Notice that $\dot{u} = \dot{x}$. As such, it is possible to rewrite the dynamics in terms of the deviation:

$$\frac{du}{dt} = \frac{\beta(u + x^*)}{K + u + x^*} - \alpha u - \alpha x^*. \tag{2.61}$$

As before, it is possible to rewrite and simplify this form as follows:

$$\frac{du}{dt} = \frac{\beta(u+x^*)}{K+x^*}\frac{1}{1+u/(K+x^*)} - \alpha u - \alpha x^* \tag{2.62}$$

$$\frac{du}{dt} \approx \frac{\beta(u+x^*)}{K+x^*}\left(1-u/(K+x^*)\right) - \alpha u - \alpha x^* \tag{2.63}$$

At this point, the equation can be simplified by retaining only first-order terms in u/K, i.e.,

$$\frac{du}{dt} \approx \frac{\beta(u+x^*)}{K+x^*} - \frac{\beta u x^*}{(K+x^*)^2} - \alpha u - \alpha x^* \tag{2.64}$$

$$\frac{du}{dt} \approx \frac{\beta u}{K+x^*} - \frac{\beta u x^*}{(K+x^*)^2} - \alpha u \tag{2.65}$$

and the last equivalence uses the fact that the definition of the steady state is that production $\beta x^*/(K+x^*)$ equals degradation αx^*. Note that because $x^* = \beta/\alpha - K$ this can be further simplified to

$$\frac{du}{dt} = \alpha u - \frac{\alpha^2 u(\beta/\alpha - K)}{\beta} - \alpha u \tag{2.66}$$

and finally to

$$\frac{du}{dt} = -\frac{\alpha^2(\beta/\alpha - K)}{\beta}u. \tag{2.67}$$

Hence, the stability of the On fixed point is based on whether β/α is greater than or less than K. When $\beta/\alpha > K$, then the sign of the coefficient is negative and small deviations decay exponentially, so the fixed point is stable. Notably, $\beta/\alpha > K$ is the condition for the existence of the fixed point. Hence, insofar as positive feedback leads to an On state, then it is invariably stable.

The analysis above is somewhat cumbersome, as it depends on expanding the non-linear dynamics and setting all higher-order terms to zero. Formally, this is equivalent to "linearizing" the nonlinear dynamics near the equilibrium points—a method that scales up to more complex systems with more than one variable. For a system whose dynamics are $\frac{dx}{dt} = f(x)$, the behavior of the system near $x = x^*$ can be described in terms of some perturbation $x(t) = x^* + u(t)$. The dynamics near the fixed point $x = x^*$ can be approximated by a Taylor approximation of the function $f(x)$ in terms of the perturbation $u = x - x^*$ such that

$$\frac{dx^*}{dt} + \frac{du(t)}{dt} = f(x) \tag{2.68}$$

$$\frac{dx^*}{dt} + \frac{du}{dt} = f(x^*) + \left.\frac{df}{dx}\right|_{x=x^*} u + \mathcal{O}\left(u^2\right) \tag{2.69}$$

$$\frac{du}{dt} \approx \left.\frac{df}{dx}\right|_{x=x^*} u \tag{2.70}$$

because $f(x^*) = 0$ and higher-order terms are ignored when u is small. The perturbation will increase or decrease depending on the change of the system dynamics with respect to changes in the population near the fixed point: $\frac{df}{dx}\big|_{x=x^*}$. When local increases in the population correspond to increases in the rate of change, the system is unstable, and vice versa. This is the algebraic equivalent of the approach illustrated in this section. In the case of the autocatalytic toggle switch, the linearized version of the system near a fixed point $x = x^*$ is

$$\frac{du}{dt} = \left[\frac{\beta K}{(x^* + K)^2} - \alpha \right] u. \tag{2.71}$$

This linearized system must be evaluated at a fixed point, i.e., $x^* = 0$ for the Off state and $x^* = \frac{\beta}{\alpha} - K$ for the On state. In both instances, the linearized system reduces to the same conditions as above. As before, the Off state is unstable when $\beta/\alpha > K$ and the On state is stable when $\beta/\alpha > K$. The linearization approach can be extended to analysis of multivariable systems.

Linear stability analysis in 2D A linear stability analysis in higher dimenions requires expressing the rate of change of populations near an equilibrium. Formally, consider the generic dynamical system

$$\frac{dx}{dt} = f(x, y) \tag{2.72}$$

$$\frac{dy}{dt} = g(x, y) \tag{2.73}$$

The population state (x^*, y^*) is an equilibrium if $f(x^*, y^*) = 0 = g(x^*, y^*)$. The local stability of an equilibrium can be evaluated by considering the perturbations u and v such that $x(t) = x^* + u(t)$ and $y(t) = y^* + v(t)$. The dynamics of these perturbations can be written as

$$\frac{du}{dt} = f(x^*, y^*) + \frac{\partial f}{\partial x}\bigg|_{x^*, y^*} u + \frac{\partial f}{\partial y}\bigg|_{x^*, y^*} v + \mathcal{O}(u^2, uv, v^2), \tag{2.74}$$

$$\frac{dv}{dt} = g(x^*, y^*) + \frac{\partial g}{\partial x}\bigg|_{x^*, y^*} u + \frac{\partial g}{\partial y}\bigg|_{x^*, y^*} v + \mathcal{O}(u^2, uv, v^2). \tag{2.75}$$

Because the local stability analysis is near the equilibrium, these two equations can be approximated as

$$\frac{du}{dt} \approx \frac{\partial f}{\partial x}\bigg|_{x^*, y^*} u + \frac{\partial f}{\partial y}\bigg|_{x^*, y^*} v, \tag{2.76}$$

$$\frac{dv}{dt} \approx \frac{\partial g}{\partial x}\bigg|_{x^*, y^*} u + \frac{\partial g}{\partial y}\bigg|_{x^*, y^*} v. \tag{2.77}$$

The local dynamics of the perturbation (u, v) can be written as

$$\frac{d}{dt} \begin{bmatrix} u \\ v \end{bmatrix} = \mathbf{J} \begin{bmatrix} u \\ v \end{bmatrix} \tag{2.78}$$

where

$$\mathbf{J} = \begin{bmatrix} \frac{\partial f}{\partial x} & \frac{\partial f}{\partial y} \\ \frac{\partial g}{\partial x} & \frac{\partial g}{\partial y} \end{bmatrix} \tag{2.79}$$

is termed the Jacobian. The Jacobian is evaluated at a specific point, corresponding to the equilibrium under investigation: (x^*, y^*).

The Jacobian encodes information on the rate of change of population dynamics with respect to changes in the population. The Jacobian can be evaluated at a given equilibrium, in this case yielding four numbers, each corresponding to a given element of a matrix. In this way, a nonlinear dynamical system in two dimensions can be linearized, yielding a linear matrix equation with coefficients that depend on the equilibria (which themselves depend on parameter values). The solution to a linear system of the form $dx/dt = Ax$ for a single variable is one of exponential growth or decay, i.e., $x(t) = x_0 e^{\lambda t}$. The same holds for two- and higher-dimensional systems. For example, consider the candidate solution $(u(t), v(t)) = (u_0, v_0) e^{\lambda t}$, in which case $du/dt = \lambda u$ and $dv/dt = \lambda v$, such that Eq. (2.78) can be rewritten as

$$\lambda \begin{bmatrix} u \\ v \end{bmatrix} = \mathbf{J} \begin{bmatrix} u \\ v \end{bmatrix}. \tag{2.80}$$

For $u_0 \neq 0$ and $v_0 \neq 0$, the only feasible values of λ are those that satisfy

$$\det (\mathbf{J} - \lambda \mathbf{I}) = 0 \tag{2.81}$$

where $\mathbf{I}$ is the 2×2 identity matrix and det refers to the determinant of the matrix.

In summary, local analysis of stability predicts the dynamics near a fixed point. More generally, the stability of a fixed point can be classified based on the sign of the real components. If one of the eigenvalues has a positive real component, then the fixed point is unstable. If none of the eigenvalues has a nonnegative real component, then the fixed point is stable. The type of stability can be further classified based on whether or not the eigenvalues have a nonzero imaginary component. When the eigenvalues are complex, this suggests that dynamics will oscillate. For example, dynamics may oscillate away from or toward a fixed point, given instability or stability, respectively. Finally, if the eigenvalues are purely imaginary (i.e., have no real components), then—to first order—deviations will not grow in magnitude, though a full evaluation of the dynamics at higher order may sometimes be necessary.

Chapter Three

Stochastic Gene Expression and Cellular Variability

3.1   LIVING WITH RANDOMNESS

A human body includes tens of trillions of cells. Humans, like other multicellular organisms, are not made up of the same cell repeated many trillion times over (in fact, each of us harbors approximately as many microbial cells as our human cells (Sender et al. 2016)). Instead, individual cells differentiate into distinct phenotypic states. These states are then heritable from one cell generation to the next, shaping the nature of the multicellular organism during development. Lineages of phenotypically convergent cells organize spatiotemporally to give rise to complex structures, functions, and behaviors. Do similar principles apply to bacteria? For example, a single bacterium grown in a 50 ml flask filled with rich media can, in the course of a day, give rise to a population of billions. A world in a bottle. Yet are all of these cells the same? Instead, could many of them have qualitatively different phenotypes than others? The answer is that cells can have markedly different phenotypes even if they are all recent descendants of a common ancestral cell, and that such phenotypic differences are possible even in genetically identical cells (for a critical early, perspective and more recent ones, see Spudich and Koshland Jr. 1976; Ozbudak et al. 2002; Balázsi et al. 2011; Eling et al. 2019).

The previous chapter already described a core concept underlying such emergent differences: bistability. In the genetic toggle switch engineered by Gardner and colleagues, bistability arose due to the nonlinear feedback between two genes and their corresponding transcription factors. Such models suggest a mechanism for incipient forms of differentiation. That is, a cell with many type A repressors will repress expression of type B repressors, thereby remaining in the A state. Similarly, a cell with many type B repressors will repress expression of type A repressors, thereby remaining in the B state. If similar processes subdivide further, it is possible to envision a process of multistability and, perhaps, differentiation in which a neuron, muscle cell, or epithelial cell have a single origin, yet vastly different fates. There is a substantive gap from engineering a synthetic toggle switch to understanding multicellular differentiation, but it is a path worth walking on, at least for a few steps (for related approaches, see Nelson 2021; Phillips et al. 2009). In doing so, the continuous model of bistability leaves a major question unresolved. Cells have a particular history, so given a colony (or lineage) arising from a single cell, in a particular state, shouldn't all the daughter cells remain in precisely the same state? Continuous models of bistability suggest that differences must be set exogeneously. But single-cell observations in *E. coli*, yeast, and viruses with both synthetically engineered and natural switches suggest otherwise (Gardner et al. 2000; Ptashne 2004; Ozbudak et al. 2004).

In the case of the synthetic toggle switch discussed in Chapter 2, the activity of repressors is controlled by small molecules, i.e., chemical "inducers." These inducers change the activity of transcription factors, enabling them to repress, or not, their targets. By varying the levels of IPTG, a commonly used inducer, cells switched gene expression from one qualitatively different state to another. Yet, at any given level of IPTG, there was a subpopulation of cells in either state. That is, bistability seemed inherent to nearly every level of inducer. To borrow the language of the prior paragraph, these observations raise a generic question: what made genetically identical cells more A-like and others, more B-like? Perhaps the very nature of transcription and translation—the processes by which information is converted from DNA to RNA into proteins—underlies these cellular-scale differences.

This is the perspective we take here: that cells can have different structures, functions, and behaviors, even if they have identical genotypes. The differences may be amplified or reinforced by nonlinear feedback in the system. Yet the differences can emerge due to noise in the system, irrespective of whether exogenous factors—and there are many—influence the particular milieu in which a cell lives. The source of noise can be divided broadly into two categories - intrinsic and extrinsic noise (Elowitz et al. 2002; Paulsson 2004; Süel et al. 2007). Intrinsic noise arises due to stochastic timing of production and decay of individual molecules. Extrinsic noise arises due to differences in the underlying reaction rates in cells—which can be caused by differences in the number of RNA polymerase molecules and ribosomes, which may be influenced by environmental context. To understand the nature of stochasticity in gene expression requires a different kind of model than the continuous models described in Chapter 2. We must turn to models that account for the *discrete* nature of molecules inside cells. For example, the process of transcription yields one new mRNA molecule in a cell at a time. Similarly, the process of translations yields one new protein molecule in a cell at a time. Such changes also modify the concentration of molecular abundances. A concentration is just the number of molecules divided by a volume. Yet dividing discrete values by a continuous volume to yield a concentration does not absolve us of a need to recognize and develop appropriate models of stochastic gene expression. A translation event does not mean that the cell now has 1.23429 more proteins. It has one more. And, if the protein were to degrade, the cell would have one less. Building models that can account for such discrete changes is the focus of this chapter.

But before we turn there, it is important to reinforce that such models are necessary to understand pressing biological challenges. To what extent will a population of bacteria be susceptible to antibiotics if the founding cell of a lineage or population is susceptible? What fraction of cells will enter dormancy given stressful conditions, e.g., starvation or exposure to toxins? Will a virus entering a cell initiate a process to kill the cell releasing new progeny or, instead, to integrate its genome with the cell, thereby continuing to divide along with the cell itself? What determines which cell or group of cells among billions or trillions will transform into a founding cancerous clone, with profound consequences for its local environment and, perhaps, the multicellular host? These questions are but a few of many ongoing challenges in modern biological sciences. The answers to each may involve theories

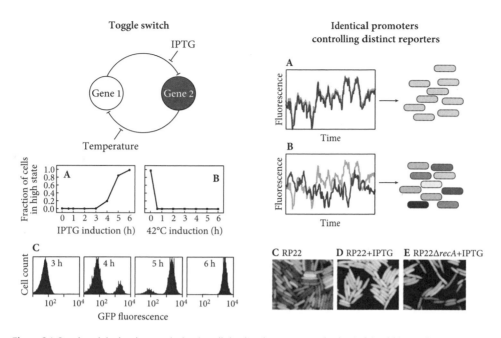

Figure 3.1: Stochasticity leads to variation in cellular-level outcomes, whether in bistable switches or in identically controlled reporters. (Left) A genetic toggle switch is a synthetically engineered bistable feedback loop where two genes mutually inhibit each other. Activities of the two genes are controlled by external factors—IPTG and temperature. When IPTG is high, the system turns on (as evidenced by the expression of GFP (green fluorescent protein); when temperatures are increased, the system turns off. However, as shown in panel c, individual cells do not move uniformly between states; instead, there are marked differences in cellular state associated with the bistability of the circuity. (Right) Gene expression variability induced by intrinsic noise. In the event that the same promoter is used to control distinct reporters that express GFP and RFP, it is possible to evaluate the extent to which the expression of a promoter is identical (as in A, in the absence of intrinsic noise) or variable (as in B, in the presence of extrinsic noise). The correlation between identical promoters is one way to measure intrinsic noise. Panels C–E show microscopy images of synthetically engineered *E. coli* with the two-color reporter system, given the absence and presence of IPTG (C and D, respectively). As is apparent, constitutive expression leads to differences that are then decreased in the presence of strong expression induced by IPTG. Images reproduced from Gardner et al. (2000) and Elowitz et al. (2002).

and principles that cut across scales, including ecological and evolutionary dynamics. Yet, as is apparent, modeling stochasticity is a key part to each.

Figure 3.1 provides two examples of intrinsic noise in microbial systems. The first example is from the synthetic toggle switch system in which the level of GFP is measured, cell by cell, as a function of increasing levels of IPTG. Increasing IPTG should induce the system—in theory—to switch from low levels of expression to high levels of expression. Yet, in reality, some of the cells remain in the low state even as some switch to the high state. This finding indicates the potential for intrinsic differences in gene expression to shape the phenotypic state of individual cells, particularly at intermediate values of an external driver. Even in the "low" and "high" states, there is still cell-to-cell variability in GFP. The extent to which the total noise in a system can be compartmentalized into intrinsic and extrinsic sources has been studied in depth (Elowitz et al. 2002). The second example in Figure 3.1 (Plate 1) shows what happens when two different (and distinguishable) fluorescent proteins are inserted into engineered *E. coli* strains, albeit under the regulatory control of two copies of the same promoter. In theory, if expression is strictly deterministic and there is not intrinsic noise, then the levels of expression of both proteins should be identical. However, intrinsic noise can lead to divergence in the stochastic trajectory of protein expression and distinct phenotypes. Images C–E (bottom right) show the variability in expression in this system given the engineered strain growing in Luria-Bertani (LB) broth with and without the inclusion of IPTG that drives the promoter toward its maximum expression. As is apparent, intrinsic noise can drive differences in expression, and the relative noise levels decrease when overall rates of expression are higher (as induced by IPTG).

This chapter introduces a framework for modeling stochastic gene expression, following upon the work of many (Thattai and Van Oudenaarden 2001, 2004). This framework provides a baseline of what to expect in systems with "simple" gene regulation, albeit including the intrinsic variation driven by the discrete nature of molecules. This framework can also be used to probe the limits of our mechanistic assumptions. In doing so, we will build toward trying to compare the predictions of simple models of stochastic gene expression with in vitro data. After establishing key formalisms for describing variation in gene expression, the second part of the chapter focuses in on the very assumptions underlying such models. Measuring gene expression at the level of molecules affords the chance to ask new questions of our quantitative models.

Notably, in the mid-2000s, Ido Golding, Edward Cox, and colleagues developed a new approach to measuring the expression of mRNA, one by one (Golding and Cox 2004; Golding et al. 2005). As we will see, doing so provides a new vista into the fundamental mechanisms of gene expression, which we have thus far considered as a memoryless or Markov process. Instead, by carefully examining the dynamics of expression, it will be possible to address whether the expression of mRNA is consistent with a model in which transcripts are continuously and stochastically produced and subsequently degraded. Or perhaps we will find we must appeal to a more nuanced model, in which the promoter alternates between On and Off states such that transcripts appear in bursts. As we will see, the value of quantitative models is often enhanced when they fail. By failing, they provide a baseline for recognizing when measurements point us toward new biological insights rather than just confirming what we thought we knew in advance.

3.2   STOCHASTICITY IN GENE REGULATION

3.2.1   Continuous and discrete paths

Rather than beginning with a model of stochastic gene regulation, it is worthwhile to revisit deterministic models that pose a useful starting point and, eventually, a relevant contrast. Consider a simple gene regulatory system in which gene X positively regulates gene Y; in other words,

$$X \to Y. \tag{3.1}$$

Assume that the inducer is present so that the transcription factor X is active. A mathematical model of the dynamics can be written in terms of continuous changes in the concentration of Y:

$$\frac{dy}{dt} = \beta(x) - \alpha y \tag{3.2}$$

where $\beta(x)$ is the concentration-dependent production rate of y proteins given concentrations x of a transcription factor. If x is present at or near saturation levels such that $\beta(x) = \beta$, then the solution is

$$y(t) = \frac{\beta}{\alpha} \left(1 - e^{-\alpha t}\right) \tag{3.3}$$

given $y(0) = 0$, as discussed in Chapter 2. In this model, every cell will eventually converge to the same limiting concentration, β/α. Moreover, small deviations from this equilibrium concentration will relax back exponentially to equilibrium with a rate constant of α. In essence, this model predicts that a target gene under control of a single transcription factor will behave identically, irrespective of initial conditions.

Yet this model assumes that the concentrations of y change continuously. Instead, to whatever extent such a model remains valid, it is important to keep in mind that it is an approximation to a process in which the number of y molecules inside each cell changes in a sequence, e.g.,

$$\text{Cell 1: } 0 \rightarrow 1 \rightarrow 2 \rightarrow 1 \rightarrow 2 \rightarrow 3. \ldots \tag{3.4}$$

This particular sequence includes four production events and one degradation event. But this is not the only path that could yield a cell with three molecules after five events, e.g.,

$$\text{Cell 2: } 0 \rightarrow 1 \rightarrow 0 \rightarrow 1 \rightarrow 2 \rightarrow 3. \ldots \tag{3.5}$$

$$\text{Cell 3: } 0 \rightarrow 1 \rightarrow 2 \rightarrow 3 \rightarrow 2 \rightarrow 3. \ldots \tag{3.6}$$

In the end, each outcome is the same, but the processes are different. In addition, whatever the underlying state may be, this is not the only final outcome possible given a series of synthesis and degradation events, e.g.,

$$\text{Cell 3: } 0 \rightarrow 1 \rightarrow 2 \rightarrow 1 \rightarrow 0 \rightarrow 1. \ldots \tag{3.7}$$

These sequences suggest a route forward: develop models that describe paths in a "molecular state space." That is, instead of quantifying cells in terms of their continuous concentration of molecules, consider quantifying cells in terms of n, the discrete number of molecules. If we further restrict ourselves to individual synthesis and degradation events, then the dynamics of cells could be represented by a discrete state space $n = \{0, 1, 2, \ldots\}$ such that degradation decreases the number of molecules and synthesis increases the number of molecules. With this in mind, the next question becomes, are sequences enough? For example, consider two cells that do not have Y proteins. Then consider what happens when an inducer is added to the system, such that the transcription factor X initiates gene expression in both. Even if the experimental conditions are tightly controlled such that both cells are exposed to exactly the same amount of inducer, the two cells need not produce the new Y protein at the same time. These differences suggest that paths differ in both the *discrete number* of molecules as well as the *timing* at which events take place. Timing of individual events constitutes the second component of stochasticity we must account for in understanding gene expression.

3.2.2   Timing between individual events

Consider an event, such as transcription, translation, or degradation, that takes place at a rate r per unit time. For example, if an event takes place at $r = 3 \text{ hr}^{-1}$, then there should be approximately three events in every hour, or one event every 20 minutes. In a large period of time T, there should be rT events on average. But what about a small period of time? For example, how many events should take place in a minute if $r = 3 \text{ hr}^{-1}$? The value of rT in this case is $3 \cdot \frac{1}{60} = \frac{1}{20}$ or 0.05. It would seem intuitive to think about this value of 0.05 as the average number of events. But in any given cell, the event either takes place or does not. If this is true for each cell, then perhaps the value of 0.05 should be interpreted as a probability of an event taking place. This interpretation becomes more evident as we consider ever smaller intervals of time, ΔT. Formally, rates of biological processes should be thought of as *probabilities per unit time*. This definition may seem counterintuitive and suggests the need for additional explanation.

Event rates, i.e., the probability per unit time, can exceed 1, just as $r = 3 \text{ hr}^{-1}$ does in this example. These rates must be combined with time to yield a probability. And, of course, probabilities of any particular event must lie between 0 and 1, inclusive. Denote the probability that the event takes place in some very small time interval dt as $P_{\text{event}} = r dt$ and the probability that the event does not take place in some very small time interval dt as $P_{\text{noevent}} = 1 - r dt$. Here $dt \ll 1/r$, so it would be appropriate to consider time intervals far less than $1/3$ of an hour given $r = 3 \text{ hr}^{-1}$. In any small increment, the chances of an event taking place will be small, and as $dt \to 0$, the chances become vanishingly small. Given these definitions, what is the probability that it will take a finite duration T before an event takes place? Formally, what we mean to ask is, what is the probability that the event takes place in a very small time interval between T and $T + dt$ and not in any small time interval before then?

To answer this question, consider dividing up the timeline between 0 and T into very small increments, each of duration dt. There are $N = T/dt$ such increments. What is the probability that the event did not place in the first increment? The answer is simply $1 - r dt$. If no events to take place in the interval $(0, T)$, then the same logic should apply for all of the N increments. This is equivalent to the probability of tossing a biased coin, in which it turns up tails (no event) N times, each with probability $1 - r dt$. The probability of such a sequence of non events is $(1 - r dt)^N$. Now the probability that the event takes place between $(T, T + dt)$ is $r dt$. The product of these two probabilities represents the probability that one must wait a time interval of T before observing an event that takes place at a rate r:

$$P_{\text{event}}(T, T + dt) = (1 - r dt)^{\frac{T}{dt}} r dt \tag{3.8}$$

$$= \left(1 - \frac{rT}{T/dt}\right)^{\frac{T}{dt}} r dt \tag{3.9}$$

$$= e^{-rT} r dt \tag{3.10}$$

where the last step utilizes the definition of an exponential

$$e^a \equiv \lim_{n \to \infty} \left(1 + \frac{a}{n}\right)^n \tag{3.11}$$

and noting that in this example $a = rT$ and $n = T/dt$. The key result is that an event that can occur, or not, at any moment in time at a fixed rate will be distributed exponentially in time. That is, events are more likely to take place earlier than later, indeed, exponentially more likely! This run counters to intuition that events should be spaced somewhere near the average interval. For example, if events occur at a rate of $r = 3$ hr^{-1}, then the average interval should be 1/3 of an hour, or 20 minutes. Yet, if we were to make many such samples of a process and then "bin" the measurements into 1-minute intervals, we would find a surprising result: the most likely waiting period is less than 1 minute! The technical appendixes in this chapter shows that the expected waiting time is $\langle T \rangle = 1/r$, even if the peak of the exponential distribution is near 0.

This result makes it evident that a model of stochastic gene expression should account for both the state of cells and the potential differences in timing for events that could change such states. Such a model would seem to be possible from an algorithmic perspective, that is, we might be able to simulate an individual cell among many. For example, here is "pseudocode" for a numerical simulation that one could plausibly envision writing that also matches the intuition of many biologists for how cells work:

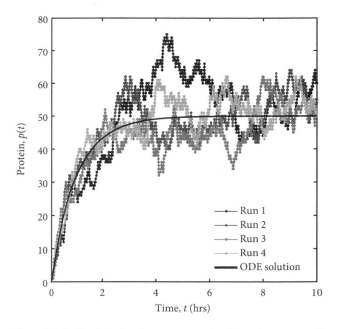

Figure 3.2: Realized stochastic gene expression dynamics compared to the ODE solutions for a fixed expression level $\beta = 50$ nM/hr and degradation rate $\alpha = 1$/hr. The ODE dynamics are given in Eq. (3.2). The stochastic dynamics utilize a Gillespie algorithm to simulate the processes of production and degradation (for more details, see the computational laboratory associated with this chapter).

```
Simulating the stochastic gene expression of a single cell

Specify initial state of cell, e.g., number of proteins and/or mRNA
While measurements are taking place, do the following:
    Calculate event rates based on the current protein count
    Randomly select the waiting time until the next event
    Identify which event type took place at the specified time
    Update the state of the cell and time based on the event
        production: add one protein
        degradation: subtract one protein
Continue
```

This pseudocode provides the starting point for the workhorse of many stochastic gene expression frameworks. In formal terms, it is knows as the Gillespie algorithm and is described in detail in the computational lab guide accompanying this chapter (Gillespie 1977). The Gillespie algorithm is a powerful tool and paradigm. Yet, for the example of $X \to Y$, a quantitative bioscientist should want to know what to expect from the simulation even before beginning. As we will find, the stochastic simulations are close to—but do not exactly coincide with—the ordinary differential equation (ODE) dynamics (Figure 3.2).

It is evident that individual cells may have different states. Hence, it is worthwhile to think about predicting the probability that a cell is in one particular state out of many. To do so, we consider the concept of an *ensemble*, that is, many systems prepared identically, but whose fates may differ precisely because of the nature of the stochastic process. Here these systems will be cells, and the processes will be the production and degradation of proteins. The stochastic simulation framework generalizes to other systems, e.g., that of individuals and the processes of birth and death (see Chapter 4). Irrespective of the application, consider the following definition: $P_n(t)$ the probability that there are exactly n molecules in the system at time t. In the case of stochastic gene expression of a protein Y, $P_{30}(10) = 0.01$ means there is a 1% probability of finding exactly 30 Y proteins in a randomly chosen cell at time 10 hrs. In practice, if the ensemble included 100,000 cells, then one would expect to find approximately 1000 cells with 30 Y proteins at time 10 hrs. Now, what happens if one of the proteins in one of these cells degraded at some point between 10 and 10.01 hrs—and no other events took place? Then $P_{30}(10.01)$ would decrease. This type of *observational* approach belies the fact that with so many events taking place, it might be possible, in advance, to predict how such probabilities should change. That is indeed our goal: to characterize the dynamics of $P_n(t)$ given our knowledge of the mechanisms taking place that shape the transitions between different states. As it turns out, characterizing the full time dependence of $P_n(t)$ is nontrivial, but we can make progress on quantifying other measurable features of such dynamics, including the mean and variance within an ensemble of cells. The mean and variance of gene expression states will become useful in comparing expectations to observations.

3.3   CHARACTERIZING DYNAMICS OF INDIVIDUAL CELLS, GIVEN STOCHASTIC GENE EXPRESSION

3.3.1   Getting to a full model of stochastic gene expression

This section outlines how to construct a principled mathematical model of stochastic gene expression. In doing so, the dynamics will modify $P_n(t)$, the probability that a cell has n proteins at time t. The two processes—protein production and degradation—change these probabilities in different ways. First, a cell in the ensemble with $m < n$ proteins may change to one with $n = m + 1$ proteins given production. Second, a cell in the ensemble with $m > n$ proteins may change to one with $n = m - 1$ proteins given degradation. Hence, production and degradation provide distinct routes to *increase* $P_n(t)$. But the same processes also provide distinct routes to *decrease* $P_m(t)$. In general, we can write these changes in probability as

$$P_n(t + \Delta t) = P_n(t) + \Sigma_{m \neq n} P_m(t) W_{mn} - \Sigma_{m \neq n} P_n(t) W_{nm} \qquad (3.12)$$

where W_{mn} (W_{nm}) denotes the rate per unit time of transitioning from a state m to n (n to m).

Eq. (3.12) is known as the Master Equation (for an extended treatment see van Kampen 2001). The Master Equation describes an infinite number of equations, one for each value of $n = 0, 1, 2, \ldots$. To make the application clear, consider the case of simple gene regulation

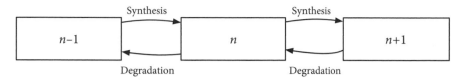

Figure 3.3: Transitions between states with n molecules depending on synthesis and degradation. The synthesis of molecules increases the value of n, and the degradation of molecules decreases the value of n. The Master Equation considers the change in probability of any particular system state n as a function of current probabilities and the rates of synthesis and degradation.

such that $X \to Y$. In this case, Eq. (3.12) can be written in terms of the 4 ways that probability "flows" in and out of $P_n(t)$ (Figure 3.3), yielding

$$P_n(t+\Delta t) = P_n(t) + \beta P_{n-1}(t)\Delta t \tag{3.13}$$
$$+ (n+1)\alpha P_{n+1}(t)\Delta t$$
$$- \beta P_n(t)\Delta t - \alpha n P_n(t)\Delta t$$

while restricting our attention to very small time differences such that we can ignore higher-order terms. Recall that synthesis occurs at a rate of β per unit time, irrespective of the current protein abundances. In addition, degradation occurs at a rate of α per unit time per protein. Hence, to get the total transition rates per unit time, α must be multiplied by the relevant number of proteins—that is, $(n+1)$ for degradation events that increase $P_n(t+\Delta t)$ and n for degradation events that decrease $P_n(t+\Delta t)$. Note that $dP_n(t)/dt = \frac{P_n(t+\Delta t) - P_n(t)}{\Delta t}$ in the limit that $\Delta t \to 0$. Hence, the dynamics for this system are

$$\frac{dP_n}{dt} = \beta P_{n-1} + (n+1)\alpha P_{n+1}(t) - \beta P_n(t) - \alpha n P_n(t). \tag{3.14}$$

If only we knew how to solve this last equation . . . . In fact, this chapter will soon explain how to solve this equation, even though generalized solutions for nonlinear feedback mechanisms lie outside the scope. To see why such a solution is meaningful, recognize that if we know $P_n(t)$ it would be possible to calculate—at any moment in time—the expected mean and variance of protein abundance within cells. The average number of proteins is defined as follows:

$$\langle n(t) \rangle = \sum_{n=0}^{\infty} n P_n(t). \tag{3.15}$$

The variance in the number of proteins is defined as follows:

$$\text{Var}[n(t)] = \left\langle \left[n(t) - \langle n(t) \rangle \right]^2 \right\rangle$$
$$= \sum_{n=0}^{\infty} \left[n - \sum_{m=0}^{\infty} m P_m(t) \right]^2 P_n(t)$$
$$= \sum_{n=0}^{\infty} n^2 P_n(t) - \left[\sum_{n=0}^{\infty} n P_n(t) \right]^2. \tag{3.16}$$

This last equation is the standard formulation of the variance written out explicitly, i.e., $\mathrm{Var}[n(t)] = \langle n^2 \rangle - \langle n \rangle^2$. Note the crucial point that we began this chapter discussing the goal of understanding the basis for stochastic gene expression. Here we are a step closer to a partial victory: a description of the expected state of cells as well as variance between them. If we knew $P_n(t)$, then calculating such dynamics would be trivial. But, as shown next, it is possible to figure out these dynamics even without knowing $P_n(t)$.

3.3.2   Deriving the mean and variance of stochastic cellular dynamics

It seems difficult to know what to do when faced with a challenging question. There are things to know about an ensemble of cells, but first and foremost, in the present context we are interested in some information about the average. That is, what is the average number of proteins found in a randomly chosen cell within an ensemble $\langle n(t) \rangle$? The mean is a dynamic variable and, in principle, should be governed by a predictable set of rules. Here, it is possible to rewrite the equation for the mean and consider its derivative in time (removing the parenthetical notation of time for P_n):

$$\frac{\mathrm{d}}{\mathrm{d}t}\left(\langle n(t) \rangle = \sum_{m=0}^{\infty} m P_m(t) \right) \tag{3.17}$$

which becomes

$$\frac{\mathrm{d}\langle n \rangle}{\mathrm{d}t} = \sum_{m=0}^{\infty} m \frac{\mathrm{d}P_m}{\mathrm{d}t} \tag{3.18}$$

because the time derivative of $mP_m(t)$ is $m\,\mathrm{d}P_m/\mathrm{d}t$. The change in probabilities is given by the Master Equation, yielding

$$\frac{\mathrm{d}\langle n \rangle}{\mathrm{d}t} = \sum_{m=0}^{\infty} \beta m P_{m-1} - \sum_{m=0}^{\infty} \beta m P_m$$
$$+ \sum_{m=0}^{\infty} \alpha m (m+1) P_{m+1} - \sum_{m=0}^{\infty} \alpha m^2 P_m \tag{3.19}$$

The technical appendix explains how to reduce these sums, term by term. This technical appendix is worth reviewing, at least once, because the dynamics of the mean number of molecules reduces to nothing other than

$$\frac{\mathrm{d}\langle n \rangle}{\mathrm{d}t} = \beta - \alpha \langle n \rangle. \tag{3.20}$$

This equation seems straightforward. Indeed, you have seen it before. It is analogous to the equation for maximal production of a target gene at a rate β with per protein degradation at a rate α, as presented in Chapter 2 on *deterministic* regulation of proteins. Yet now, instead of assuming that all cells are identical, this equation describes the dynamics of the *average* number of proteins in an ensemble of cells. The solution to this equation is

$$\langle n(t) \rangle = \frac{\beta}{\alpha} \left(1 - e^{-\alpha t} \right). \tag{3.21}$$

In this case, the average number of proteins increases from 0 at the start of addition of a signal toward an average value of β/α. Hence, despite the stochasticity inherent in the dynamics, we have not lost all sense of predictability. Rather, the mean is predicted to increase toward the same carrying capacity as predicted in a deterministic model. But, unlike the deterministic model, not all cells are predicted to have the same protein abundances. Variation remains due to the stochastic nature of gene expression. Rather than repeating the derivation for the variance, let's examine the variance of protein abundances at equilibrium.

The Master Equation for the gene regulatory network, $X \to Y$, operating at saturation is:

$$\frac{dP_n}{dt} = \beta P_{n-1} + (n+1)\alpha P_{n+1}(t) - \beta P_n(t) - \alpha n P_n(t). \tag{3.22}$$

If the ensemble of cells is at a steady state, then there should be no change in $P_n(t)$, that is, $\frac{dP_n(t)}{dt} = 0$ for all values of n. Denote P_n^* as the equilibrium distribution of probabilities such that this steady state condition is satisfied. As described in the technical appendix setting this condition yields a recursive relationship such that

$$P_n^* = \frac{\lambda^n e^{-\lambda}}{n!} \tag{3.23}$$

where $\lambda = \beta/\alpha$. Examples of the results of stochastic simulations of simple gene expression are shown in Figure 3.4, in which it is evident that proteins are in fact Poisson distributed. In other words, even in an environment with the same production rate and the same degradation rate, individual cells would nonetheless differ in their expression levels—thus leading to intrinsic noise in the system. The distribution describes probabilities that cells differ in their *discrete* number of proteins. It also has the property that the variance is equal to the mean. This then completes our description of the state of the system, at least at steady state. Note that convergence to the steady state is a more complicated dynamic, nonetheless it reinforces the central message: that stochastic dynamics elucidate features of the continuous dynamics we have already described, as well as shed light on new features that are not present in the continuous model.

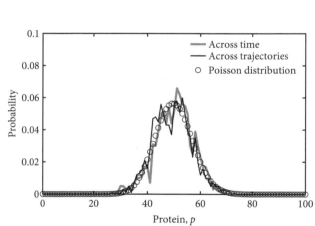

Figure 3.4: Emergence of Poisson distributions via a stochastic gene expression process, given $\beta = 50$ nM/hr and $\alpha = 1$/hr.

In essence, the equilibrium is characterized by noise, such that the variance is equal to that of the mean number of proteins and the standard deviation is equal to the square root of the mean number of proteins. The coefficient of variation is defined as the standard deviation divided by the mean. For a Poisson distribution, the variance scales like the mean, $\bar{y}$, such that the standard deviation scales like $\bar{y}^{1/2}$ and the standard error scales like $\bar{y}^{-1/2}$. For example, a system with 400 molecules has a 5% coefficient of variation compared to a system with 20 molecules, which has a 22% coefficient of variation (Figure 3.5). This finding is consistent with the example shown at the start of this chapter that intrinsic noise

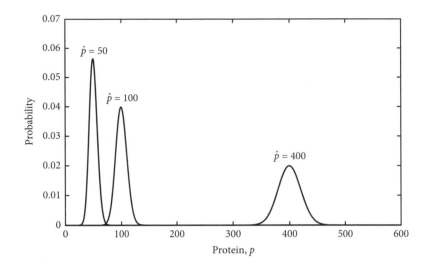

Figure 3.5: Expected Poisson distributions have variances that scale with the mean, here for $\hat{p} = 50, 100$, and 400. Hence, although the overall magnitude of variability increases, the relative variation actually decreases with increasing mean. For example, given $\hat{p} = 100$, there should be a standard deviation of 10, and therefore 10% noise (10/100) is typical. In contrast, given $\hat{p} = 400$, there should be a standard deviation of 20, and therefore 5% noise (20/400) is typical. Hence, intrinsic noise should decrease with the increasing strength of a promoter.

can lead to variability in expression and that the extent of such variability diminishes with stronger expression (e.g., when induced; see right panel of Figure 3.1). The scaling of the mean and variance is yet another hallmark of intrinsic noise and will prove critical in the next section as we explore whether gene expression can be described as a Poisson process, i.e., memoryless.

3.4   IS GENE EXPRESSION BURSTY?

Thus far, this chapter has reviewed how simple gene expression of a target gene Y controlled by a transcription factor X can lead to variation in expression, including variation in steady state values. Implicitly, such a a model also suggests that the timing between expression events (e.g., mRNA transcription or protein translation) should be exponentially distributed, which is a characteristic of a Poisson process (not to be confused with a Poisson distribution). A cell undergoing a Poisson process of mRNA transcription or degradation should lead to a Poisson distribution of mRNA in a cell. Hence, even if each translation event and degradation event were random, the resulting distribution of proteins would already be non-Poisson (Friedman et al. 2006; Cai et al. 2006). Such a model can potentially include both the variability and repeatability of gene expression in genotypically identical cells. Yet perhaps even the original assumption that mRNA transcription can be described as a Poisson process does not fully describe that feature of cellular dynamics. If not, how would we know? The advantage and utility of quantitative models of living systems fully comes to fruition when we ask models to make predictions that, when tested, show signs of failure. The failure of an otherwise consistent quantitative model may point to new principles or mechanisms that govern the underlying dynamics.

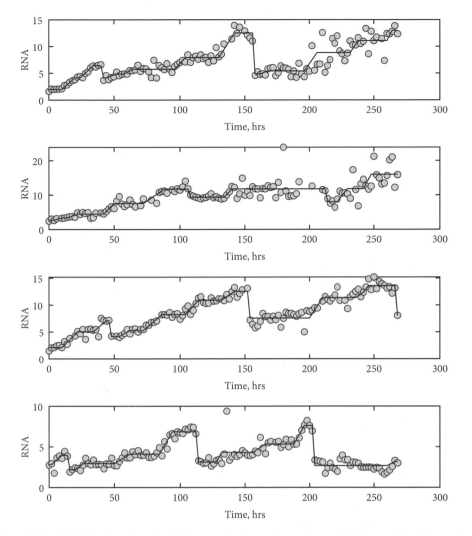

Figure 3.6: Individual trajectories of mRNA in a synthetic reporter system, exhibiting periods of transcriptional bursts and inactivity. This data provides the basis for comparing a Poisson model of gene activity with a two-state model (i.e., with both On and Off states). Data courtesy of Ido Golding; original analysis in Golding et al. (2005).

In order to evaluate variability in mRNA production, we turn to a single molecule reporter system designed to measure mRNA molecules one at a time (Golding and Cox 2004; Golding et al. 2005). This system enhances the signal associated with a single mRNA molecule so that the discrete number of mRNA can be counted at a single cell level. As explained in detail later in this chapter, the reporter system was engineered such that each mRNA molecule had an appended sequence that was not translated, but instead served as an attachment site for another protein complex that included GFP. Hence, when mRNA molecules were transcribed, a part of the mRNA could become rapidly bound with GFP proteins, leading to sharp green "foci" in the cell that could be measured directly with microscopy. This gene reporter system enables mRNA expression tracking of cells over time. Figure 3.6 shows a series of trajectories of mRNA measured one by one in a synthetic reporter system. The shape of these trajectories suggests something unusual, insofar as there appear to be periods of significant activity (the periods of steady accumulation) and other periods of relative stasis. This difference does not seem to be consistent with a memoryless (i.e., Markovian) system of gene expression, which has been the assumption used throughout this chapter.

In looking at the trajectories of Figure 3.6, it is important to keep in mind the context in the mid-2000s and try to see them with fresh eyes. The study of stochastic gene expression advanced rapidly in the late 1990s, as new methods to see and count variation at the scale of individual cells afforded new opportunities to quantify and characterize the quantitative rules of cellular life. Part of the focus of these studies was to determine features of cellular gene regulatory networks so as to identify "motifs," i.e., patterns of regulation that seemed to occur more frequently than expected by chance (Milo et al. 2002). In doing so, perhaps those motifs would also shed light on functional features of gene regulatory networks. This was a productive path, e.g., to see the work on the feed-forward loop and principles of gene expression networks more generally (Alon 2006). Other research focused on engineered systems, like the toggle switch introduced in Chapter 2 (Gardner et al. 2000). Yet a different approach altogether was to shed light on the basic, stochastic nature of the component processes, including transcription, translation, and binding (Elowitz et al. 2002).

One might think that such basic biology is best revealed by observation. Perhaps. But observation coupled to quantitative models provides an even more powerful lens. Let us return to the basic model of stochastic gene expression and consider a single cell that has not yet begun to produce a target protein. After induction, production should occur at a rate β, implying that the time before the first production event should be exponentially distributed with a mean time of $1/\beta$. Recall how Luria and Delbrück also used information from those experiments in which no resistant colonies emerged as a means to test the dependent and independent hypotheses for mutation. If experimentalists had methods to measure individual transcriptional events as they took place, then those time distributions could reveal something about production rates, β. Moreover, observing differences in transcripts between cells could also provide information on the ratio of β/α. Comparing the variance to that predicted under the "Poisson" hypothesis is analogous to the efforts of Luria and Delbrück to utilize the magnitude of variation to characterize the validity of hypotheses. Yet the evidence that Golding and colleagues observed suggested that transcription was not strictly Poisson. Instead, the cells seemed to switch between two states: Off and On. In Off states, transcription would not occur; in On states, transcription would occur, and the simplest model of transcription assumes that such events are independent of one another, i.e., are well described by the kind of model we just analyzed. Figure 3.7 provides a conceptual schematic underlying the two alternative hypotheses. The next challenge is to measure individual molecules and break down different levels of evidence to try to address the two hypotheses: is mRNA transcription bursty or memoryless?

The analysis of bursty expression was enabled by the development of a single-cell reporter system to measure mRNA one at a time (Golding and Cox 2004). A schematic of the system and expected output is shown in Figure 3.8 (Plate 2). The cell reporter system combines two constructs on different plasmids that are then introduced into *E. coli* cells. The aim of the system is to count the number of mRNA associated with a red fluorescent protein (RFP) reporter. When active, transcription will yield a concatenated mRNA

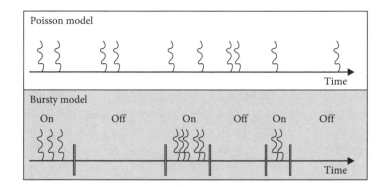

Figure 3.7: The Poisson model and "bursty" model of transcription. (Top) In the Poisson model, transcripts are generated at a constant rate with an exponential distribution between events. (Bottom) In the bursty model, no transcripts are generated in Off periods, whereas transcripts are generated in bursts during On periods. Note that these two schematics include the same number of total transcripts, albeit distributed in starkly different ways.

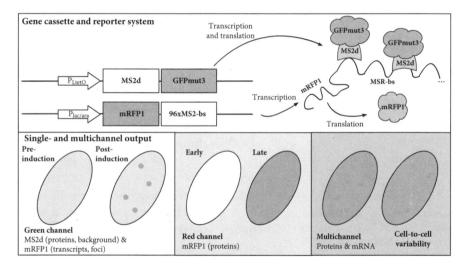

Figure 3.8: Gene reporter system for measuring transcripts per cell via an mRNA detection system in *E. coli*. (Top) The two-promoter reporter system, including a binding site for a fusion protein on RNA. (Bottom) Expected output via the green channel, red channel, and multichannel, including cell-to-cell variation. For more details, see the description in the text and refer to Golding and Cox (2004) and Golding et al. (2005).

molecule, including both RFP mRNA and a long mRNA that comprises 96 repeats of the same sequence: an MS2 binding site. The other promoter controls the production of a hybrid MS2-GFP protein. MS2 is the phage protein that recognizes the binding site. Hence, when the system is induced by a synthetic IPTG-regulated promoter, then the labeling protein will bind to the concatenated mRNA molecules. The use of 96 potential binding sites intensifies the signal, revealing green foci, ideally one per mRNA molecule. Later, after activation, the mRNA RFP signal will be translated into RFP proteins, generating a red background signal. Altogether, this system provides a means to potentially count mRNA (via the accumulation of green foci) and the relationship between mRNA and proteins (by contrasting the number of green foci with the strength of the protein signal (as measured in the red channel).

3.4.1 Evidence of individual-level mRNA molecule detection

Before evaluating evidence for burstiness, it is essential to ask: does the detection system measure individual mRNA molecules? Qualitative lines of evidence are documented in Golding et al. (2005), and one of the most critical components (described in Golding and Cox 2004) is worth reviewing. First, given the detection system, the number of green foci should provide a rough approximation of the number of mRNA. For example, if g is the GFP intensity per mRNA and G_{tot} is the total intensity measured in foci per cell, then presumably $n = G_{tot}/g$. Such intensities discount the background GFP, which is not bound to an mRNA molecule. When measured across cells, this intensity provides an estimate of the unknown number of transcripts. Figure 3.9 shows the result of estimating transcripts per cell from nearly 100 different cells from a given experiment. Notably, the estimates peak at discrete values. This suggests that the intensity is both sufficiently uniform and resolvable at *individual* mRNA molecules, and that the system can be used to discriminate a difference of a single mRNA molecule between cells. That is, frankly, quite remarkable. Beyond being a technical feat, the impact of this technology was made evident when the ability to measure RNA dynamics in living *E. coli* cells was put to the service of answering a scientific question. And to answer this question requires that we combine a quantitative model of gene expression with this new type of measurement.

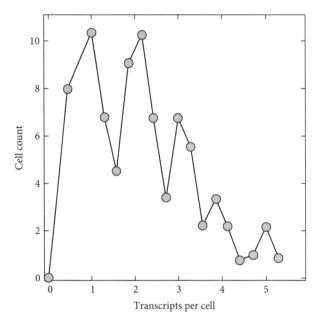

Figure 3.9: Integer-valued peaks for measurements of transcripts per cell in an mRNA detection system in *E. coli*. The *x* axis denotes the number of estimated transcripts per cell. The *y* axis denotes the number of cells with the corresponding number of estimated transcripts—note that the measurement system has evident peaks at integer values. See text for more details.

3.5 THE GEOMETRY OF BURSTS

What is a hallmark signal of burstiness? What is a hallmark signal of genes that are not bursty? Like the work of Luria and Delbrück, it is critical to ask quantitative questions of alternative mechanistic models. In this case, if transcription is not bursty, there should be an equal chance of transcription at any particular moment. This represents the default Poisson process model of transcription. As was shown earlier in this chapter, a Poisson process model exhibits exponentially distributed waiting times between events. The steady state level of mRNA should have a variance equal to that of the mean. This implies that if one could measure mRNA one at a time, then the time between each transcriptional event should be exponentially distributed with a rate equal to the average transcriptional rate.

However, in a "bursty" transcription model, there are periods with many transcripts produced in a relatively short period of time interspaced with periods in which no transcripts are produced. Moroever, if transcription is bursty, then it should be possible to characterize the size of such bursts. Just as in the case of Luria and Delbrück, understanding the predictions of the non-Poisson model takes a bit more effort.

3.5.1 Hallmark of burstiness I: Time before mRNA appears

Consider the production of mRNA in an induced cell. The average dynamics given a Poisson model of transcription can be derived via the Master equation given a production rate β and a decay rate α. As shown earlier in this chapter and in the technical appendix, the expected dynamics are

$$\langle n(t) \rangle = n^* \left(1 - e^{-\alpha t}\right) + n_0 e^{-\alpha t} \tag{3.24}$$

such that, if there are no mRNA molecules in the cell before induction, the dynamics of mRNA should follow

$$\langle n(t) \rangle = n^* \left(1 - e^{-\alpha t}\right) \tag{3.25}$$

where $n^* = \beta/\alpha$. This equation can be directly compared to an experiment. Notice that at early times $\langle n(t) \rangle \approx \beta t$, whereas at later times the deviation of the average expression from the equilibrium, $\delta(t) \equiv n^* - n(t)$, should decay exponentially given a rate α: $\delta(t) = n^* e^{-\alpha t}$. Hence, by measuring the initial production of mRNA and its relaxation to equilibrium, it is possible to estimate both the production rate and decay rate. These constants were estimated as $\hat{\beta} = 8.4$/hr and $\hat{\alpha} = 0.83$/hr (Golding et al. 2005). This implies there should be approximately a steady state of 10 mRNA molecules in the system. Such a Poisson model can in fact fit the data; but just like fitting a Poisson distribution estimate of a selection-dependent mutation rate, these estimates are insufficient to distinguish between the alternative hypotheses. But they yield a clue.

Given that the production rate is 8.4 molecules/hr, one would anticipate that it should take approximately 0.12 hr (or 7 minutes) for the first mRNA molecule to appear. Formally, this is equivalent to a waiting time problem: how long should we wait until the system generates a molecule given a single process (transcription) that takes place at a fixed rate? The waiting time for a Poisson process with rate β is simply the exponential distribution, $P_0(t) = \beta e^{-\beta t}$. Instead, Golding et al. found that the observed time of the first mRNA transcript was exponentially distributed but with the wrong characteristic time! Instead of 7 minutes, the average waiting time was approximately 30 minutes (Figure 3.10). Yet there is even more to learn from these curves.

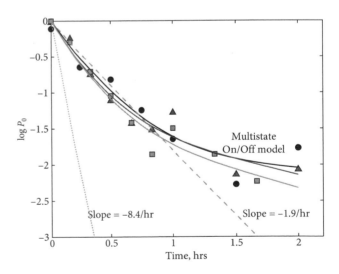

Figure 3.10: Dynamics of the time to first appearance of mRNA are inconsistent with a Poisson production hypothesis. The steeper slope with rate 8.4/hr corresponds to that derived from an estimate of the effective transcription rate. The shallower slope is the estimate reported in Golding et al. (2005). The solid lines denote the best fit to experimental data (in symbols) assuming a multistage process as described in the text. An extended description of this multistage model is included in the technical appendix.

The timing data itself seems to have longer, nonexponential tails. What could give rise to the slower dynamics?

One explanation is that not all the genes in every induced cell were ready to transcribe. In a bursty model of gene transcription, the gene switches between Off and On states, corresponding to when transcription is silenced and when it is active, respectively. Hence, instead of a stochastic system with one state, we must consider an additional expression state, which we denote as $B = \{0, 1\}$ where 0 and 1 are the transcriptionally inactive and active states, respectively. Hence, the complete model can be described in terms of (B, m) where $B = 0$ or 1 to denote the transcriptional activity and $m = \{0, 1, 2, \ldots\}$ is the nonnegative number of mRNA transcripts. The transitions in this system can be cataloged as follows:

$$\text{Transcription}: (1, m) \xrightarrow{\beta} (1, m + 1) \tag{3.26}$$

$$\text{Degradation}: (B, m) \xrightarrow{\alpha m} (B, m - 1) \tag{3.27}$$

$$\text{Gene turn on}: (0, m) \xrightarrow{k_{on}} (1, m) \tag{3.28}$$

$$\text{Gene turn off}: (1, m) \xrightarrow{k_{off}} (0, m) \tag{3.29}$$

These four events correspond to transcription in the active state, degradation of mRNA (irrespective of the expression state), switching from off to on, and switching from on to off. The inclusion of a transcriptional expression state has profound consequences on expected dynamics.

In order to expand the model, we must also specify the switching rates k_{on} and k_{off} for the transitions from 0 to 1 and from 1 to 0, respectively. As a result, a cell may initially be in the On state. If so, an mRNA molecule will be transcribed if and only if the mRNA is initiated prior to the switch to the Off state. Otherwise, the system will have to wait until a switch back to the On state before an mRNA can be transcribed. In essence, one can then catalog the sequence of gene expression states before the first mRNA transcript as follows:

1*
101*
10101*
. . .
01*
0101*
010101*

. . .

where * denotes the first measured transcriptional event, 1 denotes a transcriptional On state, and 0 denotes a transcriptional Off state. The first set of sequences denotes a situation where an On gene leads to a transcription event, then one in which the gene goes from on to off to on before a transcription, etc. The second set of sequences denote a situation where an Off gene turns on followed by a transcription event, then one in which the gene goes from off to on to off to on, and then transcription, etc. It should be apparent that these multiple events would necessarily delay the average time and potentially drive the system away from purely exponential waiting times.

Figure 3.11 reveals that a Poisson process model of transcription leads to exponentially distributed waiting times. But that is not what is observed. The waiting times seem longer than anticipated based on the maximum transcriptional rate and also nonexponential. To diagnose this observation, note that using a fast switching model delays the waiting time, even if the observed waiting times are exponential. Indeed, in the limit of very rapid switching (here 10/hr), then the gene state switches back rapidly between On and Off states. In that event, the probability of being in an On state can be denoted as p, whose dynamics must satisfy

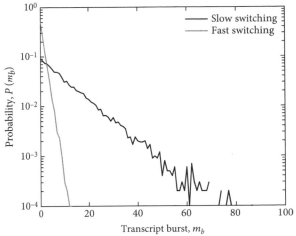

$$\frac{dp}{dt} = -pk_{off} + (1-p)k_{on} \qquad (3.30)$$

Figure 3.11: Transcript burst distributions given fast and slow switching. In both cases, $\beta = 10$/hr. In the fast switching model, $k_{off} = 10$/hr. In the slow switching model, $k_{off} = 1$/hr.

with an equilibrium probability $p^* = \frac{k_{on}}{k_{on}+k_{off}}$. For example, the equilibrium probability $p^* = 1/2$ in the event that the transition rates are equal. In the event that these rates are very fast (compared to gene expression), this finding implies that the rate of production should be slowed down by a factor of p^* such that the effective waiting period is longer than that expected in a Poisson model. In essence, rapid switching changes the waiting time but not the shape of the distribution. In contrast, when switching is very slow (i.e., slower than the time to transcribe, conditional upon the gene being in the On state), then the distribution changes shape and is characterized by short waiting times if the gene starts in the On state but very long waiting times if the gene starts in the Off state. In that case, the longer tail is a result of waiting for the gene to switch between Off and On states, corresponding to when transcription is silenced and when it is active. The technical appendix derives an approximate distribution of such a multistate model (see Figure 3.10 for the consequences). It is critical to note that the observed data seems to have a longer-than-exponential tail, which is a hallmark of a multistate model! But it's not the only hallmark; for more, we need to turn to the size of bursts.

3.5.2　Hallmark of burstiness II: Variation in numbers of mRNA produced

In a bursty model of gene transcription, the gene switches between Off and On states, corresponding to when transcription is silenced and when it is active, respectively. We have shown that, in the limit of fast switching, the system should exhibit exponentially distributed waiting times, albeit slower than that expected in a Poisson model. And, in the limit of slow switching, then waiting times should no longer be exponentially distributed, and in fact in very slow switching should be characterized by two rates: the (fast) transcriptional rate and the (slow) switching rate. However, once switched, many transcription events happen in close succession. How big are these bursts?

To understand the nature of transcriptional bursts, it is important to revisit the principles underlying which of two random events occurs first. To do so, consider a system in which there are two processes that occur at rates r_1 and r_2. If $r_1 = r_2$, then it seems apparent that both should have an equal chance. But when $r_1 \neq r_2$, then it would seem that the event with the higher rate is more likely to occur first. But how much more likely? Formally, the probability that process 1 occurs first can be broken down in terms of conditional probabilities. That is, it could be that process 1 occurred in a small interval of time $(0, dt)$ and process 2 did not. Or it could be that process 1 occurred in a small interval of time $(dt, 2dt)$ before process 2 occurred. And so on.

Formally, the probability that event 1 occurs before event 2 can be written as the following integral:

$$P_{event\ 1} = \int_0^\infty dt \ \overbrace{r_1 e^{-r_1 t}}^{\text{(event 1 occurs)}} \ \overbrace{e^{-r_2 t}}^{\text{(event 2 has not yet occurred)}} \tag{3.31}$$

As shown in the technical appendix, the result of this integral is

$$P_{event\ 1} = \frac{r_1}{r_1 + r_2}. \tag{3.32}$$

In essence, the probability of an event is equal to its relative contribution to the total rate at which events occur. This equation generalizes, so that $P_{event\ i} = \frac{r_i}{\sum_{j=1}^s r_j}$ where s is the number of process types. With this in mind, how many transcripts should be produced in a single burst? To answer this, assume that production during On states occurs at a rate β and a switch from an On to an Off state occurs at a rate k_{off}. Hence, the probability that n transcripts are produced in one burst is

$$p(n) = \overbrace{\left(\frac{\beta}{\beta + k_{off}} \right)^n}^{n\text{ production events}} \times \overbrace{\left(\frac{k_{off}}{\beta + k_{off}} \right)}^{\text{Off event}} \tag{3.33}$$

In essence, the two processes are competing. For n transcripts to be produced in one burst, the production process must take place before the Off process occurs, precisely n times. The total probability is analogous to getting the sequence

$$\overbrace{1, 1, \ldots 1, 0}^{n\text{ times}}$$

given a biased Boolean coin that comes up with 1 with a probability of $\beta/(\beta + k_{off})$ and 0 with a probability of $k_{off}/(\beta + k_{off})$. As shown in the Appendix, these burst distributions are geometric distributions. The geometric distribution has the feature that, when $\beta > k_{off}$, then it is more likely to have larger burst events. This is precisely what is seen in Figure 3.6, in which long trajectories of activity alternate with periods of inactivity. These kinds of periods also explain why there is far more variation in the stationary levels of mRNA than would be expected given a Poisson model. If the process of expression includes additional states, this variation must be included in understanding the total observed variation in mRNA levels in a cell. Quantifying the extent of that gap is a story for you to explore (in the homework).

3.6   TAKE-HOME MESSAGES

- Cellular gene expression is a stochastic process arising, in part, from the fact that individual mRNA are transcribed and then translated into proteins.
- For a Poisson process, the timing between one event and the next is exponentially distributed.
- Stochastic gene expression dynamics are expected to generate Poisson distributions at steady state under the assumption of Poisson kinetics.
- The average dynamics of a purely Poisson kinetic system converges to the deterministic prediction with a time scale given by the cellular growth rate.
- Poisson distributions have the feature that their variance is equal to their mean, implying that the relative error (i.e., the standard deviation divided by the mean) decreases like the inverse of the square root of the mean. In practice, the relative error goes down (and stochasticity becomes less important) as the mean increases.
- An engineered reporter system can be used to measure individual expression of mRNA.
- Measurements of single mRNA dynamics reveal multiple lines of evidence for bursty transcription that diverge from those expected in a Poisson process, including a longer time to the first appearance of mRNA.
- The bursty dynamics of mRNA lead to longer periods of activity and longer periods of inactivity than would be expected given a Poisson model where all the periods are exponentially distributed.
- A model of bursty transcription implies that expression switches between On and Off states, implying geometric rather than exponential distributions of expression.
- Variation in gene expression in genetically identical cells drives phenotypic differences.

3.7   HOMEWORK PROBLEMS

This problem set builds upon the developing toolkit you have accumulated, including ideas explored in the computational laboratory guide associated with this chapter. The lab guide addresses the following major concepts:

- Finding the time to the next random event given one or more concurrent Poisson processes
- Combining multiple discrete events together to simulate a stochastic trajectory governed by one or more reactions
- Modeling stochastic gene expression, including a basal level of production and a fixed decay/degradation
- Sampling a stochastic trajectory over discretely spaced time points
- Assessing the statistics of an ensemble of stochastic trajectories

With these tools in hand, the following problems are intended to deepen your understanding of stochastic gene expression, including the opportunity to analyze data from in vitro studies. The overall objective of this problem set is to explore the principles by which cellular individuality emerges, one molecule at a time.

PROBLEM 1. Autoregulation and Memory

Write a program to simulate an autoregulatory, positive feedback loop where X activates itself via Boolean logic. Assume that the protein dilution rate is α, the max production rate is β_+, the basal level is β_-, and the half-saturation concentration is K. Here you will explore the dynamics that unfold when expression is noisy. Throughout, assume that $\beta_- = 20$ nM/hr, $K = 30$ nM, and $\alpha = 1$/hr. Identify the critical value of β_+ beyond which you expect the long-term dynamics to exhibit bistability. Provide evidence in support of this critical β_+^c.

PROBLEM 2. Autoregulation and Bistability

Write a program that simulates stochastic gene expression given autoregulation—use a maximal production rate of 5 nM/hr, 10nM/hr, and 20 nM/hr above the critical value identified in Problem 1. Initialize a cell in the Off state and characterize the time it takes to move to the On state. Show sample trajectories and characterize the distribution of this time. Similarly, initialize a cell in the On state, and characterize the time it takes to move to the Off state. As before, show sample trajectories and characterize the distribution of this time. Describe differences in the time it takes to transition between states.

PROBLEM 3. Long-Term Dynamics and Bistability

Using your stochastic gene expression model developed in the computational laboratory and parameterized in Problem 2, take very long samples of a single, stochastic trajectory. Characterize the distribution of the protein concentration and discuss the notion of bistability in this model. How frequently and for how long do you have to sample cells to ensure that you "see" the bistability in action?

PROBLEM 4. Transcription to Translation to Noise

This problem focuses on the circuit $X \rightarrow Y$ where the inducer for X is present. Write a program that simulates the stochastic transcription of mRNA and translation of gene Y given a rate of $\beta = 5$ nM/hr transcription and a dilution rate of $\alpha = 1$/hr. In this program, consider a scenario where Y proteins are precisely r-fold higher than mRNA. Vary r from 20 to 40, and compare the variance-to-mean ratio for mRNA and proteins. Which, if any, seems to follow a Poisson distribution? Now revise the model so that there is both transcription and translation, using the following set of equations as a guide to the synthesis and degradation steps involved:

$$\frac{dm}{dt} = \beta - \alpha_m m$$

$$\frac{dp}{dt} = rm - \alpha_p p$$

Characterize the fluctuations in this model across the same range of r as in the first program for this question. Next, try to measure the correlations between your mRNA and protein signals. If they are not perfectly correlated, can you explain why not?

PROBLEM 5.   Evidence for Transcriptional Bursting—Theory

Develop a stochastic model of mRNA and protein dynamics including transcriptional bursting. Utilize the same parameters as in the single-cell mRNA expression study (Golding et al. 2005), but do not explicitly account for cell division. First, show a set of sequences of trajectories in your mRNA and protein counts after induction of the system by IPTG. Second, use many simulations together and present statistics on (i) the mean mRNA levels; (ii) the time until the first appearance of an mRNA. Present evidence in favor of/against a Poisson hypothesis versus that of the transcriptional bursting hypothesis.

PROBLEM 6.   Evidence for Transcriptional Bursting—Part 1

This problem uses the mRNA stochastic gene expression dataset available on the book's website. The central question is simple: provide evidence for or against the bursty gene hypothesis. At minimum, you should visualize the raw data and attempt to assess the extent to which the single-cell trajectory supports either Poisson bursting or transcriptional bursting. Notes: (i) cell division takes place at 46, 152, and 266 minutes in this trajectory; (ii) given the stochastic nature of the trajectory, you may not necessarily be able to make definitive conclusions.

PROBLEM 7.   Evidence for Transcriptional Bursting—Part 2

As before, extend the model in problem 2 to include transcriptional bursting, including transitions between On and Off states where transcription only happens when the gene is in the On state. Utilize the same parameters as in the single-cell study of Golding et al. (2005)—but do not explicitly account for cell division. Use your model to vary the On and Off rates and provide evidence for bursting when evaluating the shape of the P_0 curve, i.e., the probability that a cell has 0 mRNA molecules postinduction. Use the data available on the website and see if you can find parameter regimes that coincide with the qualitative, and event quantitative, features of the dataset.

3.8   TECHNICAL APPENDIX

Exponential distributions   The term *Poisson process* denotes a mechanism by which events take place at a rate r per unit time. The probability of a single event taking place in a very small unit of time, dt, is rdt, and the probability that no such event occurs is $1 - rdt$. Given sufficiently small values of dt, ignore the possibility of two events. Section 3.5 discussed how the waiting time distribution between events generated by a Poisson process should be exponentially distributed, i.e., $P(t) = re^{-rt}$. The mean waiting time of an exponential distributed random variable is

$$\langle t \rangle = \int_0^\infty dt\, t P(t),\tag{3.34}$$

which for the exponential distribution becomes

$$\langle t \rangle = \int_0^\infty dt\, r t e^{-rt}.\tag{3.35}$$

This equation can be rescaled in terms of dimensionless time $\tau \equiv rt$ such that $d\tau = rdt$, and therefore

$$\langle t \rangle = \frac{1}{r} \int_0^\infty d\tau\, \tau e^{-\tau}.\tag{3.36}$$

Utilizing the standard method of integration by parts and identifying $u = \tau$ and $dv = d\tau e^{-\tau}$, the integral reduces to

$$\langle t \rangle = \frac{1}{r} \left(\left[-\tau e^{-\tau} \right]_0^\infty + \int_0^\infty d\tau e^{-\tau} \right).\tag{3.37}$$

The left side is 0 due to evaluation of the extreme values of the integral while the right side reduces to 1, so that $\langle t \rangle = 1/r$. Hence, as anticipated, the mean waiting time is the inverse of the underlying rate. Note that the mode of the distribution, corresponding to the most likely time interval, is 0. This gap between the mean and mode is important to keep in mind when considering events that likely have a fixed interval; if so, then the exponential distribution is unlikely to be a suitable approximation.

Competing Poisson processes   Living systems have multiple stochastic processes operating simultaneously, each with its own rate. If these are Poisson processes, then it is possible to understand both how long it will take for *any* event to happen and the probability that at a *particular* event will occur first. To begin, consider two processes that occur at rates r_1 and r_2, respectively. The probability that the first event, of any kind, occurs between T and $T + dt$ must correspond to the probability that no event occurs between 0 and T and that either event occurs in the small unit of time $(T, T + dt)$. Following the logic in this chapter, we can write this as

$$P_{\text{either event}}(T, T + dt) = (1 - r_1 dt - r_2 dt)^{\frac{T}{dt}} (r_1 dt + r_2 dt)\tag{3.38}$$

$$= \left(1 - \frac{r_{tot} T}{T/dt} \right)^{\frac{T}{dt}} r_{tot} dt\tag{3.39}$$

$$= e^{-r_{tot} T} r_{tot} dt\tag{3.40}$$

where $r_{tot} = r_1 + r_2$ is the sum of rates. Hence, rates add rather than mean times. This argument holds for any finite number of finite rates, i.e., $r_{tot} = \sum_i^p r_i$ for p distinct processes. As is evident, the expected mean time for any event also changes:

$$\bar{t}_{\bar{r}} = \frac{1}{\sum_i r_i}. \tag{3.41}$$

Given this exponential waiting time, which process takes place first? In the chapter discussion, the probability that one event takes place before another is written as

$$P_{\text{event 1}} = \int_0^\infty dt r_1 e^{-r_1 t} e^{-r_2 t}. \tag{3.42}$$

We can nondimensionalize time by setting $u = (r_1 + r_2)t$ such that $du = (r_1 + r_2)dt$ and then rewriting the integral as

$$P_{\text{event 1}} = \frac{r_1}{r_1 + r_2} \int_0^\infty du e^{-u}. \tag{3.43}$$

The value of the integral is 1; hence,

$$P_{\text{event 1}} = \frac{r_1}{r_1 + r_2}. \tag{3.44}$$

This derivation also extends to an arbitrary number of processes, i.e.,

$$P_{\text{event i}} = \frac{r_i}{\sum_i^p r_i}. \tag{3.45}$$

There is another way to think about this problem. If we know that some event has taken place in the interval $(t, t + dt)$, then this changes the nature of the problem into one of conditional probability. That is, conditioned upon some event having taken place, which of the p events took place? The probability of each event is $r_i dt$. Hence, the probability is $P(i, event) = P(i|event)P(event)$. Therefore,

$$P(i|event) = \frac{P(i, event)}{P(event)} = \frac{r_i dt}{\sum_i^p r_i dt} \tag{3.46}$$

and after canceling out the small time interval, dt, it yields the same probability as derived above. Note that this dual interpretation of waiting and selecting underlies the simulation technique of the Gillespie algorithm, developed further in the accompanying computational laboratory.

Master Equation The Master Equation given production at a rate β and per molecule degradation at a rate α can be written as

$$\frac{dP_n}{dt} = \beta P_{n-1} + (n+1)\alpha P_{n+1}(t) - \beta P_n(t) - \alpha n P_n(t). \tag{3.47}$$

The change in the average number of proteins per cell is then

$$\frac{d\langle n \rangle}{dt} = \frac{d}{dt} \Sigma_m m P_m(t), \tag{3.48}$$

which can be expressed by moving the derivative in front of each of the $P_m(t)$s, yielding

$$\frac{d\langle n \rangle}{dt} = \sum_{m=0}^{\infty} \beta m P_{m-1} - \sum_{m=0}^{\infty} \beta m P_m$$
$$+ \sum_{m=0}^{\infty} \alpha m(m+1) P_{m+1} - \sum_{m=0}^{\infty} \alpha m^2 P_m. \tag{3.49}$$

A few key points are worth noting. First, note that $P_{-1} = 0$, that is, there cannot be a negative number of proteins in a cell. Second, note that $\sum_{m=0}^{\infty} P_m = 1$ given the definition of P_m as a probability distribution. Finally, note that both $\sum_m m P_m$ and $\sum_m (m-1) P_{m-1}$ are equal to the average number of proteins. The sums are shifted in order, but again, because $P_{-1} = 0$ there is just an extra 0 in the second version. With these points in mind, we can rewrite the sum as

$$\frac{d\langle n \rangle}{dt} = \sum_m \left(\beta(m-1) P_{m-1} + \beta P_{m-1} \right) - \sum_m \beta m P_m$$
$$+ \sum_{m=0}^{\infty} \left(\alpha(m+1)^2 P_{m+1} - \alpha(m+1) P_{m+1} \right) - \sum_{m=0}^{\infty} \alpha m^2 P_m \tag{3.50}$$

which after canceling terms yields

$$\frac{d\langle n \rangle}{dt} = \beta - \alpha \langle n \rangle. \tag{3.51}$$

The dynamics of higher-order moments can be derived in a similar fashion.

Next, at equilibrium, $dP_n^*/dt = 0$ for all values of n. The processes of degradation and synthesis must be balanced so that the probability of finding a cell with n proteins remains constant. This means that at equilibrium we should expect that

$$\beta P_{n-1} + (n+1)\alpha P_{n+1}(t) - \beta P_n(t) - \alpha n P_n(t) = 0. \tag{3.52}$$

Let us denote the expected equilibrium concentration of proteins as $\lambda = \beta/\alpha$. Dividing the equilibrium condition for P_n by α yields

$$(n+1) P_{n+1} - \lambda P_n = n P_n - \lambda P_{n-1}. \tag{3.53}$$

This is a recursive relationship, which should hold irrespective of the value of n such that eventually $f(0) = n P_n - \lambda P_{n-1} = 0$. But this means that $n P_n = \lambda P_{n-1}$ for all n. For the value $n = 1$, this means that, for P_0 to be at equilibrium, then the probability of a new translation event, with rate β, should balance the probability that the system is in a state P_1 and experiences a decay event with rate α. Furthermore, for arbitrary n,

$$P_n = \frac{\lambda}{n} P_{n-1}$$
$$= \frac{\lambda^2}{n(n-1)} P_{n-2}$$

$$= \frac{\lambda^3}{n(n-1)(n-2)} P_{n-3}$$

$$\cdots$$

$$= \frac{\lambda^n}{n!} P_0. \tag{3.54}$$

There is one more constraint to keep in mind. In order to fix the value of P_0, recall that $\sum_n P_n = 1$, which means that

$$P_0 \sum_n \frac{\lambda^n}{n!} = 1. \tag{3.55}$$

However, the sum in the equation is the definition of e^λ, which means that $P_0 e^{-\lambda} = 1$ such that $P_0 = e^{-\lambda}$ and finally

$$P_n^* = \frac{\lambda^n e^{-\lambda}}{n!}. \tag{3.56}$$

At equilibrium, the number of proteins expected in the cell follows a Poisson distribution.

Solving the deterministic equation Consider a gene product—either mRNA or protein—whose dynamics can be written as

$$\frac{dx}{dt} = \beta - \alpha x \tag{3.57}$$

where β is a production rate (nM/hr) and α is a degradation/dilution rate (1/hr). This equation can be written as

$$\frac{dx}{\beta - \alpha x} = dt \tag{3.58}$$

and integrated such that

$$\int \frac{dx}{\beta - \alpha x} = \int dt \tag{3.59}$$

$$\int \frac{dx}{\beta - \alpha x} = \int dt \tag{3.60}$$

$$\frac{\log(\beta - \alpha x)}{-\alpha} = t + \tilde{C} \tag{3.61}$$

$$\beta - \alpha x = C e^{-\alpha t} \tag{3.62}$$

Here C is an integration constant. Finally, by rearranging sides, note that

$$x(t) = \frac{\beta}{\alpha} - \frac{C}{\alpha} e^{-\alpha t}, \tag{3.63}$$

and recognizing that $\beta/\alpha = x^*$ leads to

$$x(t) = x^* - \frac{C}{\alpha} e^{-\alpha t}. \tag{3.64}$$

Hence, the system asymptotically approaches $x(t) \to x^*$. Note that when $x(t) = x_0$, then this equation reduces to

$$x(t) = x^* \left(1 - e^{-\alpha t}\right) + x_0 e^{-\alpha t}. \tag{3.65}$$

This last equation reveals that the initial conditions disappear in relevance given the decay rate and that the steady state is approached at a rate controlled by α. In the limit that $x_0 \to 0$, this reduces to

$$x(t) = x^* \left(1 - e^{-\alpha t}\right). \tag{3.66}$$

Waiting time distributions in multistage transcription Consider a gene that can be in a state of active or silenced transcription. The active state on and the silenced state is off. The gene switches between states at a rate k_{on} and k_{off} characterizing the switching from off to on and from on to off, respectively. The gene is transcribed at a bursty rate b during On states and at a rate of 0 during Off states. What then is the waiting time for the appearance of the first mRNA transcript? For a Poisson case, the waiting time is exponentially distributed, $P(t) = be^{-bt}$, such that the probability that there are no mRNA in a cell after time t is simply $1 - \text{CDF}$ (where CDF denotes the cumulative distribution function). That is, because the CDF denotes the probability that an event has taken place at or before t, $1 - \text{CDF}$ must denote the probability that an event has *not* yet taken place as of time t. For the Poisson, memoryless case, the CDF is $1 - e^{-bt}$. When $t = 0$, the CDF $= 0$, since an event can't have occurred before the start of the process. In the limit that $t \to \infty$, then the CDF $= 1$, since the event certainly must take place in finite time.

The analysis of a multistep process is more complicated. Consider an approximation where cells are in the On state with probability $p \equiv \frac{k_{on}}{k_{on} + k_{off}}$ and in the Off state with probability $1 - p$. We approximate the probability distribution before the appearance of an mRNA transcript by ignoring multiple events, i.e., considering the contributions of sequences like on-mRNA and off-on-mRNA and ignoring sequences like on-off-on-mRNA and off-on-off-on-mRNA. As such, the complete $P_0(t)$ can be derived by first approximating the probability distribution of waiting times and then using that information to estimate $P_0(t)$. Note that if the waiting time distribution is $Q(t)$, then $P_0(t) = 1 - \int_0^t Q(t')dt'$, i.e., the fraction of cells that have not yet had an mRNA transcript.

Consider that the waiting time distribution is the sum of two components: (i) the waiting time distribution for sequences of the kind on-mRNA; (ii) the waiting distribution for sequences of the kind off-on-mRNA. These two distributions are weighted by the probabilities p and $1 - p$, denoting the relative initial distribution of On and Off cells, respectively. The exponential waiting time for transcription beginning in the On state is $Q_1(t) = be^{-bt}$. The second term is the gamma distribution waiting time for bursts beginning in the Off state. This distribution is also equivalent to an Erlang distribution because two events are required—first a transformation from off to on and then a burst event:

$$Q_2(t) \approx \int_0^t dt_b\, be^{-bt_b} k_{on} e^{-(t-t_b)/k_{on}}. \tag{3.67}$$

Note that we do not include the impact of k_{off} in the waiting time distribution but use it only to approximate the initial distribution of cells. The second term is an integral over all potential times to transcribe t_b, noting that if t is the waiting time and t_b is the time before transcription in the On state, then $t - t_b$ denotes the waiting time in the Off state. The following lines show the step-by-step integral of the off-on-mRNA contribution to the waiting time:

$$Q_2(t) = \int_0^t dt_b \, be^{-bt_b} k_{on} e^{-(t-t_b)/k_o n} \tag{3.68}$$

$$= bk_{on} e^{-tk_{on}} \int_0^t dt_b e^{-(b-k_{on})t_b} \tag{3.69}$$

$$= \frac{bk_{on}}{b - k_{on}} e^{-tk_{on}} \left[1 - e^{-(b-k_{on})t} \right] \tag{3.70}$$

$$= \frac{bk_{on}}{b - k_{on}} \left[e^{-tk_{on}} - e^{-bt} \right] \tag{3.71}$$

such that, given $Q_1(t) = be^{-bt}$, we can now write the full equation for $P_0(t)$:

$$P_0(t) = p \left(1 - \int_0^t Q_1(t) \right) + (1 - p) \left(1 - \int_0^t Q_2(t) \right). \tag{3.72}$$

This equation connects the waiting time distribution to the probability that an event has not happened (i.e., 1 minus the probability the event happened between 0 and t). This reduces to

$$P_0(t) = pe^{-bt} + (1 - p) \left(1 + \frac{b}{b - k_{on}} \left(e^{-tk_{on}} - 1 \right) - \frac{k_{on}}{b - k_{on}} \left(e^{-bt} - 1 \right) \right). \tag{3.73}$$

In the limit that $t \to 0$, this reduces to $P_0(t) = p + 1 - p = 1$, i.e., the probability that no mRNA has been released after induction is 1 and then reduces with time. In the limit that $t \to \infty$, this reduces to $P(t \to \infty) = p \times 0 + (1 - p) \times (1 - 1) = 0$. This theoretical approximation is used in the chapter. The homework problems include the challenge to develop a multistage transcription computational model to connect mechanism, theory, and data.

Geometric distributions As in the preceding technical section, consider a gene that can be in a state of active or silenced transcription, i.e., on or off, respectively. Here we will ask: what is the probability that n transcripts are produced in one burst? To answer this question, we first reframe it in terms of a conditional probability, $Q(n|t_{on})$; that is to say, the probability that n transcripts are produced during an On burst that lasts t_{on} amount of time. Then the probability that n transcripts are generated in one burst should be the integral over all potential burst durations, multiplied by the conditional probability of burst sizes with time:

$$P_{burst}(n) = \int_0^\infty dt_{on} P(t_{on}) Q(n|t_{on}). \tag{3.74}$$

Because each On state is a Poisson process, then $Q(n|t_{on}) = \frac{\lambda^n e^{-\lambda}}{n!}$ where $\lambda = bt_{on}$ is the expected number of transcripts generated. Similarly, we expect that On states have an exponential distribution of duration such that $P(t_{on}) = k_{off} e^{-k_{off} t_{on}}$.

Combining terms, we can now derive the expected distribution of burst sizes:

$$P_{burst}(n) = \int_0^\infty dt_{on} P(t_{on}) Q(n|t_{on}) \tag{3.75}$$

$$= \int_0^\infty dt_{on} k_{off} e^{-k_{off}t_{on}} \left(\frac{(bt_{on})^n e^{-bt_{on}}}{n!} \right) \tag{3.76}$$

$$= \frac{k_{off} b^n}{n!} \int_0^\infty dt_{on} e^{-(b+k_{off})t_{on}} t^n. \tag{3.77}$$

It is useful to convert to dimensionless units and ignore the subscript notation for time such that $\tau = (b + k_{off})t$ and $d\tau = (b + k_{off})dt$, so that

$$P_{burst}(n) = \frac{k_{off} b^n}{(b+k_{off})^{n+1}} \frac{1}{n!} \int_0^\infty d\tau \tau^n e^{-\tau} \tag{3.78}$$

where the integral is equivalent to the gamma function, i.e., $n!$ such that

$$P_{burst}(n) = \frac{k_{off}}{b+k_{off}} \times \frac{b^n}{(b+k_{off})^n}. \tag{3.79}$$

Eq. (3.79) is equivalent to a geometric distribution. As noted in the chapter, it can be interpreted as the probability that n events corresponding to the process with rate b (i.e., transcription) occur before the process with rate k_{off} (i.e., switching to the Off state).

Geometric distributions have longer tails than do exponential distributions when $b > k_{off}$. For example, when $b = k_{off}$, then $P_{burst}(n) = 2^{-(n+1)}$, or 50%, 25%, 12.5%, etc., for events of size 0, 1, 2, and so on. When $b = 10k_{off}$, then the probability $P_{burst}(n) = (1/11) \times (10/11)^n$, or approximately 9%, 8% and 7% for events of size 0, 1, 2. In contrast, when $b \ll k_{off}$, then the system begins to resemble a typical memoryless process because it is unlikely to have long bursts interspersed with periods of stasis given fast switching between On and Off states. For example, when $b = k_{off}/10$, then the decay of bursts appears exponential, e.g., $P_{burst}(n) = (10/11) \times (1/11)^n$, or ~91% and ~8% for events of size 0 and 1, respectively, with all other events being highly unlikely. The geometric distribution has a mean equal to the ratio of the events, i.e., b/k_{off}. Because $1/k_{off}$ is the mean duration of the On state, another way to think of this is $b\bar{t}_{on}$, i.e., a rate of transcription multiplied by the duration of the On state. The variance is $(b + k_{off})/k_{off}$ and so is higher due to the additional variance of the duration of the On state.

Chapter Four

Evolutionary Dynamics: Mutations, Selection, and Diversity

4.1 EVOLUTION IN ACTION

The modern study of evolution via natural selection is now more than 150 years old. Whereas the process of evolution is many billions of years old. Evolution has shaped the emergence of life and the diversification of a multitude of forms, and is essential to understanding global-scale challenges, ranging from adaptation to climate change to the spread of antibiotic-resistant pathogens. As Thomas Dobzhansky (1973) famously remarked, "Nothing in biology makes sense except in light of evolution." Yet the core mechanisms of how evolution works remain controversial to some and poorly understood by others—both within and without the scientific community. Hence, this chapter aims to explain how evolution works and how evolution can be modeled, as a "historical" as well as an ongoing process.

To begin, it is worth recalling a few key definitions:

Evolution Any change in the genetic makeup of individuals in a population over subsequent generations

Evolution via natural selection Any *nonrandom* changes in the genetic makeup of a population due to the *differential reproduction/survival* of the individuals

The differences are important. Organisms have genomes that contain the code of life. These genomes are usually DNA based (for cells and viruses) and sometimes RNA based (for viruses). The sequence of chemical letters that make up genomes are copied during reproduction. The copying process is done with high fidelity, but it is not perfect. The small error rate of genome replication varies between organisms. Hence, there is a chance that the genome of an offspring will have a different genome than that of their parent. For haploid organisms (i.e., those with a single copy of their genome), single-nucleotide polymorphisms alone are sufficient to generate differences between mother and daughters. There are many other factors that influence the genetic makeup of organisms at the molecular level. For example, large sections of genomes can be deleted, inserted, and swapped with other genomes. Moreover, for organisms with more than one copy of each genome—like diploid organisms—the offspring genomes reflect site-by-site differences in the combinations of copies passed on from the two parents. Yet it is equally important to note that the theory of evolution via natural selection was developed many decades before the identification of the genetic basis for heritability. Evolution can unfold whenever there is some form of heritable variation.

Although nothing in biology makes sense except in light of evolution, the converse is not necessarily true. Simply knowing that evolution happens does not mean that everything in biology makes sense. Indeed, the list of processes above is partial and there are many unknowns to understanding joint changes in genotypes and phenotypes. But this chapter is not meant to review all the ways in which genomes can change upon reproduction. Irrespective of the mechanism, the question that is central to this chapter is simply: how do changes in a genome translate into changes in the *frequencies* or *numbers* of distinct "genotypes" in a population? This question can be addressed in many ways. This chapter introduces mathematical models in concert with *experimental* approaches to address the problem of assessing evolutionary dynamics. The rationale for doing so is twofold. First, direct tests of evolution in action provide the context to probe both changes and consequences over time. How fast does evolution operate, what are the shape of fitness landscapes, and are populations limited by mutations or, instead, by the competition between many paths to adaptation? Second, by archiving samples in evolution experiments, it is also possible to probe the genomic mechanisms underlying observed changes, a process that has become both increasingly accessible and indeed common given the ongoing revolution of sequencing in the past 30 years.

The popularization and dissemination of the concept that studies of evolution at the population scale could be an experimental science is due, in no small part, to the seminal work of Richard Lenski (Lenski et al. 1991; Elena and Lenski 2003; Sniegowski et al. 1997; Lenski and Travisano 1994; Wiser et al. 2013; Good et al. 2017). Lenski and his team have now studied the evolutionary dynamics of bacteria for over 35 years (Fox and Lenski 2015)—partially interrupted only by the COVID-19 pandemic. At the outset, Lenski's key insight was to recognize that many of the principles underlying evolution via natural selection could be probed, in a forward sense, by growing an organism—*Escherichia coli*—in a relatively simple condition: a shaken flask. The bacteria were inoculated in a DM25 growth medium—a medium in which the principal glucose carbon source is available at relatively low concentrations. The medium also contained small amounts of citrate—a point that will have relevance in a moment. The bacteria grow for approximately 6–7 generations per day, and then 1% of the population is transferred to fresh media the subsequent day. In addition, the population is archived on a regular basis. This process of growth and dilution has continued day after day, month after month, and now year after year for more than 73,000 generations—a span of over 1 million human years. If we are to translate, somewhat dubiously, this time scale into anthropogenic terms, then one might argue that this experiment has lasted more than three times longer than the earliest documented evidence of the emergence of *Homo sapiens* in Africa (Hublin et al. 2017). The "long-term evolution experiment" (or LTEE) has enabled many levels of insights into how evolution works in practice, as well as stimulated new methodological developments to probe genomic variation among individual cells in a population.

The result has been remarkable in many ways. First, as was expected, the bacteria adapted to their new environmental conditions. By "adapted," we mean the bacteria grew

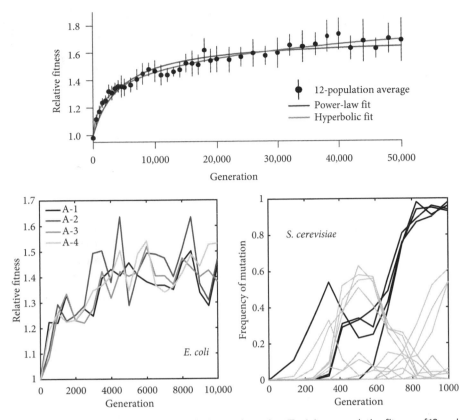

Figure 4.1: Reproducibility and chance in evolutionary dynamics. (Top) Average relative fitness of 12 evolved *E. coli* strains compared to ancestral strain in the first 50,000 generations of the long-term evolution experiment (Wiser et al. 2013, 2014). (Bottom left) Individual resolution of relative fitness of evolved *E. coli* strains compared to ancestral strain in the first 10,000 generations of the long-term evolution experiment (Lenski and Travisano 1994). (Bottom right) Relative frequency of different mutations in the first 1200 generations of an evolving *S. cerevisiae* population (Lang et al. 2013).

faster and evolved to have higher fitness than their ancestors (Figure 4.1). In essence, the bacteria changed over time, often in rapid transitions. As is apparent, initial results over 10,000 generations (bottom left panel) might give the impression that the bacteria could only transiently adapt to the environment, leading to a plateau in fitness. This was not the case. In fact, the population continued to exhibit increased relative fitness (top panel, comparing population averages to alternative fitness models). Second, the advantage of performing experimental evolution is that the genomic basis for such evolution can be probed systematically, by sequencing and comparing changes in genotypes across thousands of generations. In doing so, it is also possible to ask: what is the shape of fitness landscapes, locally at least? That is to say, given a particular genotype, how many nearby mutations lead to organismal death, to increased fitness, or to organisms with slightly worse fitness than the "wildtype"? A long line of papers and reviews of the LTEE are available elsewhere (e.g., Kawecki et al. 2012; Wiser et al. 2013; Good et al. 2017; Blount et al. 2018), and hopefully this brief introduction whets your appetite to learn more.

Over time, the emergent understanding from the LTEE was that a bacterial genotype would adapt increasingly slowly to present conditions. The adaptations are enabled by mutations that enhance fitness, e.g., improving uptake mechanisms, shortening the lag period, and other means of growing in the same, fixed environment. In the same vein, the work of many others have branched out from this initial premise. That is, take a microbe, choose a condition, let the microbe grow again and again in this condition, and then use the changes in the genotype and perhaps even the diversifying community to understand how evolution works now and perhaps even some of the conditions of how evolution has worked in the past given the appropriate choice of conditions. This gets particularly interesting insofar as the selective condition enables the experiment to recapitulate a major evolutionary transition, e.g., the evolution of multicellularity in a test tube (Ratcliff et al. 2012).

At this point, you have may have the impression that the space of genotypes is replete with beneficial adaptations despite the fact that there are many ways for mutations to have a deleterious effect on fitness. How easy it is to adapt lies in the eyes of the beholder. But the visualization of increases in fitness over 10,000 generations belies something else: punctuated evolution. That is, when viewed at higher resolution, it is evident that there are *discrete* jumps in fitness related directly to cell size. The bacteria cells evolved larger cell sizes concurrent with increases in fitness. It could be that cell size was incidental to fitness. For various reasons, Santiago Elena, Rich Lenski, and colleagues favored the hypothesis that cell size was of direct fitness benefit (probably related to increased nutrient uptake). It was also possible that increasing cell size was related to traits that conferred a fitness benefit (Elena et al. 1996), i.e., many genetic changes can lead to changes in more then one phenotype—this is termed *pleiotropy*. In this interpretation, individual bacteria would, on occasion, mutate their genotype and modify their cell size, thereby increasing their growth rate (Figure 4.2). This is, in essence, evolution via natural selection—a concept that applies to Darwin's finches and their evolution of beak size in relationship to environmental disturbances and changing food availability (Grant and Grant 2014), just as well as to bacteria growing in shaken flasks given a stable food source. But is that all there is to the process—stasis interrupted by occasional selective sweeps?

This chapter introduces the key features of models that bridge the gap between mechanisms of evolutionary change and the resulting changes in populations, particularly with a focus on asexual organisms. In doing so, the text is motivated by a number of questions. How fast should such a mutation "fix," i.e., go from a single individual to (nearly) the entire population? How does

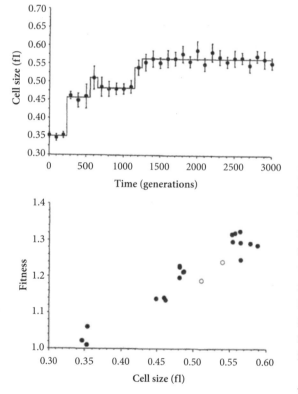

Figure 4.2: The punctuated growth of cell size in the LTEE experiment over 3000 generations. (Top) Evidence of punctuated changes in cell size over 3000 generations, where 1 fl = 10^{-15} L. (Bottom) Evidence that increases in cell size are related to increases in fitness. Reproduced from Elena et al. (1996).

the rate of adaptation depend on mutation rate and other features, e.g., background levels of mutants? And are populations "beneficial mutation limited," i.e., only occasionally experiencing beneficial mutations? Or, instead, are such beneficial mutations in ample supply such that genotypes with different mutations compete with each other, perhaps even with other mutations that have no benefit one way or another but nonetheless hitchhike along for the ride?

This last question also raises a puzzle framed by the bottom right panel in Figure 4.1. This panel shows changes in the frequency of mutations measured in an entirely different set of evolution experiments in yeast (Lang et al. 2013). This study tracked evolving populations of yeast, albeit those that had been sequenced with sufficiently high time resolution to document the spread of mutations in a population. Notably, the timing of these increases suggests a problem: is it really possible that multiple chance mutations swept through the population at nearly precisely the same time (and speed)? (See the darker lines representing mutations that reach nearly 100% frequency). Or, instead, is it possible that the sweep of multiple concurrent mutations represents a balance between selection and chance? The mechanistic basis for evolution via natural selection suggests that reproducibility and chance both play a role in evolution. But how big a role do these two factors play and how would we recognize the balance from experiments? In other words, if you were to measure the changes in frequencies in Figure 4.1 in the case of evolving yeast, how would you interpret it? Does the rise of many mutations at or near the same moment in time reflect a hallmark of reproducible evolution or, instead, a hybrid, in which certain reproducible features are mixed with chance? This chapter introduces models and methods to approach and perhaps even answer this question.

4.2 SELECTION AND THE DISAPPEARANCE OF DIVERSITY

The phrase *survival of the fittest* has long been invoked to describe dynamics in which a subset of individuals outcompetes others. The "more fit" subset is made up of those with traits that increase their competitiveness, fecundity, and/or ability to survive. In some cases, differences in traits can lead to changes in the relative abundance of populations, particularly meaningful in microevoluationary settings (Holt 2009). These traits depend strongly on the organism of interest. For Darwin's finches, both body size and beak size critically shape access to limited seed resources (Grant and Grant 2014). For bacteria, relevant traits may include physical changes to surface moieties that confer resistance to viral infection or the effectiveness of efflux pumps that pump otherwise lethal antimicrobials out of the cell and back into the environment (Madigan et al. 2009). In other cases, adaptation may lead to novel traits that allow organisms and their offspring to explore empty "niches" and potentially catalyze macroevolutionary adaptation (Hutchinson 1959). For fish like the ancient *Tiktaalik*, key adaptive traits include protoappendages that enable them to walk on land; features that could, over time, become the genesis of an entirely new form of life (Shubin 2008). These examples are so disparate that it would seem hard to invoke a single or even limited set of modeling frameworks that could accommodate each. This diversity often poses what seems like an insurmountable obstacle to model development. Perhaps every detail must be included. Indeed, these examples include highly disparate phenotypes and associated genotypes. Yet what is common is that, irrespective of size, scale, or evolutionary epoch, all of these organisms exhibit heritable traits that vary at the population level, the organisms must reproduce, and the degree of variation changes over time. If abstracted in a certain way, the change in the number and type of individuals over time may be predictable, and comparable, across systems, even if highly detailed models may be necessary to explain the changes in both phenotypes and growth rates. In that sense, the path of model building in evolutionary dynamics resembles an argument made by Oliver Wendell Holmes Jr. on the (ir)relevance of certain details when trying to move from legal cases to a "general form":

> The reason why a lawyer does not mention that his client wore a white hat when he made a contract . . . is that he foresees that the public force will act in the same way whatever his client had upon his head. It is to make the prophecies easier to be remembered and to be understood that the teachings of the decisions of the past are put into general propositions and gathered into textbooks, or that statutes are passed in a general form.   —"The Path of the Law", 1897

Let us then move to these general propositions and their prophecies, keeping in mind that they make certain assumptions, often strong assumptions, about the drivers of evolution. And we will soon explore the many ways that conditions may well violate these assumptions, leading to altogether new conclusions. First, we consider perhaps the simplest model of evolution, in which the size of the population remains fixed.

4.2.1   Replicator Dynamics

Consider two populations, A and B, growing at rates r_A and r_B. For a continuous process, one would expect that the population of A would grow like $n_A(0)e^{r_A t}$ and that of B would grow like $n_B(0)e^{r_B t}$. Although both are seemingly better off from an abundance standpoint, the relative amount of an exponentially growing population may nonetheless decline exponentially fast. Note that this paradoxical idea is of practical relevance; it explains why variants of SARS-CoV-2 can take over a population, in a logistic sense, even sometimes when both the variant and wild-type strains increase in abundance (Davies et al. 2021). If we term x_A and x_B as the relative proportion of the two types, then after a long period of time one would expect $x_B \sim e^{(r_B - r_A)t}$, which means that the relative population of B would decrease exponentially whenever $r_B < r_A$. Moreover, such exponential growth cannot continue unabated. Hence, it is worthwhile to develop a simple model of replicators that continuously reproduce in a system in which the total population remains constrained to be finite and less than some total carrying capacity.

To do so, let us begin with the logistic model of population growth:

$$\frac{dN}{dt} = N\left(r - \frac{rN}{K}\right) \tag{4.1}$$

where r is the growth rate and K is the carrying capacity. This model can be extended to multiple types each with abundance N_i growing at a maximal rate r_i, equivalent to

Malthusian fitness. The dynamics of each population can be written as

$$\frac{dN_i}{dt} = N_i \left(r_i - \frac{\sum_{j=1}^{s} r_j N_j}{K} \right) \tag{4.2}$$

where there are s distinct species. In this case, the total population at any point is $N_T = \sum_{j=1}^{s} N_j$ and the average growth rate (or mean fitness) is

$$\bar{r} = \frac{\sum r_j N_j}{\sum N_j} \tag{4.3}$$

where the indices have been suppressed for convenience. This multispecies logistic model is known as the Verhulst equation (see Schuster 2011 for more details). Given that the growth rate of any individual population growing on its own is constrained to be less than the carrying capacity, it seems intuitive that the total population would be similarly constrained. This intuition is correct. As shown in the technical appendix, the dynamics of the total population can be rewritten as

$$\frac{dN_T}{dt} = \bar{r}(t) N_T \left(1 - \frac{N_T}{K} \right) \tag{4.4}$$

such that the total population converges to K in the long-term limit with the caveat that the mean fitness changes over time.

As a result, a set of s populations that can each grow with a fitness r_i in the absence of density-dependent limitation will lead to an increase in the total population toward the carrying capacity K. In doing so, the relative abundances of each of the species will change—but how? To explore these changes, define the relative abundance of each species as $x_i = N_i/N_T$. The dynamics of the relative abundances can be expanded using the quotient rule while using the shorthand notation $\dot{N} = dN/dt$ to denote a derivative with respect to time, such that

$$\frac{dx_i}{dt} = \frac{\dot{N}_i}{N_T} - \frac{N_i}{N_T} \frac{\dot{N}_T}{N_T} \tag{4.5}$$

$$= \frac{N_i}{N_T} \left(r_i - \frac{N_T}{K} \bar{r} \right) - \frac{N_i}{N_T} \left(1 - \frac{N_T}{K} \right) \bar{r} \tag{4.6}$$

By replacing the relative abundances N_i/N_T with x_i and removing canceled terms, we find

$$\frac{dx_i}{dt} = x_i \left(r_i - \bar{r} \right). \tag{4.7}$$

This is the standard replicator equation from evolutionary dynamics. It says that each population grows in relative abundance at a rate equal to the difference between its fitness and that of the average fitness. In other words—relative fitness matters. This equation is typically interpreted to represent the changes in the relative abundances of different species (or replicators) given a constant population. However, we can also interpret the replicator equation as describing the dynamics of the relative abundances of s populations each with

Case: $r_A > r_B$ Case: $r_A > r_B$

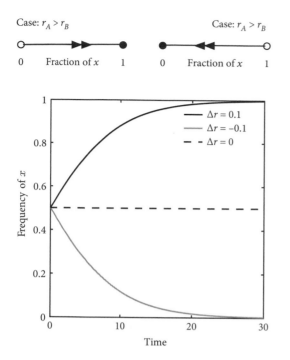

Figure 4.3: Long-term dynamics in replicator dynamics. (Top) One-dimensional state space representation with two cases. When $r_a > r_b$, then the system converges to $x = 1$ (associated with the type A replicator) and the converse holds when $r_a < r_b$. (Bottom) Simulation of replicator dynamics with $\Delta r = 0.1, -0.1$, and 0, using the standard replicator dynamics and initializing the system with $x_0 = 0.5$ (an equal balance of type A and type B replicators).

different maximal growth rate and growing toward a fixed carrying capacity. Note that there are other ways to derive the replicator equation, and the technical Appendix focuses on the example of replication of two populations at a fixed population size.

Indeed, for two populations it is possible to describe the dynamics in terms of $x \equiv x_A$, from which x_B can be deduced, i.e., $x_B(t) = 1 - x_A(t)$. In that case, we now have

$$\frac{dx}{dt} = r_A x - \left(r_A x + r_B(1-x) \right) x \qquad (4.8)$$

$$= x \left(r_A - r_A x - r_B(1-x) \right) \qquad (4.9)$$

$$= x \left(r_A(1-x) - r_B(1-x) \right) \qquad (4.10)$$

$$= x(1-x) \quad \overbrace{\left[r_A - r_B \right]}^{\text{selection coefficient}} \qquad (4.11)$$

This last equation is the logistic growth equation where the selection coefficient, i.e., $\Delta r = r_A - r_B$. The solution to the logistic growth equation is

$$x(t) = x_0 \left[\frac{e^{\Delta r t}}{1 + x_0 \left(e^{\Delta r t} - 1 \right)} \right]. \qquad (4.12)$$

As is apparent, the fitness difference determines which of the two types grows to dominate (Figure 4.3). This growth equation is also used in practice—the technical appendix describes how to use changes in frequency among competing strains to estimate the selection coefficient from experiments (a process used in Lang et al. 2013).

However, it is often the case that many strains are limited by a shared carrying capacity. Given dynamics that are strictly determined by Malthusian growth rates, then it seems the fittest should indeed survive, i.e., $x_i \to 1$ if r_i is the maximum of all the growth rates $\{r_1, r_2, \ldots, r_S\}$. As a consequence, the other populations should go to zero given pure Malthusian growth competition among diverse replicates. Indeed, a solution of the analogous s species version of the logistic equation model is

$$x_i(t) = \frac{x_i(0)e^{r_i t}}{\sum_{j=1}^{s} x_j(0)e^{r_j t}}. \qquad (4.13)$$

One consequence is that the mean fitness continually increases with time. As a result, only those species whose fitness exceeds the mean fitness increase in abundance, and the other species decrease in abundance. Over time, fewer and fewer species have fitness values above the mean, until only one does—the species with the highest Malthusian fitness.

In essence, selection in the replicator model eliminates diversity. The model also suggests that, if $\Delta r = 0$, then the proportions will remain constant. Proportions may remain constant given the absence of fitness differences in a theoretical sense for an infinitely large population. However, the process of genetic drift in a finite population leads to changes in the frequency of types even when there are no intrinsic fitness differences. Yet, for cases where fitness differences are present, the finding that a subset will dominate raises a question: are there any general conclusions we can make about what drives the rate at which selection can, in theory, eliminate diversity? This question is at the heart of the premise underlying Fisher's fundamental theorem of natural section.

4.2.2 Fisher's Fundamental Theorem

Ronald Fisher, a statistician and evolutionary biologist, proposed the following theorem: the average rate of increase in fitness is equal to the variance of fitness in a population (Fisher 1930). In essence, more variation can somehow drive faster rates of change, presumably accelerating the rise of the most fit. One way to demonstrate this "fundamental theorem of natural selection" is to consider the case of S replicators, each with a population fraction x_i, whose dynamics can be described as

$$\frac{dx_i}{dt} = r_i x_i - \langle r \rangle x_i \tag{4.14}$$

where $\langle r \rangle = \sum_{i=1}^{S} r_i x_i$ is the average fitness. The change in the average fitness is then

$$\frac{d\langle r \rangle}{dt} = \frac{d}{dt}\left(\sum_{i}^{S} r_i x_i\right) \tag{4.15}$$

$$= \sum_{i}^{S} r_i \frac{dx_i}{dt} \tag{4.16}$$

$$= \sum_{i}^{S} r_i \left(r_i - \langle r \rangle\right) x_i \tag{4.17}$$

$$= \left(\sum_{i}^{S} r_i^2\right) - \left(\langle r \rangle \sum_{i}^{s} r_i x_i\right) \tag{4.18}$$

$$= \langle r^2 \rangle - \langle r \rangle^2 \tag{4.19}$$

where the last line shows that the change in mean fitness is equal to the difference between the average squared fitness and the average fitness squared. This is precisely the definition of the variance, i.e.,

$$\frac{d\langle r \rangle}{dt} = \text{Var}(r). \tag{4.20}$$

The variance is positive by definition so that the mean fitness seems to always increase. But this concept seems incompatible with our earlier finding that selection purges diversity. The resolution is found by realizing that the very process of selection operating to increase mean

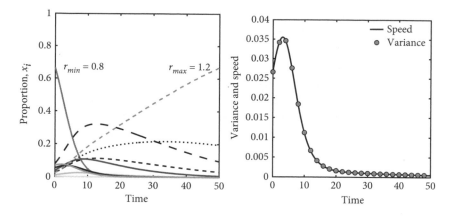

Figure 4.4: Dynamics of $S = 10$ populations competing via replicator dynamics. (Left) The faster replicator dominates the system; all other populations will eventually go to zero. (Right) Variance-speed relationship with time; the values coincide at all times. Note that the system increases in mean fitness at its maximal value precisely when there is the greatest variance in fitness, and that both decay to zero as the system is dominated by the fastest replicator.

fitness also purges diversity, thereby slowing down the increase of mean fitness. Asymptotically, one expects that $\text{Var}(r) \to 0$, i.e., the system increases in fitness while purging itself of variation—the grist for the acceleration in fitness.

These results can be seen in a model of 10 competing populations, shown in Figure 4.4. As is apparent, the replicator with the highest growth rate eventually dominates the system and the replicator with the lower growth rate (despite having the highest initial frequency) is rapidly purged from the system. This is the essence of frequency-independent selection, insofar as a faster replicator always outgrows its competitors (in relative terms). In doing so, the speed of adaptation increases, at least at the beginning, until it starts to approach 0 (Figure 4.4, right). The trajectory of adaptation precisely matches that of the measured variance during the simulation. Hence, Fisher's theorem is true in a mathematical sense, but is also of limited value in a biological sense, as it has many requirements that continue to limit its utility in understanding all but the most constrained of evolutionary dynamics. The key insight is that variation provides the basis for evolution to modulate diversity and average fitness. The other insight is that selection is relative, but as we will see later, fitness need not always go up in an absolute sense.

4.3   MECHANISMS THAT RESTORE DIVERSITY

There are many mechanisms to restore diversity. But rather than list/discuss all of them, this chapter examines two: mutation-selection balance and negative frequency-dependent selection. This first mechanism is important because it shows the dueling tension between selection, which can act to purge diversity, and the process of mutation, which increases diversity. The second mechanism is often invoked when context matters, i.e., when ecology or feedback between organisms (or the environment) means that there is not a unique, global optimum. Instead, the very same reasons that might explain the rise of a type that is rare may also explain its failure to eliminate competitors when abundant.

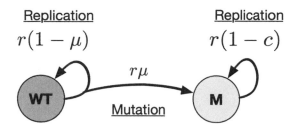

Figure 4.5: Mutation-selection balance in which the wild type (WT) replicates at a rate r with mutation rate μ, and the mutant M replicates at a rate $r(1-c)$ where c denotes the fitness cost of the mutation.

4.3.1 Mutation-selection balance

Consider two variants: an ancestral population, which we term the wild type, and a mutant population, that has a deleterious mutation. The wild type replicates at a rate r, and the mutant replicates at a rate $r(1-c)$ where c is the fitness cost of the mutation. This model of wild types and mutants is inspired by the dynamics of RNA viruses for which mutations occur frequently. Some of these mutations may be lethal such that the RNA virus is not able to infect or replicate a target host cell. Such offspring cannot survive. Yet what about those daughter viruses that bear a partial cost? How can they continue to persist? To answer this question, consider the following replicator dynamics describing changes in the frequencies of wild-type individuals, x, and mutants, y:

$$\frac{dx}{dt} = r(1-\mu)x - \langle r \rangle x \tag{4.21}$$

$$\frac{dy}{dt} = r\mu x + r(1-c)y - \langle r \rangle y \tag{4.22}$$

where μ is the mutation probability per replication and $\langle r \rangle = rx + r(1-c)y$ is the average growth rate (Figure 4.5). Because replicator dynamics describe changes in the relative growth rate, we can rescale these equations in terms of a dimensionless time, $\tau \equiv rt$, where a change of 1 in τ represents a typical generation time. The rescaled dynamics become

$$\frac{dx}{d\tau} = (1-\mu)x - \tilde{r}x \tag{4.23}$$

$$\frac{dy}{d\tau} = \mu x + (1-c)y - \tilde{r}y \tag{4.24}$$

where $\tilde{r} = x + (1-c)(1-x)$ given that frequencies must add to 1, i.e., $y = 1 - x$. Given that the dynamics of the wild type completely specify the mutant fraction, this system of equations has an equilibrium when $\frac{dx}{dt} = 0$ or when $x^* = 0$ and $x^* = 1 - \frac{\mu}{c}$—only one of which is stable given the ratio of mutation rate to fitness cost of mutants. The former is stable when $\mu > c$ and the latter is stable when $\mu < c$. Hence, the wild type should persist insofar as the fitness cost of mutations exceeds the mutation probability. When this occurs, then $y^* = \frac{\mu}{c}$ and the system coexists due to mutation-selection balance (Figure 4.6, left). In a mutation-selection

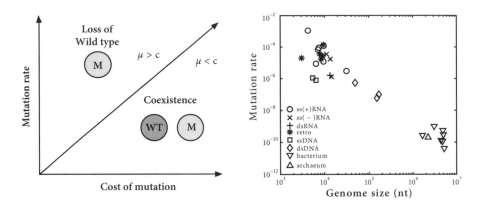

Figure 4.6: Mutation-selection balance. (Left) Expected dynamics in the $c - \mu$ phase plane. Wild types and mutants coexist when $\mu < c$, i.e., when the levels of mutation are lower than the relative fitness costs of the mutant compared to the wild type. However, when $\mu > c$, then the mutant population grows and eventually eliminates the wild type. (Right) Relationship between mutation rate per site and genome length in different viral and microbial organisms. Adapted from Sanjuan (2010).

balance scenario, the more fit wild type generates a constant flux of new mutants, which themselves replicate, albeit at a lower rate. For RNA viruses, the mutation probability per replication is approximately 1/1000 (Figure 4.6, right). Hence, mutants will not be able to outcompete wild-type individuals so long as costs exceed more than a 0.1% reduction in fitness. Yet the wild-type individuals will not be able to completely displace the mutants, given that their success continues to represent the source of new variants. Note that in effect the system is driven to a lower fitness in terms of replicator speed, but to a higher fitness when accounting for the mutational costs on fitness.

4.3.2 Frequency-dependent selection

Mutation-selection balance can give rise to coexistence of types. However, such coexistence seems fragile, as the continual persistence of the mutant type requires the continual renewal of its population via new mutations. Instead, consider two replicators, each of whose frequency changes according to the following:

$$\frac{dx_i}{dt} = r_i x_i - \langle r \rangle x_i \tag{4.25}$$

where $\langle r \rangle = \sum_{i=1}^{S} r_i x_i$ is the average fitness. When r_i is fixed, this leads to the elimination of diversity, which is rapidly purged by evolution. However, what happens if the fitness of each strain is context dependent? A strain may produce some form of public good, like the bacteria *Pseudomonas aeruginosa*, which secretes an extracellular enyzme called a siderophore that binds to nonsoluble iron, thereby enabling it to be taken up—at least potentially—by cells. Yet, when abundant, a producer strain might be less fit than a cheater strain that can take advantage of the iron without incurring the costs of producing and secreting siderophores. In this situation, then, the frequency of the producer strain $r_1(x_1)$

might decrease with its own frequency x_1. If that is the case, then the strain with the larger relative fitness will shift with frequency. As one conceptual example, the curves in Figure 4.7 depict a case where type 1 has a higher fitness than does type 2 when rare, but has a lower fitness when abundant. Recall that for two populations where $x \equiv x_1$, $y \equiv x_2$, and $x + y = 1$, the replicator dynamics reduce to

$$\frac{dx}{dt} = x(1 - x)\left(r_1(x) - r_2(x)\right). \qquad (4.26)$$

Hence, the critical value x_c satisfying $(r_1(x_c) = r_2(x_c))$ will represent a stable equilibrium irrespective of the particular shape of r_1, insofar as $\partial r_1(x)/\partial x < 0$ and $\partial r_2(x)/\partial(x) > 0$. This implies that frequency should increase for $x < x_c$ and decrease for $x > x_c$, leading to a stable fixed point of the system at $x^* = x_c$, corresponding to a mixed system with two strains. Yet, notably, this simple example also illustrates that this fixed point $x^* = 0.5$ need not be the maximum average fitness, which occurs for a value $x < x^*$. This example reinforces the point that the selection process need not act to maximize the fitness for the entire population. Moreover, it also reveals the limitation of Fisher's fundamental theorem of natural selection: it does not apply to cases where fitness depends on context (e.g., frequency).

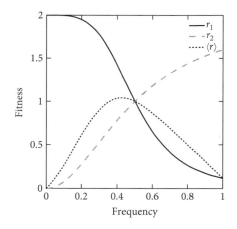

Figure 4.7: Illustration of negative frequency dependence for two frequency-dependent fitness functions; here $r_1(x)$ decreases with x and likewise $r_2(y)$ decreases with $y \equiv 1 - x$. The average fitness is equal to $\langle r \rangle = r_1 x + r_2(1 - x)$. The fixed point of this system occurs when $r_1 = r_2$ (denoted by the intersection of curves). However, note that this equilibrium does not correspond to the largest average fitness, which is at a smaller value of x. In this example, $r_1(x) = \frac{2}{1 + x^4/0.5^4}$ and $r_2(x) = \frac{2x^2}{x^2 + 0.5^2}$.

4.4 STOCHASTICITY IN THE EVOLUTION OF POPULATIONS—BASELINE EXPECTATIONS

4.4.1 Reproduction and survival—a recipe for evolution without natural selection

Evolution denotes the heritable change in the frequency of genotypes from one generation to the next. Hence, any description of evolution must include at least two genotypes and some way of transmitting, with fidelity, the genotype from parent to offspring (i.e., from one generation to the next). Put concretely, consider a population with N individuals in which there can be one or more types (or genotypes). In this model, asexual organisms produce n_i offspring each, such that there is a pool of new individuals of size $N_{off} = \sum_{i=1}^{S} n_i$ where S denotes the number of types in a population. Further, there need not be any difference in the expected number of offspring between types such that the average number of offspring $b \equiv \langle n_i \rangle$ is the same for all genotypes. A key approximation underlying a large class of evolutionary dynamics models is that the number of individuals remains constant over time, i.e., from one generation to the next. Hence, in this model, all N of the "mothers" die while N of the N_{off} offspring survive to become the next generation of mothers. In such a case, differences in genotype frequencies will emerge even if there are no inherent differences in the expected number of offspring or the survivorship of offspring, The notion that the stochasticity of reproduction and survival can lead to "neutral" evolutionary dynamics is,

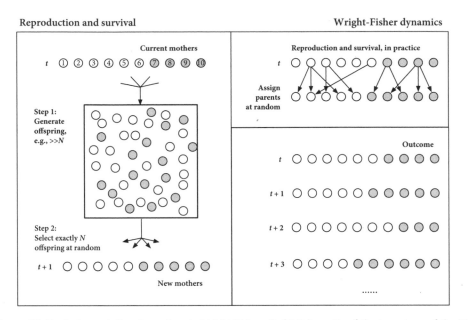

Figure 4.8: Neutral population dynamics via Wright-Fisher. (Left) Schematic of the two steps of the WF model given non-overlapping generations. Step 1: Random production of offspring. Step 2: Selection of *N* offspring for the next generation. (Right) Expected outcome—parent-offspring relationships define differences in genotype frequency that change stochastically over time, even if the average frequency is expected to remain unchanged.

in essence, the central assumption of the Wright-Fisher (WF) model of population genetics (Figure 4.8).

There are different ways to think about these dynamics. For example, one could envision the state of the system as being a set of N unique identifiers, each referring to the genotype identity of the individuals:

$$\overbrace{t: \text{A E B B A A A B B D} \ldots \text{M J B}}^{N \text{ in total}} \tag{4.27}$$

However, if there is no difference between individuals of the same genotype, then the composition of a population can be described with a set of pairs: the genotype ID and its corresponding abundance. For example, with three types in a population of 100, this might be

ID	A	B	C
Number	60	10	30

Hence, the "state space" is a set of S numbers, referring to the S genotypes, such that $\sum_{i=1}^{S} n_i = N$. In the case that there are only two types, we can refer to the types as A and B such that $n_A + n_B = N$. Because the sum is fixed, we only need one number, $n \equiv n_A$. This implies that the Wright-Fisher process has the effect of changing the number n from one generation to the next. This also means that the Wright-Fisher model is a form of Markov chain. The word *Markov* denotes the memoryless nature of the stochastic process (depending only on the current state), and the word *chain* denotes the fact that there are proscribed sequences of events that are possible. What then are the expected dynamics of n_t and how much difference do we expect between trajectories, i.e., what is the expected variance?

4.4.2   Population genetics models of non-overlapping generations

A system in state n at generation t can become a system in state n' in the next generation $t + 1$. For example, if $n = 50$ now, then it is possible that there may be 50, or perhaps more or less, individuals of type A, at the next generation. Formalizing this requires considering the WF process in terms of *conditional probabilities*. That is, conditional upon the fact that the system has precisely n individuals of type A, then $T(n'|n)$ denotes the probability that the system will have precisely n' individuals of type A at the next generation. One restriction is that the system must go somewhere, i.e., $n' \in 0, 1, 2, \ldots, N$:

$$\sum_{n'=0}^{N} T(n'|n) = 1. \tag{4.28}$$

The structure of the transition matrix can be seen in the following transition matrix for $N = 8$ (Figure 4.9) and then for $N = 80$ and then for $N = 800$. As is apparent, as N increases,

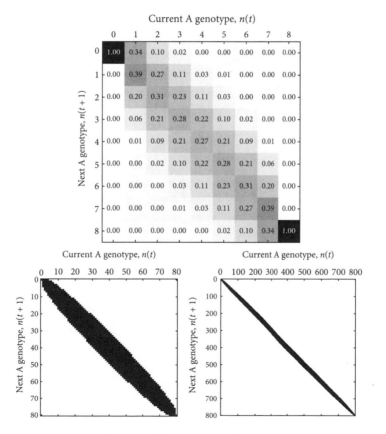

Figure 4.9: Transmission matrix in the neutral WF model given the current state n in columns and the next state $n(t+1)$ in rows. (Top) Transition values given a population of size $N = 8$. (Bottom left) Transition values $T > 0.01$ for $N = 80$. (Bottom right) Transition values $T > 0.01$ for $N = 800$.

the probability that the next state will be close to the current state becomes more tightly centered (in relative terms) on the current value—a point that is more readily seen when considering those states that might occur next with a probability exceeding 0.01. In essence, $T(n'|n)$ is the probability that a type A individual is selected n' out of the N trials, one for each individual in the next generation. This should remind you of something: flipping coins, albeit biased coins. The bias comes in because the probability of choosing a type A individual is n/N and not necessarily 1/2. Therefore, given the probability of "success" of n/N (type A) and the probability of "failure" of $1 - n/N$ (type B), this probability is equivalent to a binomial distribution,

$$T(n'|n) = \binom{N}{n'} p_A^{n'} p_B^{N-n'} \tag{4.29}$$

or

$$T(n'|n) = \binom{N}{n'} \left(\frac{n}{N}\right)^{n'} \left(1 - \frac{n}{N}\right)^{N-n'}. \tag{4.30}$$

The expected mean of a binomial process is equal to the number of trials, N, multiplied by the probability of success, i.e., n/N—the fraction of type A individuals such that the expected number of type A individuals remains the same from one generation to the next, i.e., $\langle n \rangle = n_0$. Formally speaking, this means that the expected value of the Markov chain does not change with time (Feller 1971). This property becomes essential to resolve a crucial question concerning WF dynamics. Given that $n(t=0) = n_0$, what is the probability of fixation, i.e., that the system will converge to being entirely dominated by type A individuals, $n \to N$? The potential paradox is that the expected number of type A individuals should not change, even though it seems that variation may drive the system to one of two extremal outcomes.

4.4.3 Variation and fixation

One way to think about fixation is to note that the WF model represents a Markov chain with two absorbing states. An *absorbing state* is one in which $T(n'|n_a) = \delta_{n',n_a}$ such that the only transition is to remain in the same place. For an evolutionary dynamics model without mutation or immigration, there are two absorbing states, $n = 0$ and $n = N$, corresponding to the elimination and fixation, respectively, of type A. Although the mean value of n is expected to remain constant, the variance in the next generation is nonzero, i.e., $\text{Var}(n'|n) = n(1 - n/N)$ for a binomial process. Hence, different realizations of the WF process are often different than each other, maximally so when they originate from the point $n = N/2$ where the types are in equal proportion. How much variation should be expected over time across evolutionary trajectories given that the initial conditions were shared? In other words: how repeatable is evolution given drift?

One answer can be approached by calculating the probability that two randomly chosen individuals in the population are of different genotypes. Formally, this is termed *heterozygosity*, or

$$H = 2\frac{n}{N}\left(1 - \frac{n}{N}\right) \tag{4.31}$$

where n is the number of A genotypes, $N - n$ is the number of B genotypes, and therefore n/N is the fraction of A genotypes and $1 - n/N$ is the fraction of B genotypes. The origin of the 2 may appear mysterious, until one considers the different ways of choosing two individuals: (A, A), (A, B), (B, A), and (B, B). Out of these four options, only the sequences (A, B) and (B, A) contribute to heterozygosity. The probability of selecting (A, B) or (B, A) is the same: $\frac{n}{N}\left(1-\frac{n}{N}\right)$, but both contribute to the total probability of choosing two different genotypes in a population. Over time, the expected dynamics of heterozygosity, $H(t)$, can be computed by calculating the value one generation ahead, where $E(H)' \equiv \langle H(t+1)\rangle$:

$$E(H') = E\left(2\frac{n'}{N}\left(1-\frac{n'}{N}\right)\right). \qquad (4.32)$$

By expanding these expectations, it is possible to show (see technical appendix) that for large populations, $N \gg 1$,

$$H(t) \approx H_0 e^{-t/N}. \qquad (4.33)$$

Hence, heterozygosity decays exponentially, leading to the rapid elimination of variation. This equation implies that the WF model of neutral drift will eventually lead to a state in which there is no variation between individuals, i.e., to one of the two absorbing states, either local extinction or fixation. The question now is, which one and with which probability?

To answer this question, we must turn back to the feature of this Markov chain—that its expected value remains constant over time. Because of the two absorbing states, the ensemble probability distribution should converge to

$$P^*(n) = \pi\delta_{n,N} + (1-\pi)\delta_{n,0} \qquad (4.34)$$

where π is the fixation probability (and not the number 3.14159…) and the δ denotes the Kronecker delta, which is equal to 1 when the two arguments in its subscripts are equal and zero otherwise. Therefore, it should be possible to calculate the expected value of n, or

$$E^*(n) = \Sigma_{n=0}^{N}P^*(n) = \pi N + 0 \times (1-\pi) = \pi N. \qquad (4.35)$$

However, given that the expected mean value in the ensemble should remain constant, this value πN must equal the initial value n_0, such that $\pi = n_0/N$. In other words, the probability of fixation is equal to the initial fraction of type A genotypes in the population. Likewise, the probability of extinction is equal to the initial fraction of type B genotypes in the population. Another way to think about this is that eventually one of the current individuals will be the ancestor of all extant individuals. Since each individual has an equal chance of being that progenitor, then the probability that the progenitor is type A is n_0/N, or the probability of randomly choosing a type A individual. Hence, in the absence of selection and/or mutation, even the case of replication with equivalent fitness values will eventually lead to the extinction of diversity, albeit over time scales set by the size of the population.

4.5   EVOLUTIONARY DYNAMICS WITH STOCHASTICITY AND SELECTION

This section combines principles developed thus far that suggest that drift can lead to the local fixation or extinction of a population and that selection is a powerful force to drive one population to fixation at the expense of others. Both drift and selection seem likely to make it harder for coexistence to emerge, yet unlike drift, the process of selection provides the basis for understanding the extent to which the outcomes of evolution are repeatable and predictable. Building simple evolutionary models that incorporate both selection and drift is a first step to parsing patterns emerging from experimental evolutionary data, whether in bacteria, yeast, or beyond.

4.5.1 The Moran model of population genetics

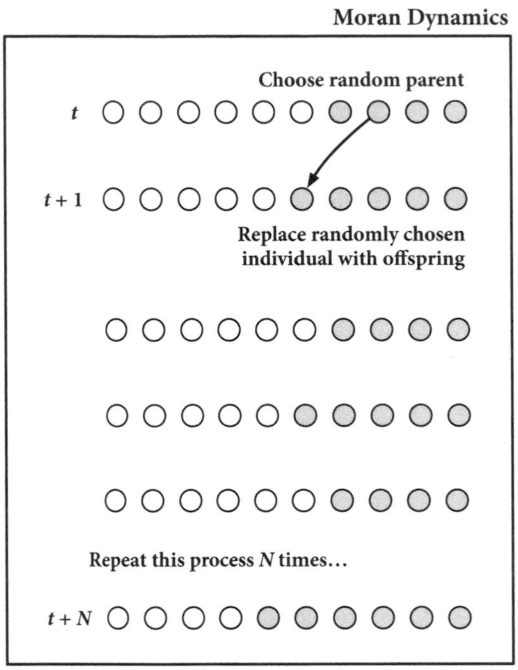

Figure 4.10: Schematic of the Moran model, in which a randomly selected parent reproduces and replaces any one of the N current individuals with one of its offspring. A total of ~ N such events are equivalent to a single generation of the WF model.

Here we consider a variant of evolutionary dynamics in which there are exactly N individuals, but the dynamical steps involve single replacement events only. The Moran model includes two steps. In the first step, a random individual is chosen to give birth, i.e., to replicate. In the second step, a randomly chosen individual is selected to die, and is replaced by the offspring of the first individual, as is shown in Figure 4.10. In this case, the transition matrix is different, in that given a system in state n at time t, there can be either $n - 1$, n, or $n + 1$ individuals of type A at time $t + 1$. Here it appears evolution moves more slowly than it does in the WF case. Another way to think about this is that it takes more steps of the Moran model to be equivalent to a single generation in the Wright-Fisher model—formally speaking, this is on the order of N steps. In the event that all individuals have the same fitness, then

$$T(n'|n) = \begin{cases} \frac{n}{N}\left(1 - \frac{n}{N}\right) & n' = n+1 \\ \frac{n}{N}\left(1 - \frac{n}{N}\right) & n' = n-1 \\ 1 - T(n-1|n) - T(n+1|n) & n' = n \\ 0 & otherwise \end{cases}$$

$$(4.36)$$

The transition matrix for the Moran model is shown in Figure 4.11, noting that the only nonzero values lie on or adjacent to the diagonal.

The properties of the Moran model are quite similar to that of Wright-Fisher, namely, that the expected number remains constant and the heterozygosity declines exponentially (see technical appendix). The major difference is that the time scales are distinct and it takes N Moran steps to equal a single generation step in the Wright-Fisher model (again, see technical appendix). The other key point to recall is that the transition matrix for the Moran model has positive values either on or just adjacent to the diagonal. This is unlike the Wright-Fisher model, whose probabilities are centered on the diagonal, but have nonzero weight (nearly) everywhere. The Moran model is also useful as a means to consider the combination of selection and drift—and how both converge to yield realized, evolutionary dynamics.

4.5.2 Selection in light of stochasticity

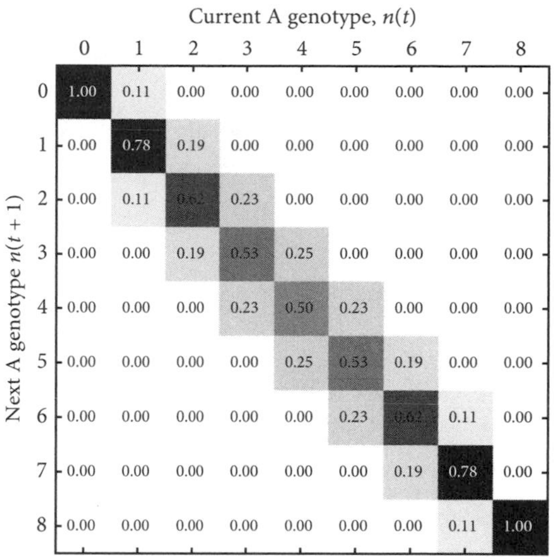

There are multiple mechanisms by which selection can change the state of populations. Advantageous changes in genotypes can increase fitness relative to the background wild type. In a deterministic model, the more fit the organism is, the faster it takes over the population. But that is merely a quantitative issue. The deeper implication of deterministic models is that a better-fit mutant *always* outperforms the wild type. In the real world, that is not necessarily the case. Consider one better-fit individual that has a 10% growth rate advantage compared to all other individuals in a population. This mutant may, by chance, die before reproducing. Hence, it is obvious that selective advantages, in and of themselves, do not guarantee success. Moreover, it is also possible for the converse to occur. That is, a mutant with a fitness disadvantage may grow in number and reach fixation in a small population. A model will help us conceptualize the relative importance of selective advantages, initial sizes of mutant populations, and the total size of the population, while estimating the chance of fixation as well as the time scale over which fixation occurs.

To explore the impact of selection requires modifying the Moran model. Instead of assuming strict neutrality, we will assume that type A individuals have a slightly higher chance of being chosen for replication, i.e., by a factor of $(1+s)$ relative to that of type B individuals. The value s denotes the fitness difference such that $s>0$ represents a selective advantage and $s<0$ represents a selective disadvantage. The rescaled transmission matrix can be written as

$$T(n'|n) = \begin{cases} \frac{n(1+s)}{N+sn}\left(1-\frac{n}{N}\right) & n'=n+1 \\ \left(\frac{N-n}{N+sn}\right)\frac{n}{N} & n'=n-1 \\ 1-T(n-1|n)-T(n+1|n) & n'=n \\ 0 & \text{otherwise} \end{cases} \tag{4.37}$$

where the weak selection limits correspond to $0\leq|s|\ll 1$. Note that the ratio of transition rates $T(n+1|n)/T(n-1|n)=1+s$—unlike the neutral case where the ratio is equal. Given this transmission matrix, what is the probability that type A fixes if there are n_0 individuals of type A initially—or perhaps just a single one?

To answer this question, denote π_n as the probability of fixation given that there are n individuals of type A. Because the Moran process is a Markov process, π_n is independent of time, e.g., the future chance of fixation will be the same whether or not the system has $n=20$ individuals at time 0, 5, or 500. We can leverage this Markovian process to write down a recurrence relationship among fixation probabilities:

Figure 4.11: Transmission matrix in the neutral Moran model given the current state n in columns and the next state $n(t+1)$ in the rows. Note that the only non-zero terms are those that are on the diagonal or adjacent to the diagonal. Note also that in this population the total size is 8, such that when $n=0$ or $n=8$, then the only nonzero term is $T(0|0)$ and $T(8|8)$ respectively, given that such systems have reached local extinction or fixation.

$$\pi_n = T(n+1|n)\pi_{n+1} + T(n-1|n)\pi_{n-1} + T(n|n)\pi_n. \tag{4.38}$$

In essence, this says that the probability of fixation given n individuals can be decomposed into the probability of fixation given the three possible outcomes at the next moment ($n-1$, n, and $n+1$) multiplied by their corresponding fixation probabilities.

After some algebra (see the technical appendix), it is possible to conclude that

$$\left(\pi_{n+1} - \pi_n\right) = \frac{\pi_n - \pi_{n-1}}{1+s}. \tag{4.39}$$

This recurrence relationship provides a route to find π_n. It helps that we already know two values, irrespective of s, that is, $\pi_0 = 0$ and $\pi_N = 1$. Hence, the recurrence relationship can help once we realize that if $\pi_1 = C$, then we can use the "boundary conditions" to write

$$\pi_2 - \pi_1 = \frac{C}{1+s} \tag{4.40}$$

and by extension

$$\pi_{n+1} - \pi_n = \frac{C}{\left(1+s\right)^n}. \tag{4.41}$$

By using the fixation boundary condition and solving for C, we can arrive at the final result,

$$\pi_n = \frac{1 - \left(1+s\right)^{-n}}{1 - \left(1+s\right)^{-N'}} \tag{4.42}$$

which in the limit of very large populations becomes

$$\pi_n = 1 - \left(1+s\right)^{-n}. \tag{4.43}$$

This allows us to ask: what is the chance that a new mutant in a very large population reaches fixation? The answer should be π_1, or

$$\pi_1 = \frac{s}{1+s}. \tag{4.44}$$

Moreover, if the fitness is much less than 1, then the relative fitness advantage s denotes the approximate probability of fixation. That is a 10% fitness advantage yields an approximately 10% chance of fixation in an infinite population. Likewise, for deleterious mutations, then

$$\pi_1 = \left(1+s\right)^{N-1} \tag{4.45}$$

so that, when $N|s| \gg 1$, it becomes virtually impossible for a deleterious mutation to fix. Hence, with a 10% fitness disadvantage, populations beyond 10 or so individuals rarely have such a mutant fix, whereas a very weakly deleterious mutation, e.g., with a 0.01% fitness disadvantage, might reach fixation in a population of 1000. That is to say, fitness need not always increase in real populations, no matter what Fisher's fundamental theorem might tell you about replicators in the limit of infinite populations and deterministic dynamics.

4.6 SWEEPS OR HITCHHIKING OR BOTH?

4.6.1 Apparent punctuated equilibrium, sweeps, and the LTEE

This chapter began with a puzzle: are evolutionary dynamics repeatable and, if so, are they repeatable at the scale of genotypes or phenotypes (or both)? The work of Richard Lenski and colleagues has shown that evolutionary dynamics is repeatable, and that organisms (like *E. coli*) that face the same selection pressure can evolve, independently, leading to systematic changes in fitness. In the case of the LTEE, these changes also coincided with changes in cellular physiology: cells got bigger—though, surprisingly, Lenski and Travisano (1994) were unable to prove a direct mechanistic relationship between cell size and fitness. Instead, it seemed fitness increased in each lineage in a seemingly continuous fashion. However, that continuous increase belied a mechanism of rapid

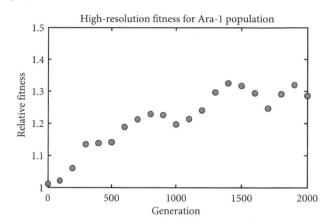

Figure 4.12: High-resolution sampling of fitness in 2000 generations of *E.coli* in the LTEE reveals stepwise changes in fitness. Data from Elena et al. (1996).

sweeps when the system was resequenced at higher resolution. Figure 4.12 shows that the fitness of the Ara-1 population was characterized by plateaus and then rapid changes in fitness (Elena et al. 1996). These rapid changes appear to support Stephen Jay Gould and Niles Eldredge's proposal of punctuated equilibrium (Gould and Eldredge 1993). The population remains in an evolutionary stasis (from the perspective of fitness) until a mutation that confers a benefit appears and then rapidly spreads in the population, leading to a relatively rapid shift in fitness. Indeed, this work suggests that *E. coli* are limited—in a mutational sense—such that only when beneficial mutations appear do they rapidly spread in the system. Yet, as limited, it would also suggest that the timing of these jumps will vary across experiments and associated populations.

One of the reasons why there can be variation in evolutionary dynamics is that, as we have shown, not all beneficial mutants fix. A beneficial mutant with fitness benefit s will invade a fraction $s/(s+1)$ of the time in the infinite population limit. Moreover, there can be extended periods where the more fit mutant persists at relatively low frequencies. It is only once the number of mutants with a fitness advantage exceeds some critical level, usually on the order of $1/s$, that they increase exponentially. That is, if there is a fitness advantage of 10%, then it requires 10 individuals (irrespective of the large, final population size) to likely lead to fixation, whereas it would take 100 individuals if the fitness advantage is only 1%. Hence, even a selectively beneficial gene will predominantly experience the stochastic effects of drift until it reaches a critical subpopulation size. These insights lead to the open question of whether or not mutants with beneficial alleles (or genotypes) appear frequently or rarely. The use of such adjectives implies a time scale of reference. Hence, for reference, consider the time scale to be the fixation or "sweep" time—this should be on the order of $\tau_s = \log N/s$. In that event, the following operational definitions will be helpful:

Limiting beneficial mutation regime: New beneficial mutations appear slowly (e.g., more slowly than the time over which such mutations can sweep).

Overlapping beneficial mutation regime: New beneficial mutations appear rapidly (e.g., faster than the time over which such mutations can sweep).

Returning to the first dataset from Lang et al. (2013), the question becomes what evidence, if any, is there to suggest that real populations may be in the overlapping beneficial mutation regime—a regime that may not necessarily have been self-evident when looking at evolutionary dynamics after long intervals in measurement? The early data from the LTEE suggests that, for this particular experiment, *E. coli* growing in a limited carbon resource environment may be mutationally limited. But, thankfully for the study of evolutionary biology, sequencing capabilities have increased so that it is possible to delve in greater detail—not just in terms of the temporal resolution of fitness but also in terms of complete genomes. It turns out there were many mutations, some of small effect, some of large effect, and some that surprised even the experimental team. In one case, *E.coli* evolved the ability to use citrate under oxic conditions—indeed, this is deeply surprising since *E. coli* is unable to use citrate in normal circumstances (Blount et al. 2008, 2012). But to evaluate evidence that sheds light on the question of the operative mutational regime for microbes, we turn to another lab workhorse—baker's yeast, i.e., *Saccharomyces cerevisiae*.

4.6.2   Evidence for clonal interference and hitchhiking in microbial populations

Lang et al. (2013) used an experimental evolutionary approach to analyze competition of yeast in a serial dilution framework. Each day a fraction of the flask was resuspended in fresh media, allowed to grow, resuspended in fresh media, and so on. Then samples were preserved approximately every 80 generations over a 1000-generation experiment, i.e., at generations 0, 140, 240, 335, 415, 505, 585, 665, 745, 825, 910, and 1000. At each time point, the entire population was sequenced using a whole genome–based approach that could use reference genomes to identify mutations across 40 distinct populations. The team identified 1020 mutations. As such, each particular experiment yielded a unique set of mutational trajectories (Figure 4.13 [Plate 3]). Using a high-resolution approach, it is apparent that the system is not mutationally limited. That is, multiple examples reveal the emergence of mutations that appear to be on their way to sweeping when another set of mutations appear and start to increase, and then another, and so on. The challenge is that, because the experiment could identify mutations and their frequency, it is not immediately apparent how (or whether) these mutations are linked. Yet there are clues.

As is apparent, there are coincident increases in the frequency of multiple mutations in a population (see the overlapping lines in both the small and large population figures in Figure 4.13). This could mean that the yeast genome luckily experienced multiple concurrent beneficial mutations, or perhaps many mutations all of a similar effect size. But, if so, how come so many mutations appeared all at once? Instead, it may be that some of these mutations were not beneficial but were actually neutral, and therefore that real evolutionary dynamics must be expanded from the simplified concept of sweeps and stasis to one in which there is a bubbling up of mutation after mutation that sometimes sweeps in

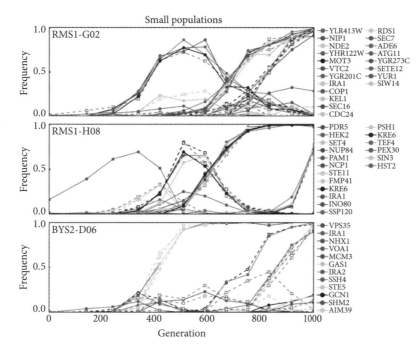

Figure 4.13: Trajectories of spontaneous mutations in evolving yeast populations. Solid lines denote synonymous mutations and dashed lines denote nonsynonymous or intergenic mutations. The labels on the right-hand side denote the mutation identity. Reproduced from Lang et al. (2013).

cohorts to partial fixation. This is the subject of one of the homework problems, but at least some additional evidence is worthwhile to consider.

At first, a beneficial mutation appears and begins to sweep through the population. Call this genome A. But then another mutation appears on a different genome (genome u), potentially with neutral mutations in the background, and genome u begins to outcompete the original sweeping genome. However, then a new mutation appears in the A background—call this genome AB—and it (and its neutral mutations) recovers and outcompetes u and its mutations. This process—called *clonal interference*—can repeat many times over. Clonal interference denotes the fact that different "clones" with different beneficial mutations can compete (i.e., interfere) with what might otherwise be the selective sweep of a focal lineage. One hallmark is that nonneutral mutations recur time and again in the evolutionary history (i.e., repeatedly) even as neutral

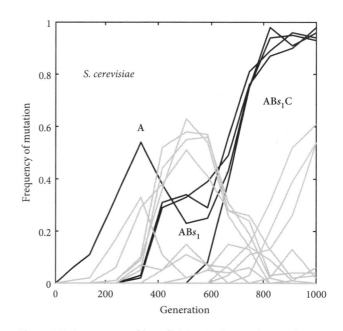

Figure 4.14: A sequence of beneficial mutations and neutral mutations that hitchhike offer a parsimonious explanation of the concurrent rise of multiple mutations in experimentally evolving yeast populations. Reproduced from Lang et al. (2013).

mutations appear by chance and sweep (i.e., contingently). The sweeping of neutral mutations is called *genetic hitchhiking*. Figure 4.14 shows precisely this point, in which groups of mutations appear at nearly precisely the same time and change in frequency in nearly precisely the same shape. Many of these mutations are neutral. One way to verify the presence of clonal interference is to evaluate if the genomic sites of nonneutral mutations repeat. (Hint: They do.) Likewise one way to verify the presence of genetic hitchiking is to evaluate whether or not the sites of synonymous mutations repeat. (Hint: They do not.) In the end, this experiment revealed novel modes of evolutionary dynamics by looking with higher resolution—finding that sweeps were really made up of near-sweeps, hitchhiking, and reproducibility despite the apparent randomness of many of the sweeping mutations. Like the old adage "a watched pot never boils," perhaps a carefully observed evolution experiment never really sweeps!

4.7 TAKE-HOME MESSAGES

- Evolutionary dynamics arise due to a combination of deterministic and stochastic processes.
- There are multiple mechanisms that can lead to the purging of diversity, insofar as pure replication implies that the type that grows the fastest should outcompete competing replicators.
- Although Fisher's fundamental theorem suggests that fitness must always increase, this theorem has severe restrictions and is of limited value in practice.
- There are multiple evolutionary mechanisms that can restore diversity, including mutation-selection balance and negative frequency-dependent selection. Other processes, including sexual reproduction and heterozygote fitness advantages, are outside the scope of this chapter.
- The study of how stochasticity impacts evolutionary dynamics is usually framed in terms of models of neutral drift, e.g., the Wright-Fisher or Moran model.
- In both cases, stochastic drift leads to either local extinction or fixation in finite populations over time scales set by the size of the population.
- In the event of selective benefits, then even rare mutants can lead to fixation, but such fixation is not inevitable and depends on the size of the population and the selective advantage.
- Recent observations of contemporary evolutionary dynamics suggest that, unlike assumptions of slow evolution and limiting beneficial mutations, natural systems often operate with many mutations competing at once (i.e., clonal interference) even as many neutral mutations are linked to beneficial mutations as part of multimutation sweeps (i.e., genetic hitchhiking).

4.8 HOMEWORK PROBLEMS

These problems build upon a growing set of computational tools, including methods elaborated on in the associated lab guide. In order to work on these problems you should be prepared to

- Translate a Markov chain into a numerical representation
- Simulate an ensemble of trajectories
- Compare individual trajectories to that expected given the transition probabilities
- Apply basic principles of Markov chain theory to the problem of population genetics
- Observe the fixation of neutral alleles due to genetic drift

These problems leverage these skills and the theory in this chapter to deepen your understanding of population genetics and evolutionary dynamics, including a chance to analyze recent data. The overall objective is to explore the role key factors, including population size and selection coefficients, play in shaping time scales over which population structure changes.

(Hint: For the Moran model, you will want to implement stochastic birth-death events, one at a time; that is, one random individual is chosen to die and one random individual replaces it. Hence, you need about N Moran events to be equivalent to one WF step. The key here is to think about how to incorporate differences in the probability of reproduction (e.g., birth) of some individuals relative to others.)

PROBLEM 1. How Fast Is Drift? Part 1

Here you will build on the computational laboratory analysis of the Wright-Fisher (WF) model to examine variability in the speed of fixation/extinction given neutral genetic drift. First, using the WF model with $N = 200$ individuals, begin with 100 type A individuals and show three representative samples that reach either fixation or extinction. What is the approximate number of time steps you expect the process to take? Next, run an ensemble of stochastic trajectories. (Hint: Use 100 to have sufficient statistics.) Do the same starting with 25, 50, 100, 150, and 175 individuals of type A. What is the relationship between the fraction of A individuals and the probability of fixation—and how does it compare to theoretical expectations?

PROBLEM 2. How Fast Is Drift? Part 2

Using $N = 100$ individuals of which 50 are type A, run a WF model using an ensemble of at least 200 trials. In each case, track how long it takes for A either to go extinct or to fix. What do you find? If you were to decrease/increase the initial fraction of individuals type A to either 20 or 80, how long would the process take? Finally, using the Markov chain transition matrix, find the theoretical expectation for the time-varying probability of extinction and fixation. Explain your rationale and comparisons to the results of stochastic realizations among the ensemble.

PROBLEM 3. A Sense of Scale—Part 1

Develop a Moran model where type A individuals have an enhanced fitness of $(1 + s)$ relative to type B individuals. That is, their reproductive chances are enhanced by a factor of $(1 + s)$ though their death rates are the same. Starting with a single individual in a population of $N = 500$, simulate the Moran process until the mutant population goes extinct or fixes. (Hint: Don't simulate every individual, but rather the cohort of As and Bs all at once.) Use $s = 0.05$ and find three examples in which the population fixes. How many simulations did it take to get these three examples? Explain your findings and what it says about the invasibility of beneficial mutations of (relatively) small effect.

PROBLEM 4. A Sense of Scale—Part 2

Using your Moran model, modulate the magnitude of the beneficial mutation s from 0.05 to 0.5 in increments of 0.05. Use ≈ 100 ensembles and answer this question: how often does a beneficial mutation fix? How does this compare to theory? Finally, recall that we expected a critical number of individuals with a beneficial mutation so that the mutation would almost certainly fix. We identified this scale as $1/s$. Focus on the case $s = 0.05$ and redo your ensemble analysis by varying the initial number of mutants and its effect on the probability of fixation. What do you conclude about the critical establishment size in this model?

PROBLEM 5. Overlapping or Limiting Beneficial Mutations

The Lang et al. (2013) analysis measures the change in mutations with time. Explore these patterns, focusing on one population, G7. First, load the data and plot the frequency of mutations over time. Using two lines of quantitative evidence, make an argument as to why this population experiment is in the limiting beneficial mutation regime or the overlapping beneficial mutation regime. This problem is intentionally open ended.

Additional Challenge Problems

- How does the time to fixation/extinction scale with N, the size of the population; s, the fitness benefit of the mutation; and x, the initial fraction of the population that has a beneficial mutation? Can you identify any potential crossover regimes based on the benefit of the mutation and the population size?
- The paths in the Moran model look diffusive. Can you develop a scaling formalism for the change in the relative frequency of a neutral allele in a population of size N given there is x fraction of the neutral allele present?
- Develop a clustering algorithm to automatically group mutants together that are likely on the same genome using data from Lang et al. (2013).

4.9 TECHNICAL APPENDIX

From the multispecies logistic model to the replicator equation The replicator equation can be derived from the multispecies logistic model as described in the chapter. A key step is to derive the dynamics of the total population, $N_T \equiv \sum_{i=1}^{s} N_i$, from the dynamics of individual populations, N_i. To do so, consider that

$$\frac{dN_T}{dt} = \sum_{i=1}^{s} \frac{dN_i}{dt}, \tag{4.46}$$

in which each of the populations is governed by the same multispecies logistic model, i.e.,

$$\frac{dN_i}{dt} = N_i \left(r_i - \frac{\sum_{j=1}^{s} r_j N_j}{K} \right). \tag{4.47}$$

Substituting the species-level dynamics into the total population dynamics yields

$$\frac{dN_T}{dt} = \sum_{i=1}^{s} N_i \left(r_i - \frac{\sum_{j=1}^{s} r_j N_j}{K} \right) \tag{4.48}$$

$$= N_T \frac{\sum_{i=1}^{s} r_i N_i}{N_T} - \sum_{i=1}^{s} N_i \left(\frac{\bar{r} N_T}{K} \right) \tag{4.49}$$

$$= N_T \bar{r} - \frac{N_T^2 \bar{r}}{K} \tag{4.50}$$

$$= \bar{r} N_T \left(1 - \frac{N_T}{K} \right) \tag{4.51}$$

This is the logistic equation, albeit for the population as a whole where the average fitness $\bar{r}$ is a time-dependent average fitness guaranteed to be positive. Hence, in the long-term limit, the total population N_T converges to K.

Replicator dynamics at a fixed population size This section provides an alternative approach to derive the replicator dynamics, focusing on two populations growing at rates r_A and r_B with the condition that their total population remains fixed. This means that any exponential growth of individual populations must be compensated by a time-dependent loss rate $\omega(t)$, i.e.,

$$\frac{dn_A}{dt} = r_A n_A - \omega(t) n_A \tag{4.52}$$

$$\frac{dn_B}{dt} = r_B n_B - \omega(t) n_B \tag{4.53}$$

Hence, we need to identify which choice of $\omega(t)$ would guarantee that $n_A(t) + n_B(t) = N$, a fixed target population size. If the total population is fixed, then by definition $\frac{dN}{dt} = 0$. Adding together the changes dn_A/dt and dn_B/dt leads to

$$r_A n_A + r_B n_B - \omega(t) \left(n_A(t) + n_B(t) \right) = 0. \tag{4.54}$$

Identifying $N(t) = n_A + n_B$ and ignoring the time dependencies on the right-hand side, this equation becomes

$$\omega(t)N = r_A n_A + r_B n_B \tag{4.55}$$

or

$$\omega(t) = r_A x_A + r_B x_B \tag{4.56}$$

where $x_A = n_A/N$ and $x_B = n_B/N$ represent the fraction of types A and B in the population. Hence, the loss rate must equal the average growth rate, or what we term the average fitness, i.e., $\langle r(t) \rangle = \omega(t)$. Altogether, this yields a model that is widely known as replicator dynamics:

$$\frac{dx_A}{dt} = r_A x_A - \langle r \rangle x_A \tag{4.57}$$

$$\frac{dx_B}{dt} = r_B x_B - \langle r \rangle x_B \tag{4.58}$$

and as should be evident, $\dot{x}_A + \dot{x}_B = 0$. These equations imply that populations will increase in proportion insofar as their growth rates at a given moment in time exceed that of the average fitness.

Dynamics of heterozygosity The heterozygosity of a population is equal to the probability that two randomly chosen individuals in a population of size N have different genotypes (or alleles). As noted in the chapter, this is defined as

$$H = 2\frac{n}{N}\left(1 - \frac{n}{N}\right) \tag{4.59}$$

where n is the number of A genotypes and $N - n$ is the number of B genotypes, and therefore n/N is the fraction of A genotypes and $1 - n/N$ is the fraction of B genotypes. The expected dynamics of heterozygosity, $H(t)$, can be computed by calculating the value one generation ahead, where $E(H)' \equiv \langle H(t+1) \rangle$:

$$E(H') = E\left(2\frac{n'}{N}\left(1 - \frac{n'}{N}\right)\right). \tag{4.60}$$

This is formally equal to selecting A then B or B then A when randomly selecting two individuals in the population; either way the genotypes are different. The expected value of H' depends on the frequency and variance of genotypes at the next generation.

Heterozygosity in the Wright-Fisher model In the WF model, the dynamics of heterozygosity are

$$E(H') = \frac{2}{N}E(n') - \frac{2}{N^2}E\left((n')^2\right)) \tag{4.61}$$

$$= \frac{2n}{N} - \frac{2}{N^2}\left[Var(n) + (E(n'))^2\right] \tag{4.62}$$

$$= \frac{2n}{N} - \frac{2}{N^2}\left[n(1-n) + n^2\right] \tag{4.63}$$

$$= \frac{2n}{N} - \frac{2n}{N^2} + \frac{2n^2}{N^3} - \frac{2n^2}{N^2} \tag{4.64}$$

$$= \frac{2n}{N} \left[1 - \frac{1}{N} + \frac{n}{N^2} - \frac{n}{N} \right] \tag{4.65}$$

$$= \frac{2n}{N} \left(1 - \frac{n}{N} \right) \left(1 - \frac{1}{N} \right) \tag{4.66}$$

$$= H \left(1 - \frac{1}{N} \right) \tag{4.67}$$

The final result implies that $E(H(t+1)) = H(t) \left(1 - \frac{1}{N} \right)$ or more generally

$$E(H(t)) = H_0 \left(1 - \frac{1}{N} \right)^t, \tag{4.68}$$

i.e., heterozygosity is expected to decay exponentially such that when $N \gg 1$ then

$$E(H(t)) = H_0 e^{-t/N} \tag{4.69}$$

with a time scale of decay on the order of N generations. This result and its implications are discussed in the chapter.

Heterozygosity in the Moran model In the Moran model, the dynamics of heterozygosity are

$$E(H') = E \left(2 \frac{n'}{N} \left(1 - \frac{n'}{N} \right) \right) \tag{4.70}$$

where $E(n') = n$ for all time and

$$E(n'^2) = \Sigma (n')^2 P(n') \tag{4.71}$$

$$= (n-1)^2 T(n-1|n) + n^2 T(n|n) + (n+1)^2 T(n+1|n) \tag{4.72}$$

$$= (n^2 - 2n + 1) T(n-1|n) + n^2 (1 - T(n-1|n) - T(n+1|n))$$
$$+ (n^2 + 2n + 1) T(n+1|n) \tag{4.73}$$

$$= n^2 + 2 \frac{n}{N} \left(1 - \frac{n}{N} \right) \tag{4.74}$$

given that $T(n-1|n) = T(n+1|n) = \frac{n}{N} \left(1 - \frac{n}{N} \right)$ for all $n > 0$ and $n < N$. Combining terms yields

$$E(H') = \frac{2n}{N} - \frac{2}{N^2} \left[n^2 + \frac{2n}{N} \left(1 - \frac{n}{N} \right) \right] \tag{4.75}$$

$$= \frac{2n}{N} - \frac{2n^2}{N^2} - \frac{4n}{N^3} + \frac{4n^2}{N^4} \tag{4.76}$$

$$= \frac{2n}{N} \left(1 - \frac{2}{N^2} \right) - \frac{2n^2}{N^2} \left(1 - \frac{2}{N^2} \right) \tag{4.77}$$

$$= \frac{2n}{N} \left(1 - \frac{n}{N} \right) \left(1 - \frac{2}{N^2} \right) \tag{4.78}$$

which is equivalent to

$$E(H') = H(n)\left(1 - \frac{2}{N^2}\right) \tag{4.79}$$

such that for $N \gg 1$

$$E(H(t)) = H_0 e^{-2t/N^2}. \tag{4.80}$$

Indeed, if we recognize that generations in Moran are measured as $G = t/N$, then heterozygosity in both the WF and Moran models decays exponentially with generation times G—the difference is that the factor of 2 comes from the variance of the one-step dynamics.

Fixation in the Moran model with selection   Consider the following recurrence relationship among fixation probabilities

$$\pi_n = T(n+1|n)\pi_{n+1} + T(n-1|n)\pi_{n-1} + T(n|n)\pi_n \tag{4.81}$$

for the Moran model with selection. This can be written as

$$\begin{aligned} \pi_n &= \frac{T(n+1|n)\pi_{n+1} + T(n-1|n)\pi_{n-1}}{1 - T(n|n)} \\ &= \frac{T(n+1|n)\pi_{n+1} + T(n-1|n)\pi_{n-1}}{T(n-1|n) + T(n+1|n)} \end{aligned}$$

Recall that the difference between $T(n-1|n)$ and $T(n+1|n)$ is a factor of $(1+s)$. Hence, we can rewrite this as

$$\pi_n = \frac{1}{2+s}\pi_{n-1} + \frac{1+s}{2+s}\pi_{n+1} \tag{4.82}$$

or by rearranging terms

$$(2+s)\pi_n = \pi_{n-1} + \pi_{n+1}(1+s) \tag{4.83}$$

and again by rearranging

$$\pi_n - \pi_{n-1} = (1+s)\left(\pi_{n+1} - \pi_n\right) \tag{4.84}$$

and finally

$$\left(\pi_{n+1} - \pi_n\right) = \frac{\pi_n - \pi_{n-1}}{1+s} \tag{4.85}$$

as examined in the chapter.

Measuring the strength of selection given a change in the frequency of variants   Consider two variants growing in a population such that the frequency of type A grows from an initial value of x_0 over time as $x(t)$ given a selective growth advantage of s:

$$x(t) = x_0\left[\frac{e^{st}}{1 + x_0\left(e^{st} - 1\right)}\right]. \tag{4.86}$$

Given a measurement time t_1 and an observation of $x_1 = x(t_1)$, what is the estimated selective growth advantage? This framework is the standard approach to estimate fitness advantages in direct competition between two types. In this case, Eq. (4.86) can be rewritten as

$$x_1 \left(1 + x_0 \left(e^{st} - 1\right)\right) = x_0 e^{st} \tag{4.87}$$

$$x_1 (1 - x_0) + x_1 x_0 e^{st} = x_0 e^{st} \tag{4.88}$$

$$x_1 (1 - x_0) = x_0 (1 - x_1) e^{st} \tag{4.89}$$

$$e^{st} = \frac{x_1}{1 - x_1} \frac{1 - x_0}{x_0} \tag{4.90}$$

such that, finally, the estimated selection coefficient is

$$\hat{s} = \frac{1}{t_1 - t_0} \log \left[\frac{x_1}{1 - x_1} \frac{1 - x_0}{x_0} \right]. \tag{4.91}$$

This estimate is used in practice to convert frequency changes to estimates of the selection coefficient.

Part II

Organismal Behavior and Physiology

Chapter Five

Robust Sensing and Chemotaxis

5.1 ON TAXIS

Organisms exhibit a wide range of behaviors. Of these, motility is essential to nearly all forms of life. There are many reasons to move, e.g., to acquire food, find mates and reproduce, defend territory, and avoid danger. Evolution can favor the emergence of *directed* movement when the benefits of moving in what appears to be the right direction outweigh the costs. However, moving in the right direction requires sensing the right direction in the first place. Typically, such sensing is imperfect. The basis for detecting the right direction may come from many different kinds of signals, including chemical cues, visual information, sounds, magnetic fields, physical pressure, and olfactory signals. When an organism moves preferentially in response to a particular kind of cue, we refer to this as a variant of *taxis*. For example, chemotaxis denotes directed movement in response to chemical cues, magnetotaxis denotes directed movement in response to electromagnetic fields, phototaxis denotes directed movement in response to light, and so on. Although the benefits seem myriad, there are costs. Sensory apparatus are energetically expensive to maintain. Moreover, sensing takes time; and decisions that follow sensing can take even more time. As leaders faced with decision after decision, one strategy is simply to make fast decisions, not necessarily the right decisions. Indeed, if time is of the essence, sometimes fast reactions (however inaccurate) can outperform slow, deliberate reactions (however accurate). Finally, sensing must operate in noisy environments and need not be foolproof. As Lee Segel (1984) quipped, "There is nothing certain in life except for death and taxis." Resolving the costs and benefits of sensing helps provide insights into evolutionary questions: why does sensing evolve and in what contexts? Yet, given that the focus of Part II is organismal physiology, we instead address this question: how can organisms sense and how do they manage to modify their behavior in response to changes in the environment?

Although there are many sensory systems that could serve as a model to answer this question, we will focus on the chemotactic machinery of *E. coli*. *E. coli* is able to detect and respond to changes in the concentration of chemical cues, termed *chemoattractants* and *chemorepellants*, or attractants and repellants for short. As implied by their name, individual cells can sense and move toward attractants as well as sense and move away from repellants. Yet in doing so, they will also manage to accomplish something quite profound—responding to gradients in concentration and not just levels of background concentration. This difference is subtle but important. The ability to sense gradients implies that the *E. coli* sensory system can robustly adapt to different environmental contexts, i.e., by somehow comparing differences rather than responding to absolute concentrations. By focusing on *E. coli*, we will explore how external signals are sensed by organisms, and then how information is transmitted and represented via internal dynamical systems and propagated into physical actions that lead to changes in behavior.

This kind of organismal behavior may seem limited for those accustomed to studying charismatic megafauna (or even more charismatic microfauna). Dennis Bray (2002) encapsulates one perspective of why this is important:

> What does it matter that the response of a tiny microbe has this or that numerical value?
>
> The answer, surely, is that the chemotaxis pathway of coliform bacteria is so well documented it has become a benchmark for the responses of cells in general to chemical stimuli. It is, in a sense, the bellwether of signal transduction.
>
> If we are unable to understand how this simple network of proteins functions as an integrated system, then what hope have we of understanding the complex pathways in eukaryotic cells?

Yet there are even more reasons to be interested in this system beyond thinking of *E. coli* as a "model" for more interesting forms of chemotaxis. Instead, it is worthwhile to explore questions of sensing, signal transduction, and organismal behavior at the microbial scale. Bacteria are everywhere. They are organisms too. Their behavior has lessons to teach us about eukaryotes but also about life at the microbial scale. Figure 5.1 shows a series of traces of *E. coli* movement. These images were taken in the early 1970s by Howard Berg (1993), author of the timeless *Random Walks in Biology*. As is apparent, individual cells appear to move in quite nonrandom ways, with long stretches of relatively straight movement and then squiggly little stopping points where they seem not to move at all. The former are called *runs* and the latter *tumbles*. The story of runs and tumbles has quite a lot to teach us about behavior as a whole. How *E. coli* modulates these runs and tumbles lies at the very core of their chemotactic machinery and forms the basis for this chapter.

Hence, this chapter is motivated by the following question: how does *E. coli* respond robustly to chemical gradients? The robust response includes two salient features: an ability to move toward attractants and away from repellants. At the crux of this question is another puzzle: why aren't *E. coli* tricked by changes in background concentrations? That is to say, if the environment suddenly experiences a uniform increase in attractant concentration, an organism might perceive that it is traveling in a better direction, in the near term. But, over time, this new normal with a changed background does not provide actionable information on preferred directions. How does *E. coli* adjust itself to shifted baselines so as to respond to gradients and not the background level itself? We will take a gradual approach to answer this question; first by addressing the benefits of swimming from a biophysical perspective, then reviewing the basics of how *E. coli* do in fact swim, and finally examining the chemotactic circuit to understand precisely how such robust gradient sensing is possible.

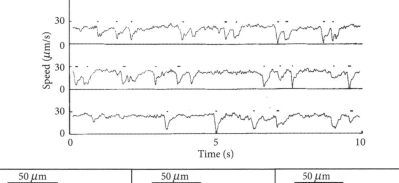

Figure 5.1: Runs and tumbles of *E. coli*; reproduced from Berg and Brown (1972). (Top) Measurement of velocity over time reveals periods of extended motion interrupted by cessation of movement. (Bottom) Individual tracks are over ≈ 30 sections of track data, with a mean speed of ≈ 21 μm/sec. Notice the long, straight segments, which are referred to as *runs*, broken up by what appears to be densely aggregated points denoting the cessation of movement, a period termed a *tumble*. As illustrated in the top and bottom panels, the movement of *E. coli* can be characterized in terms of runs and tumbles, and modulation of these two modes underlies its chemotactic ability.

5.2   WHY SWIM?

Individual *E. coli* swim to get food, escape predators, and move in fluctuating chemical environments. Other bacteria can use other forms of motility, including gliding and twitching, to translocate across surfaces. But many microbes, including ubiquitous ocean microbes, don't swim. This may come as a surprise to scientists trained to think of the world of bacteria through the lens of *E. coli*. Indeed, before discussing the chemotactic machinery of bacteria, it is worthwhile to focus on the conditions when swimming is not favored, as a means to understand the conditions when it is. This is not just a theoretical question. As it turns out, the ocean's most abundant photoautotroph—the cyanobacterium *Prochlorococcus*—does not swim. These small microbes, approximately 0.25 μm in radius, form the base of the oceanic food web (Chisholm et al. 1988; Moore et al. 1998; Chisholm 2017). They are present in concentrations of ≈ 10^5/ml in tropic and temperate waters globally. *Prochlorococcus* cells use light-driven photosynthesis to fix organic carbon from CO_2 and are estimated to be responsible for approximately 25% of the oxygen we respire each and every day. Yet, for all that these invisible plants of the sea do, they do not swim for their food. Instead, *Prochlorococcus* cells passively diffuse, taking up nutrients, and utilizing light energy as a means to fix CO_2 into organic carbon.

One way to begin to think about the problem of why such a ubiquitous ocean microbe does not swim is to consider the advantage of swimming. By swimming at a velocity v, a microbe can traverse a distance l in a time l/v. This is termed the *stirring time*, i.e., the time it takes to stir up the environment so as to access molecules a distance l away. In contrast, diffusing molecules will reach a nonmotile microbe in a characteristic time l^2/D where D is the diffusion constant of the molecule. Hence, understanding whether ocean microbes should swim as a strategy to access essential molecules depends on comparing the stirring versus diffusion times.

Consider a small ocean microbe, like a cyanobacterium, that must uptake extracellular nutrients. Iron is one such nutrient. If iron is available at a concentration of 0.5 nM, then individual iron molecules are approximately $l = 1.1$ microns away from a focal bacteria. If an ocean microbe were to swim at a velocity $v = 10$ μm/sec toward the nearest molecule, then it would take $l/v \approx 100$ msec to arrive. In contrast, diffusion on these time scales can be even faster. Diffusion of small nutrients in aqueous media at 20 degrees Celsius is approximately $D = 10^3$ μm^2/sec. Hence, in this example it would take $l^2/D \approx 1$ msec for the iron to diffuse to the microbe. In other words, waiting is 100 times more effective than swimming. This may appear counterintuitive, but it reflects a long-standing quantitative argument on the benefits and disadvantages of swimming for food in certain limits.

This argument can be generalized to other molecules given biologically reasonable concentrations. A general argument was first proposed by Edward Purcell (1977) 40 years ago. Purcell concluded that "local stirring accomplishes nothing" when the distance to swim is small, speeds are low, and the "food" diffuses rapidly. These conditions are particularly relevant for submicron-sized ocean microbes if they are to uptake iron, nitrogen, phosphorus, and carbon at realistic concentrations in the euphotic or near-euphotic zones (Zehr et al. 2017). The relevance of swimming can be quantified by taking the ratio of swimming to diffusion times—Purcell termed this the *stirring number*, $S = lv/D$. For a molecular substrate at concentration ρ, in units of nM, given swimming speeds and diffusion rates as previously noted, the relevant dimensionless stirring number is $S = 1.7 \times 10^{-2} \rho^{-1/3}$. In order for swimming to be effective, then $S > 1$. Rearranging the equation for S, we conclude that swimming at 10μm/sec is effective in accessing small nutrients when the molecular concentration of target nutrients ρ is less than 4.6×10^{-6} nM. Hence, substrates would have to be present at femtomolar concentrations before swimming is more effective than waiting for very small molecules to diffuse. And, at such low concentrations, identifying the correct direction to swim is likely difficult.

These arguments provide a biophysical rationale for certain cells to wait, rather than swim, to their food. Waiting raises another set of questions. Do enough nutrient molecules diffuse inward over a time scale relevant to growth and reproduction? For example, in the course of a day, a nonmotile microbe would diffuse and, as a consequence, explore a volume characterized by the distance $d_{cell} = \left(D_{cell} T_{day}\right)^{1/2}$ where d_{cell} denotes the distance in centimeters, $D_{cell} = 10^{-8}$ cm^2/sec assuming the organism is approximately 0.5 μm in size, and $T_{day} = 8.64 \times 10^4$ is the number of seconds in a day. This characteristic distance is approximately 0.03 cm. Hence, the characteristic volume that such a nonmotile microbe can explore is on the order of 10^{-4} cm^3 or approximately 100 nL. We term this 100 nL volume the *cellular diffusion zone*. For a nutrient at 10 nM concentration, there are approximately 6×10^8 molecules available in the cellular diffusion zone. A cell that requires 1.5×10^{10} molecules of nitrogen to reproduce would only acquire 4% of its elemental needs. The cell would be unable to divide on anywhere near a diurnal time scale if it utilized only these atoms. But small nutrients are not static. Far from it. Small-nutrient molecules diffuse 1000 times faster than microbes. Hence, the characteristic volume that a non-motile microbe explores passively is much greater, on the order of 10^5 nL. This nutrient diffusion zone contains 6×10^{11} molecules given 10 nM concentrations. Hence, in theory, there are more than enough molecules that diffuse close to cells. Cells explore this greater volume because of the diffusion of molecules rather than the diffusion of cells.

Cells cannot take up every molecule, neither in the cellular diffusion zone nor in the passive diffusion zone. The theory of chemoreception developed by Berg and Purcell (1977) sets the upper limit on the uptake rate of small molecules at $4\pi aND$/sec. This formula can be anticipated from dimensional analysis. The interaction size should be set by the radius of the cell a (μm, rather than the much smaller molecules) whereas the relevant diffusion rate should be set by the much smaller molecules that diffuse with a constant D μm^2/sec. Hence, a parsimonious combination of factors must be on the order of aND where N is the concentration of molecules. This combination has units of an inverse time, corresponding to the number of molecules taken up per cell per unit time. The interested reader may find it helpful to refer to Barenblatt (2003) for more context on how to use dimensional analysis in practice as a guide to developing expectations for the scaling of certain biological phenomena. The comparisons of nutrient uptake via passive diffusion versus growth requirements points out an important caveat. As microbes get bigger, their requirements increase as the cubed power of their characteristic size. This means that, for bacteria greater than approximately 1 micron, it pays to swim for one's dinner. And marine microbes do just that, using chemotaxis to respond to ephemeral nutrient patches in a fluctuating environment (Stocker et al. 2008; Stocker 2012). In some cases, chemotaxis gradients may lead heterotrophic microbes to move closer to autotrophs, forming the basis for emergent symbioses (Raina et al. 2019). Likewise, it also pays for microbial predators to swim for their dinner, when their dinner is about the size of *Prochlorococcus* (Talmy et al. 2019). But for our purposes, we focus on chemotaxis in the most well-studied case, that of *E. coli*, while pointing out that understanding the basic principles of chemotaxis continues to influence studies spanning biophysics to ecosystems science.

5.3   THE BEHAVIOR OF SWIMMING *E. coli*

There are many ways to swim, including *advective motion*, *random motion*, and *biased random motion*. Advective motion is also known as *ballistic motion* in that the organism swims or moves in a "straight" line with velocity v, such that the expected distance traversed in a time t is $x = vt$. *Self-propelled random motion* is a form of random motion in which the organism moves in a random direction potentially even ballistically on short time scales, and then changes direction and again moves ballistically. On long time scales, the effective behavior converges to that of passive diffusive behavior, albeit with a much larger diffusion constant, D. The average distance traveled is expected to be zero (because of the random nature of movement), but the average distance squared is expected to increase like $x^2 \sim Dt$ where D has units of length squared per time. Finally, there can be intermediates between these paradigms, given biased random motion such that $x^2 \sim t^{1+\epsilon}$, where $-1 \leq \epsilon \leq 1$ represents a continuum of classes, from anticorrelated motion $\epsilon = -1$, diffusion $\epsilon = 0$, and ballistic

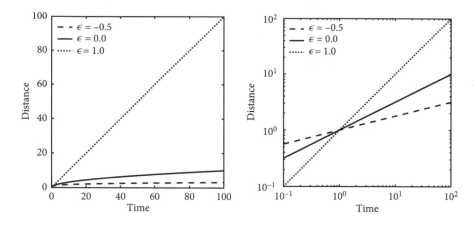

Figure 5.2: Scaling of subdiffusive, diffusive, and ballistic motion. The two plots show $x \sim t^{(1+\epsilon)/2}$ for $\epsilon = -0.5$, 0, and 1, corresponding to subdiffusive, diffusive, and ballistic trajectories, respectively. The left plot shows $x(t)$ on linear axes and the right plot shows $x(t)$ on log-scaled axes. Note that the right plot reveals that ballistic motion is faster on longer time scales but that diffusion and even subdiffusive motion can be faster on short time scales.

motion $\epsilon = 1$. Figure 5.2 visualizes the expected distance traversed in each of these cases, both in linear and log axes, revealing how ballistic motion is the most efficient over long time scales but not necessarily over short time scales.

E. coli modifies its intrinsic behavior of self-propelled random swimming to preferentially move in the direction of a chemical gradient. The random template is made of runs and tumbles. Runs are extended periods of swimming in (nearly) a straight line at speeds of approximately $20 - 35 \mu$m/sec. These runs last approximately 1 second. The swimming is made possible by the counter-clockwise motion of the cell's flagella (Berg and Anderson 1973)—there are typically four to six flagella located uniformly on the surface of the cell. In contrast, the transmembrane receptors of the chemotactic machinery are localized at the front. These runs are interrupted by tumbles. Tumbles are briefer periods of approximately 0.1 second in which at least one flagellum begins to turn in the clockwise direction, becomes unbound, and no longer generates a propulsion force. During this period, the cell has virtually zero speed. *E. coli* is swimming at low Reynolds numbers, a fancy way of saying that it has no inertia. Unlike a human swimmer or ice skater, *E. coli* will not move if it is not actively propelling. After a brief tumble, the new direction will be oriented slightly away from the prior direction as a result of torque induced by the flagella (Darnton et al. 2007), with a mode of approximately 60 degrees. Repeated again and again, these runs and tumbles generate a self-propelled random motion. However, they also raise a question: how could *E. coli* modify this behavior to respond to chemical gradients?

Before describing the chemotaxis machinery, it is worth asking which might be easier to control differentially when traversing chemical gradients:

- Speed of runs
- Duration of runs
- Duration of tumbles
- Reorientation direction

The speed of runs is set by the propulsion force generated by tethered flagella. Although it may be true that modulating their rotation speed could modulate force (or even the number of flagella that rotate counterclockwise), such a system might be a harder target for control. The number of flagella involved in a tumble and the variance in reorientation direction are coupled (Turner et al. 2000). It is challenging to control the new run direction—especially given the influence of stochastic forces induced by which and how many of the flagella become unbundled. Instead, *E. coli* modulates the duration of runs, extending runs when moving up a chemoattractant (or down a chemorepellant) gradient and shortening runs when moving down a chemoattractant (or up a chemorepellant) gradient. Hence, cells continue the same biophysical motion, albeit for a shorter or longer time. How cells control the duration of runs is examined next.

5.4   CHEMOTAXIS MACHINERY

A simplified chemotaxis circuit in *E. coli* is presented in Figure 5.3. At its core, the sensing machinery is composed of a clustered suite of transmembrane receptors that couple external signals to internal changes in state, ultimately leading to a change in motion from runs to tumbles. This chemotactic sensing machinery is located in the front of the cell, amidst a cluster of such receptors (whose numbers range from thousands to approximately 10,000 in total). This large number of receptors can reduce impacts of noise and enhance responsiveness to gradients in chemical ligands (Zhang et al. 2007; Cassidy et al. 2020). This sensing machinery is part of a complex circuit with multiple interactions and feedback. The circuit includes protein complexes, such as CheW, CheA, CheB, CheY, CheR, and CheZ. There are many proteins, a number of which can be chemically modified, thereby altering their enzymatic activity. However, it is sufficient to study the role of a few components and their feedback to understand how a bacterium responds to adsorption of a ligand. For various reasons, it is perhaps more instructive to begin this explanation by evaluating the response given the introduction of chemorepellants. In such a situation, the repellant will bind, outside the cell, to the ligand binding pocket of a transmembrane chemotactic receptor. Notably, there are five types of transmembrane chemotactic receptors, providing a repertoire of targets and sensitivity (Parkinson et al. 2015). Binding of the ligand induces a chemical conformation of the receptor complex. Two potential chemical modifications are most relevant: methylation and phosphorylation. *Phosphorylation* refers to the chemical process in which a phosphate group (PO_3) is chemically bonded to another biomolecule, in this case CheA. Binding will lead to the phosphorylation of the CheA complex. Yet the receptor may either be methylated or not. *Methylation* denotes a chemical process in which a methyl group (CH_3) is chemically bonded to another biomolecule, in this case the receptor. Here the number of bound ligand molecules and the receptor methylation state determine the receptor and CheA activity. CheA acts as a kinase if active; as such, it can phosphorylate other proteins in the system, including the CheY protein which when activated acts as a diffusible response regulator inside the cell's cytoplasm. The logic underlying the chemical-dependent activity is summarized in Table 5.1.

At this point, the logic of the chemotactic sensing circuit extends beyond the front of the cell and begins to transmit information via diffusible proteins. The phosphorylated CheA,

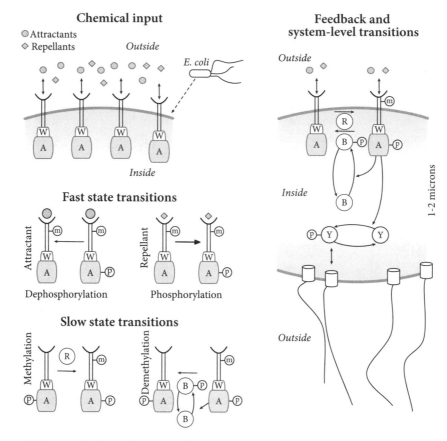

Figure 5.3: Schematic of the chemotaxis circuit, adapted from Uri Alon's *Introduction to Systems Biology* (Alon 2006). (Left panels) Components of the chemotaxis system, including binding of the transmembrane sensing receptor to both chemoattractants and chemorepellants. Fast transitions are controlled by binding events associated with (de)phosphorylation, while slow transitions are controlled by (de)methylation enzymes. (Right panel) Feedback between the chemoreceptor and the flagella is mediated by a signaling cascade. The active form of CheA phosphorylates the diffusible response regulator protein, CheY. The phosphorylated and active form of CheY reverses the chirality of the flagellar motor at the far end of the *E. coli* cell, leading to a cessation of movement and initiation of tumbling. The frequency of these tumbling events is controlled by the levels of active CheY in the system.

in turn phosphorylates CheY, a diffusible response regulator that can then rapidly travel back to the rear of the cell. The phosphorylated (i.e., active) form of CheY can bind to the flagellar motor switch and change the handedness of rotation of the transmembrane motor from counter-clockwise to clockwise. This change in chirality can lead to an unbundling of individual flagellar filaments and the temporay cessation of motion, i.e., a tumble. Diffusion times of small molecules take less than 0.1 second (see Chapter 2 on gene regulation), so that changes in the binding of repellants at the front of the cell can cause a nearly immediate consequence to cellular behavior. This description provides a glimpse into the chemotactic machinery, but it does not yet address the central problem. If the entire background level suddenly changes (as in an experiment where the system is flooded with either chemoattractant or chemorepellant), will the new level of tumbling remain permanently low (for

Table 5.1: Summary of the chemical logic underlying CheA activity

Repellant binding	Attractant binding	Methylated receptor	CheA activity
Low	High	No	Low
Low	High	Yes	Low (typically)
High	Low	No	Low (typically)
High	Low	Yes	High

Note: Chemical logic for the "activity state" of CheA depends on the type and level of ligand binding to the chemoreceptor and the methylation state of the receptor. If CheA is active, it will act as a kinase and phosphorylate CheY, a diffusible response regulator inside the cell's cytoplasm. Note that the highest output states are those in which the receptor is methylated and there are either high repellant binding levels or low attractant binding levels. The actual response can be nonlinear, hence "typically" is meant to indicate a transition toward low levels of CheA activity in the limit of low levels of repellant ligands or high levels of attractant ligands.

attractants), permanently high (for repellants), or relax back to its baseline level? The answer to this question requires a dynamical model that does more than account for rapid changes along a kinase cascade and slower changes involving methylation, demethylation, and activity. To explore this in depth requires an introduction to the basic mathematical principles of signaling cascades.

5.5 SIGNALING CASCADES

5.5.1 Basics

The function of the chemotaxis circuit depends on a process known as signal transduction. That is, information (or signals) from outside the cell are transduced into the cell to alter structure and functional properties—*transduction* refers to both the transmission of information and the transformation of that information into other states. A key component of signal transduction is the enzymatic modification of the structure and activity of proteins. (Note: There are many other processes that are essential to signal transduction, particularly allosteric regulation, involving cooperative binding of nonactive sites that can nonetheless increase activity at the distal, active site (Changeux and Edelstein 2005; Sourjik and Wingreen 2012).) Two key steps in initiating the signal transduction process are protein phosphorylation and dephosphorylation. These two processes refer to the addition or removal of a phosphate group from a protein. The enzymes that phosphorylate are termed kinases. The enzymes that dephosphorylate are termed phosphatases. The addition of a phosphate group can lead to a conformation change in the protein, which in turn can enable new activity, like binding to DNA or enabling the protein itself to phosphorylate other molecules, as depicted in Figure 5.4.

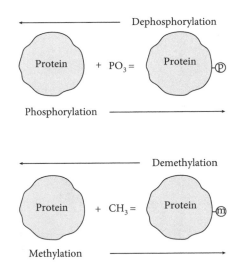

Figure 5.4: Basics of phosphorylation and methylation, involving the chemical modification of proteins (including enzymes). In both cases, the chemical modification also comes with a change in activity such that phosphorylated and/or methylated forms of the protein can have different activity.

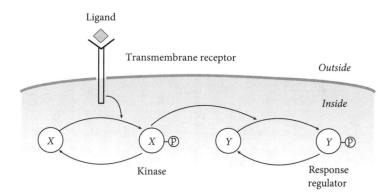

Figure 5.5: Schematic of a signaling cascade, in which a signal initiates the phosphorylation of downstream proteins. Enzymes that phosphorylate other molecules (including proteins) are called kinases. Here CheA is represented generically via the kinase X, which in the active form X-p can phosphorylate Y. In this example, the phosphorylated form of Y-p can also modify the activity of another component—the flagellar motor switch.

How does a cell sense a signal from outside and then initiate new behaviors inside as a result of a *signaling cascade*? Figure 5.5 shows a simple example of a signaling cascade. In this example, the binding of a ligand to a transmembrane receptor phosphorylates a kinase. This kinase (X) can then phosphorylate another protein (Y). In order to retain a generic form, we refer to the CheA as X and X-p for the unphosphrylated (inactive) and phosphorylated (active) forms. The steady state concentration of X_1 (active) is a function of the signal strength and the methylation state of the receptor. The signal strength is a function of the binding of the ligand to the receptor, which is itself a product of the ligand concentration and its binding affinity. To understand the dynamics of the response regulator protein Y, we should decompose the total concentration Y in terms of unphosphorylated Y_0 and phosphorylated Y_1 concentrations, such that $Y = Y_0 + Y_1$. The dynamics of phosphorylated proteins follows:

$$\frac{\mathrm{d}Y_1}{\mathrm{d}t} = \overbrace{v_1 X_1 Y_0}^{\text{phosphorylation}} - \overbrace{\alpha Y_1}^{\text{dephosphorylation}} \tag{5.1}$$

in which v_1 is a first-order kinetic constant, α is a decay constant, and X_1 is the kinase concentration. The use of a first-order approximation to enzyme kinetics will be relaxed as we consider the more complex chemotaxis circuit.

At steady state, $\mathrm{d}Y_1/\mathrm{d}t = 0$ such that

$$Y_1 = \frac{v_1 X_1 Y_0}{\alpha}$$

$$Y_1 = \frac{v_1 X_1 (Y - Y_1)}{\alpha}$$

$$Y_1 (1 + \omega_1 X_1) = \omega_1 X_1 Y$$

$$\frac{Y_1}{Y} = \frac{\omega_1 X_1}{1 + \omega_1 X_1} \tag{5.2}$$

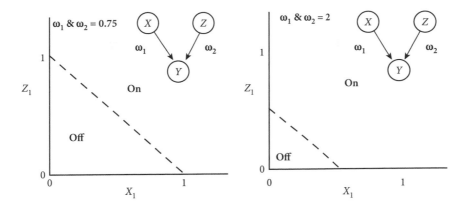

Figure 5.6: Signal cascade leading to input-output relationships in signal space. (Left)—Approaching an AND gate. (Right)—approaching an OR gate. If the inputs (X or Z) are either 0 or 1, then the output will be on (left) only when $X = 1 = Z$, and the output will be on (right) if either $X = 1$ or $Z = 1$.

where $\omega_1 = v_1/\alpha$. Here the fraction of phosphorylated output is a saturating function of the kinase strength, i.e., $\omega_1 X_1$. This input-output relationship represents a mechanism to convert signal into function. For example, consider that the fraction of Y_1 needs to cross some critical level to enable a specific biological function (like tumbling). This might be the case if Y_1 were a transcription factor that must pass a critical density to bind and activate transcription. In this event, there would be no function for weak kinase activity and there would be a function for strong kinase activity. Hence, a kinase cascade can turn a signal into an on-off switch.

Even more complex calculations can be made via kinase cascades. Imagine, for example, that two kinases (X and Z) can both phosphorylate a target protein such that

$$\frac{\mathrm{d}Y_1}{\mathrm{d}t} = v_1 X_1 Y_0 + v_2 Z_1 Y_0 - \alpha Y_1 \tag{5.3}$$

and at steady state

$$\frac{Y_1}{Y} = \frac{\omega_1 X_1 + \omega_2 Z_1}{1 + \omega_1 X_1 + \omega_2 Z_1}. \tag{5.4}$$

In essence, when $\omega_1 X_1 + \omega_2 Z_1$ exceeds some critical level, then the target function can be on. The On conditions depend on both the binding affinity and degradation (via ω) and the external signals, (via X and Z). Moreover, asymmetries between the weights can also lead to a transition from OR to AND gates. It is also worth noting that by including multiple layers of interactions as well as *negative* effects, e.g., via dephosphorylation, it is possible for kinase cascades to generate new functions, such as targeted responses to interior regions of a signal (Figure 5.6). For example, if the weights of signals are weak, then both kinases would need to be on to enable a sufficiently strong output, whereas if the weights of signals are strong, then either kinase (but not necessarily both) would need to be on to enable a sufficiently strong signal. This difference underlies the gap between an AND versus an OR gate. Even more elaborate coupling is possible, enabling computation at the intracellular or collective levels.

5.5.2 Enzyme kinetics

Kinases and phosphates are enzymes; that is, they catalyze chemical reactions involving the bonding of phosphate groups and the removal of phosphate groups, respectively, from proteins. Simplifications of complex enzyme kinetics can often be represented in terms of the rate of production, which depends on the availability of both substrate and enzyme. These molecules can bind, forming a complex that can reversibly release both substrate and enzyme or generate a product leading to the loss of substrate. In mathematical terms, this can be written as

$$\frac{dS}{dt} = -k_+ SE + k_- C \tag{5.5}$$

$$\frac{dE}{dt} = -k_+ SE + (k_- + k_f)C \tag{5.6}$$

$$\frac{dC}{dt} = k_+ SE - (k_- + k_f)C \tag{5.7}$$

$$\frac{dP}{dt} = k_f C \tag{5.8}$$

However, for many circumstances it is possible to simplify these dynamics and relate the rate of production directly to the substrate. The technical appendix show how given a substrate S and a total enzyme concentration E the rate of production can be approximated as

$$\frac{dP}{dt} = \frac{vS}{K_M + S} \tag{5.9}$$

where $v = k_f E$ is the maximum rate of the reaction and K_M is the half-saturation constant. It is important to specify two limits. First, when substrate is limited, i.e., $S \ll K$, then reaction rates can be approximated as

$$\frac{dP}{dt} \approx \frac{vS}{K_M}, \tag{5.10}$$

which is termed a "first-order" approximation to Michaelis-Menten enzyme kinetics (see the technical appendix for a detailed derivation of this approximation). Second, when substrate is abundant, i.e., $S \gg K$, then reaction rates can be approximated as

$$\frac{dP}{dt} \approx v \tag{5.11}$$

such that the generation of product is limited by the turnover rate of the enzyme and no longer by substrate availability. This second limit is equivalent to enzyme kinetics "at saturation." These concepts are central to kinase cascades and the difference between fine-tuned and robust adaptation in the case of *E. coli* response to changes in external chemical concentrations.

5.6 FINE-TUNED ADAPTATION

Our discussion of how *E. coli* responds to changes in the extracellular chemical environment can begin by avoiding some of the complexities of the full chemotactic machinery and focusing on the process of signal transduction. Consider a simple model of a signaling cascade to represent the chemotaxis machinery of *E. coli*. If CheA's activity is modulated by ligand binding, then the dynamics of CheY are controlled by the dual processes of phosphorylation and dephosphorylation. This is a single input-output reaction, which can be expressed mathematically using the notation introduced in this chapter as

$$\frac{dY_p}{dt} = v_1 X_p Y_0 - \alpha Y_p \tag{5.12}$$

such that at steady state

$$\frac{Y_p}{Y} = \frac{\omega_1 X_p}{1 + \omega_1 X_p}. \tag{5.13}$$

Hence, when attractant is added, we expect X_p to decrease and in turn for the fraction of Y_p to decrease, thereby decreasing the frequency of tumbling. Yet, in this model, there is no way for that frequency to change further even if this elevated level of attractant represents a new background concentration rather than a gradient. Similarly, when repellant is added, we expect X_p to increase and in turn for the fraction of Y_p to increase, thereby increasing the frequency of tumbling. Again, in this model, there is no way for that frequency to change further even if this elevated level of repellant represents a new background concentration rather than a gradient. Instead, we expect *E. coli* to be confused.

The response of the chemotactic circuit suggests that *E. coli* goes beyond a one-directional signaling cascade that propagates extracellular input directly to intracellular output. The full chemotaxis circuit includes feedback and can be represented—albeit in a simplified form—as a nonlinear system of differential equations that describe an aggregate receptor-CheA complex. This complex can have distinct methylation and phosphorylation states, i.e., the receptor can be methylated and CheA can be phosphorylated (as shown in Figure 5.3). The following set of equations represents an approximation to the internal dynamics involved in changing the tumbling frequency of the bacterial flagellum (Barkai and Leibler 1997; Alon et al. 1999; Alon 2006):

$$\frac{dX_m}{dt} = \frac{v_r R X_0}{K_r + X_0} + v_{deact} X_{m,p} - v_{act}(A) X_m \tag{5.14}$$

$$\frac{dX_{m,p}}{dt} = -\frac{v_b B X_m}{K_b + X_{m,p}} - v_{deact} X_{m,p} + v_{act}(A) X_m \tag{5.15}$$

In this equation, the total density of chemoreceptor complexes, X_{tot}, is fixed and equal to $X_{tot} = X_0 + X_m + X_{m,p}$ where X_0 is the number of unmethylated (and inactive) complexes, X_m is the number of methylated (and inactive) complexes, and $X_{m,p}$ is the number of methylated and phosphorylated (and hence active) complexes. In addition, here the phosphorylation rate is explicitly modeled as

$$v_{act}(A) = \frac{v_{act}^{max}}{1 + A} \tag{5.16}$$

where A is the attractant concentration in units of mM. Hence, phosphorylation events rapidly change the fraction of CheA receptors between X_m and $X_{m,p}$. This features suggests that the phosphorylation state can be modeled as in a quasi-steady state with respect to the ligand concentration, compared to slower changes in the methylation state. An increase

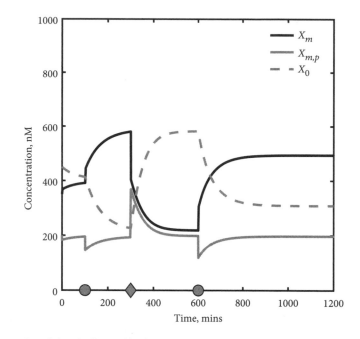

Figure 5.7: Dynamics of chemically modified variants of the chemoreceptor enzyme CheA (or A for short) such that X_0 is neither methylated nor phosphorylated, X_m is methylated but not phosphorylated, and $X_{m,p}$ is both methylated and phosphorylated. The chemical background is modified via a series of pulses that reset the system's external chemical state. The simulation is that of Eqs. (5.14)–(5.15). The simulation includes the following parameters: $v_r = 1$ (1/min), $R = 20$ (nM), $K_r = 2$ (nM), $K_b = 200$ (nM), $v_b = 1$ (1/min), $b = 40$ (nM), and $X_{tot} = 1000$ (nM), as well as $v_{act}^{max} = 100$ (mM/min), $v_{deact} = 100$ (1/min), and A (mM). Here v_{act} has units of 1/min, and initially $X_m = 350$ and $X_m^* = 200$ in units of nM. Note that the choice of total concentration is meant to be illustrative.

in attractant (or decrease in chemorepellant) acts to decrease v_{act}, whereas a decrease in attractant (or increase in chemorepellant) acts to increase v_{act}.

Figure 5.7 depicts the results of the inclusion of a two-state chemical modification system for CheA. In this figure, pulses of attractants (circles) and repellants (diamonds) are added to the system, and after a rapid deviation, the system returns to one in which the active state, $X_{m,p}$, relaxes back to its original set point of 200 nM. As such, we expect that the cell can go back to swimming "normally" (i.e., self-propelled random motion) in multiple chemical backgrounds. Instead, if the cell were to slow down or speed up with each successive pulse and remain locked in, then the frequency of tumbling would be fine-tuned to background levels. Indeed, as shown before, directly responding to the signal level of an input as part of a one-directional signaling cascade leads to input-output relationships that respond to background changes in the input rather than to gradients in the input.

Although we have used a simplified represention of the chemotaxis circuit, the corresponding dynamics respond robustly to changes in signal (i.e., ligand) concentrations. First, the circuit rapidly responds to changes in chemoattractants or chemorepellants by changing the phosphorylation state of CheA and therefore its downstream activity. This rapid response, like the fine-tuned principle, illustrates how a signaling cascade can convert input into output. Second (and critically), the circuit relaxes back to the baseline level of activity despite a long-term change in background concentration of ligands. This ensures that chemotaxis exhibits robust adaptation, and can filter out shifts in baseline levels over time. In doing so, adaptation requires sensing of the *rate of change* of ligands. Figure 5.7 shows that robust adaptation is possible. To understand this phenomenon, we need to re-examine the additional states of chemoreceptor molecules and understand precisely how this additional state space enables robust, exact adaptation.

5.7   BUFFERING AND ROBUST CELLULAR ADAPTATION

The state of the chemoreceptor varies with chemical modifications. Eqs. (5.14)–(5.15) represent a simplified model, following previous studies (Barkai and Leibler 1997; Alon et al. 1999; Alon 2006), in which there are three potential states: (i) unmethylated; (ii) methylated; (iii) methylated and phosphorylated. The last state denotes a chemoreceptor complex that is methylated and for which CheA is increasingly phosphorylated—this is the active state. This state enables a kinase activity such that CheA phosphorylates CheY, and CheY in turn changes the frequency of tumbling by binding to the flagellar motor switch. Therefore, as repellant is added (or chemoattractant decreases), CheA is increasingly phosphorylated. This newly active complex can in turn phosphorylate CheY, thereby increasing the frequency of tumbling. This output enables *E. coli* to stop its runs in a shorter period of time after moving from low to high repellant (or from high to low chemoattractant). Yet how does the cell relax back to a baseline tumbling rate if the new background level of repellant (or attractant) is in fact the norm?

We denote the overall state of the receptor-CheA complex as X_0, X_m, and $X_{m,p}$ for unmethylated, methylated, and methylated/phosphorylated, respectively. In addition, we assume that only the $X_{m,p}$ state is active in the sense that it has kinase activity on CheY. To recall, the dynamics of this system can be written as follows:

$$\frac{\mathrm{d}X_m}{\mathrm{d}t} = \frac{v_R R X_0}{K_R + X_0} + g(A, X_m, X_{m,p}) \tag{5.17}$$

$$\frac{\mathrm{d}X_{m,p}}{\mathrm{d}t} = -\frac{v_B B X_{m,p}}{K_B + X_{m,p}} - g(A, X_m, X_{m,p}) \tag{5.18}$$

where the generic function $g(A, X_m, X_{m,p})$ denotes fast changes in the phosphorylation state of chemoreceptors. This model assumes that phosphorylation is relatively fast (on the order of 0.1 sec) compared to slower processes of methylation (on the order of 10 sec). The model also assumes that methylation is required for phosphorylation (or alternatively that transitions from unmethylated to methylated states also involve dephosphorylation).

Mathematically, the consequence of this two-component system is striking. Note that on slower methylation scales

$$\frac{\mathrm{d}X_m}{\mathrm{d}t} + \frac{\mathrm{d}X_{m,p}}{\mathrm{d}t} = \frac{v_R R X_0}{K_R + X_0} - \frac{v_B B X_{m,p}}{K_B + X_{m,p}} \tag{5.19}$$

$$\approx v_R R - \frac{v_B B X_{m,p}}{K_B + X_{m,p}} \tag{5.20}$$

where the approximation in the second line is due to the assumption that CheR works at saturation, i.e., $R \gg K_R$. At steady state when $dX_m/dt + dX_{m,p}/dt = 0$, this defines an equilibrium of $X_{m,p}^*$. Hence, before adding repellant, the system begins with the expected concentration of

$$X_{m,p}^* = \frac{K_B v_R R}{v_B B - v_R R} \tag{5.21}$$

and the output should be $\omega_Y X_{m,p}^*$ where ω_Y is the prefactor related to the strength of the kinase. As is apparent, this steady state is unrelated to the input concentration of ligand! Or at least that is how it appears. Rather, the concentration may vary with parameters and baseline levels of CheB and CheR (given by the concentrations of B and R). The steady state implies that the system can relax back to equilibrium robustly. In essence, the output level is *fine-tuned* but the exact adaptation is *robust*. Here "fine-tuned" denotes the fact that the output levels have evolved to a particular set of values that could vary between cells and lineages. Nonetheless, irrespective of this output level, there are a broad range of parameters in which the circuit's dynamical state returns to this fine-tuned level after shifts in the background concentration—this is robust, exact adaptation.

The phenomenon of robust, exact adaptation leverages the extra dimension of the multi-component states of the receptor complex. You may be wondering how a system can return to the same output value after the perturbation if the fraction of phosphorylated states is directly related to the background concentration. The key is to recognize that something has changed before and after the robust adaptation. Although $X_{m,p}(0) = X_{m,p}^*$, this is not true for the concentrations of X_m. There is a chemical buffer for the system to modulate, i.e., the level of X_m can vary between 0 and $X_{tot} - X_{m,p}$. Hence, given the sum of three concentrations fixed at X_{tot}, the sensing machinery has an extra degree of freedom even if $X_{m,p}$ returns robustly and exactly to the fine-tuned value.

First, if repellant is added, this rapidly shifts molecules from $X_m \rightarrow X_{m,p}$, but since phosphorylation is fast, then the total number of methylated complexes $X_m + X_{m,p}$ should not change. Figure 5.8 shows the dynamics of the system in the two-dimensional phase plane. The phase plane shows repellant dynamics as moving in a line *parallel* to that of $X_m + X_{m,p} = X_{tot}$. Next, given this new background level, the ratio of $X_{m,p}/X_m$ will be constant, so methylation dynamics may change the total number of methylated complexes but not the ratio of phosphorylated to unphosphorylated types. Hence, dynamics will move on a *ray* whose origin is at $(0, 0)$. Because the system must return to the same fixed value, $X_{m,p}^*$, this implies that there will be fewer dephosphorylated and methylated receptors and, in fact, fewer total methylated receptors than before the addition of repellant. Similar logic applies to the addition of attractant or to the removal of repellant. It is this use of an extra dimension in state space that allows bacterial chemotaxis machinery to sense and respond to gradients, while also avoiding the pitfalls of overly fine-tuned sensitivity to changes in background concentrations. This insight reveals that a form of chemical buffering can enable the response to gradients, moving the active component of the system to a set point, even as other components shift in response to the changing baseline.

As Dennis Bray suggested, there are many reasons to strive to understand the *E. coli* chemotactic sensing circuit—perhaps as a model for examining signal transduction and sensing in "higher" organisms. But, as this chapter has shown, the insights and relevance of the chemotactic circuit and system-level responses should be evaluated on their own merits.

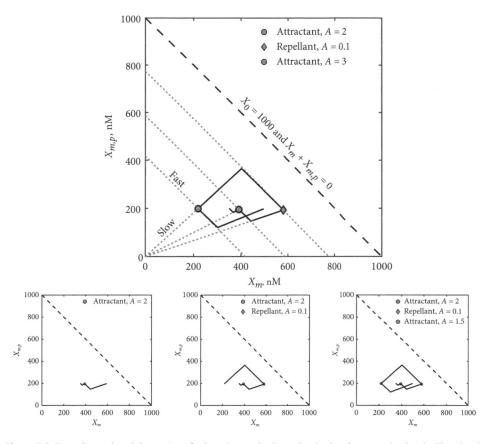

Figure 5.8: Two-dimensional dynamics of relaxation and adaptation in the chemotaxis circuit. The circuit includes rate parameters $\nu_R = 1$ and $\nu_B = 1$ (inverse minutes) and concentration constants $K_r = 2$, $K_b = 200$, $R = 20$, and $B = 40$ (in units of nM). Full details of the pulse dynamics and the phosphorylation changes are provided in the chapter. (Top) Dynamics in the $X_m - X_{m,p}$ plane, where the total concentration of receptors in this example is $X_{tot} = 1000$ nM (note there are on the order of 10,000 total receptors in the complex (Webre et al. 2003)). The dashed line denotes the domain of possibility given that $X_m + X_{m,p} \leq X_{tot}$. The dotted lines that connect the boundaries denote states of equal total methylated receptors irrespective of their phosphorylation states. The dotted lines that connect from the origin denote states where the ratio of $X_{m,p}/X_m$ remains fixed, even as the total number of methylated receptors changes. (Bottom) Visualization of phase space dynamics over time given the introduction of a sequence of input, background pulses of attractant and repellant (in units of μM).

Understanding how single-celled microbes—including bacteria and archaea—respond to complex signaling environments is likely to have far-reaching impacts.

5.8 TAKE-HOME MESSAGES

- Bacteria like *E.coli* can display a range of motilities, including flagella-enabled swimming.
- Although the benefits of swimming seem apparent, for very small microbes (like the ubiquitous ocean cyanobacteria *Prochlorococcus* living in dilute nutrient

environments) swimming is not required to uptake sufficient nutrients to survive and reproduce.

- *E. coli*'s motility is characterized by a series of runs (1 sec of nearly constant velocity) and tumbles (short periods of 0.1 sec in which movement is temporarily halted).
- The modulation of run duration can lead to chemotaxis toward chemoattractants and away from chemorepellants.
- The chemotactic sensing machinery includes a signaling cascade that enables robust, exact adaptation—the return of activity to the same fixed level after changes in background concentration of an input.
- Robust, exact adaptation enables a cell to adjust to changes in concentration and then reset its baseline level in the absence of a gradient.
- Robust, exact adaptation is enabled by a molecular buffer, given that the activity of a key kinase CheA is dependent on multiple modification states, including the methylation status of a transmembrane receptor.
- The nonlinear dynamics of robust adaptation are a paradigmatic example of advancing a systems-level understanding of bacterial physiology, and also provide insights for understanding signal cascades in other organisms.

5.9   HOMEWORK PROBLEMS

The objective of this problem set is to help deepen your understanding of bacterial chemotaxis—spanning both the phenomenology of diffusive random walks and the mechanisms by which a run-and-tumble motion is regulated. In doing so, the problem set leverages methods developed in the accompanying computational lab guide including

- Incorporating time-varying signals into a nonlinear dynamic model
- Sampling from probability distributions
- Simulating random and self-propelled diffusion in an "off-lattice" model

---

PROBLEM 1.   Scaling of Random Walks

Develop a computational model of a random walk such that the position of the cell is updated at each point by a random deviate, drawn from a normal distribution with mean 0 and standard deviation 1. Then simulate the process 1000 times, each to a maximum of 100 time steps. What is the *scaling* of the average displacement as a function of time? Repeat this problem using a model in which both the x and y positions are updated in the same way. Is the average displacement linear or nonlinear with time?

PROBLEM 2. Scaling and Details

Repeat the analysis of Problem 1, except this time compare the scaling results given an alternative choice of random deviates. Option A: The x or y deviate is either +1 or −1, i.e., the deviation is a delta function, $p(\delta x) = 0.5$ if and only if $\delta x = \pm 1$; the same holds for the y displacements. Option B: Choose a uniform distribution between −1 and 1. Compare and contrast the scaling of random walks given these alternative mechanisms; do they have the same scaling behavior? If there are differences, explain how they impact the scaling and/or the prefactor. Repeat this problem using a model in which both the x and y positions are changed.

PROBLEM 3. Robust Adaptation at the Cellular Scale—Part 1

Consider the following set of equations meant to represent the internal dynamics involved in changing the tumbling frequency of the bacterial flagellum:

$$\frac{dX_m}{dt} = \frac{v_r R X_0}{K_r + X_0} + v_{deact}X_m^* - v_{act}(A)X_m$$

$$\frac{dX_m^*}{dt} = -\frac{v_b B X_m^*}{K_b + X_m^*} - v_{deact}X_m^* + v_{act}(A)X_m$$

In this equation, the total density of chemoreceptor complexes, X_{tot}, is fixed and equal to

$$X_{tot} = X_0 + X_m + X_m^*$$

where X_0 is the number of unmethylated (and inactive) complexes, X_m is the number of methylated (and inactive) complexes, and X_m^* is the number of methylated and phosphorylated (and hence active) complexes. In addition, the phosphorylation rate is

$$v_{act}(A) = \frac{v_{act}^{max}}{1 + A}$$

where A is the attractant concentration in units of mM. Here you will (i) develop simulations of this model that include dynamic pulses; (ii) use your model to evaluate the properties of robust, exact adaptation.

  a. Demonstrate mathematically that the total number of methylated complexes is unchanged by ligand binding events.
  b. Write a script and a function program that simulates the equations. Further, include the ability to simulate a series of attractant pulses. Explain your choice of methods to simulate attractant pulses and include the code in the submission. In doing so, use the parameters $v_r = 1$ (1/min), $R = 20$ (nM), $K_r = 2$ (nM), $K_b = 200$ (nM), $v_b = 1$ (1/min), $B = 40$ (nM), and $X_{tot} = 1000$ (nM), as well as $v_{act}^{max} = 100$ (mM/min), $v_{deact} = 100$ (1/min), and A (mM). Because of this choice of units, v_{act} has units of 1/min. Assume that initially $X_m = 350$ and $X_m^* = 200$ in units of nM.

c. Using your program, simulate a series of attractant pulses such that

- $A = 1$ mM for $0 \leq t \leq 200$
- $A = 0.5$ mM for $200 \leq t \leq 400$
- $A = 2$ mM for $400 \leq t \leq 600$
- $A = 0.25$ mM for $600 \leq t \leq 1000$

Run the program using the default parameters, include a plot, and describe your findings. Is robust adaptation satisfied? Why or why not?

PROBLEM 4. Robust Adaptation at the Cellular Scale—Part 2

a. Modify the program developed in Problem 3 using the default parameters so that there is only a single pulse at $t = 200$ and let the program run until $t = 500$, i.e.,

- $A = 1$ mM for $0 \leq t \leq 200$
- $A_{attract}$ mM for $200 \leq t \leq 500$

Set the attracting pulse to be $A_{attract} = 0.25, 0.5, 1.5$, and 2. Find the relaxation time in each case. Does it vary with $A_{attract}$? Is exact adaptation satisfied? Contrast the relaxation profiles with that of Figure 1 in Alon et al. (1999).

b. Modify the program using the default parameters as in Problem 3. Set the attracting pulse to be $A_{attract} = 2$ using the following values of R: $R = 5, 10, 15$, and 20 nM. Find the adaptation time and steady state X_m^* in each case. Compare and contrast your result to Figure 2 of Alon et al. (1999).

c. Modify the program using the default parameters as in Problem 3. Here consider a single pulse at $t = 200$ of $A_{attract} = 4$ or $A_{attract} = 8$ and let the program run until $t = 500$. Is robust, exact adaptation satisfied? If not, describe the cause of the violation.

PROBLEM 5. Self-Propelled Diffusion as Search

This problem extends the computational laboratory model of self-propelled diffusion to one in which *E. coli* modifies its direction after each tumble randomly with respect to the prior run. Begin with your lab code that simulates self-propelled diffusion in a 2D spatial domain. In that code, the direction of each "run" was random. Here modify the code so that the direction of each run is skewed toward "smaller angles," i.e., closer to going straight than choosing a strictly random direction. In doing so, use the following data as input (Berg and Brown 1972) (or refer to the website):

$\Delta\theta$	5	15	25	35	45	55	65	75	85	95
Events	18	55	71	120	140	138	135	93	89	58

$\Delta\theta$	105	115	125	135	145	155	165	175
Events	52	53	39	41	18	11	10	3

a. First, utilizing the turning data, find the best-fit beta distribution. What are the shape parameters that you identify? Show an overlay of the data versus that of the beta.

b. Second, modify your code so that the direction of the next run, θ', is chosen as $\theta' = \theta + \beta$ where θ is the direction of the prior run and β is a randomly distributed variable drawn from a beta distribution with shape parameters determined in part a. Note that in doing so you should also manipulate the sign, either positive or negative, so that the direction encompasses the entire 360 degrees. Include your code snippet and provide evidence that your new directions are in fact oriented slightly closer to the current direction.

c. Integrate this reorientation function into the self-propelled diffusion model. Use a run speed of 25 μm/sec and exponentially distributed runs of 1 sec. In this model, provide three sample visualizations over a period of 100 seconds. Compare and contrast to the dynamics in which the new direction is chosen completely independently of the prior direction.

d. Simulate an ensemble of realization of self-propelled diffusion in the absence of a chemotactic signal. Calculate the root mean square displacement for a 1000-second interval for your ensemble. Provide evidence in favor of/against a hypothesis that the bacteria have effectively "diffusive" behavior.

Additional Challenge Problems

The following are additional challenge problems that can potentially form the basis for in-class work and perhaps even final projects.

- Consider ways to couple an *intracellular model* of adaptation with the *organismal* behavior of movement. For example, build a model that describes each bacterium in terms of its position, y, and its internal state variables (x_m, x_m^*), representing the concentration of methylated-inactive receptors and methylated-active receptors. Assume that when $x_m^* = 200nM$ the frequency of tumbling is 1/sec, one-half of its maximum frequency when x_m^* is at its highest value possible.

- Extend your code from Problem 5 so that individual bacteria undergo runs and tumbles; however, the tumbling frequency should depend on their internal state— which itself is changing over time. Explain the logic of your code and show evidence that both the position and internal state change concurrently in the way that you expect.

- Include the pulses as explained in Problem 3c. Provide evidence that the adaptation undergone at the cellular scale has an effect on the resulting swimming behavior of *E. coli*. Does the *E. coli* change behavior when moving in and out of areas with different background levels of chemoattractant? Does it change its behavior when traveling up/down a gradient?

5.10   TECHNICAL APPENDIX

Expected distance scaling of random walks   Consider a random walk in 1D in which an individual organism moves forward or backward with equal probability. The position, $x(t)$, after t steps is a random variable. What is the expected distance moved from the origin, $x(0) = 0$? By definition, $x(t) = \Sigma_{i=1}^{t} \chi_i$ where χ_i is a random number that takes one of two values, $D^{1/2}$ and $-D^{1/2}$, such that the expected values $E(\chi_i) = 0.5D^{1/2} - 0.5D^{1/2} = 0$. Here the value of $D^{1/2}$ is an expected step size, where the choice of a square root is done for reasons that will become apparent in a moment. The expected distance traveled in t steps is

$$E(x(t)) = E\left(\Sigma_{i=1}^{t} \chi_i\right) \tag{5.22}$$

$$= \Sigma_{i=1}^{t} E(\chi_i) \tag{5.23}$$

which because of the unbiased nature of the walk yields

$$E(x(t)) = 0. \tag{5.24}$$

This calculation reveals that a random walk doesn't get anywhere, on average. However, there is a subtlety. Even if the sign-weighted average is 0, the random walk will appear to get somewhere on long trajectories, when calculated in an absolute sense, e.g., irrespective of whether the random walk has moved 20 steps to the left or right or 40 steps to the left or right. We can evaluate the average *squared* distance of such a random walk as follows:

$$E(x^2(t)) = E\left(\left[\Sigma_{i=1}^{t} \chi_i\right] \times \left[\Sigma_{j=1}^{t} \chi_j\right]\right) \tag{5.25}$$

$$= E(\Sigma_{i=1}^{t} \chi_i^2) + E(\Sigma_{i,j,i \neq j}^{t} \chi_i \chi_j) \tag{5.26}$$

$$= \Sigma_{i=1}^{t} E(\chi_i^2) + E(\Sigma_{i,j,i \neq j}^{t} \chi_i \chi_j) \tag{5.27}$$

$$= Dt + E(\Sigma_{i,j,i \neq j}^{t} \chi_i \chi_j) \tag{5.28}$$

$$= Dt + \Sigma_{i,j,i \neq j}^{t} E(\chi_i \chi_j)) \tag{5.29}$$

$$= Dt \tag{5.30}$$

This result arises because each step is independent, so the expected value of the product of two random steps is 0. As a result, the expected distance squared is linear with the number of steps (or time). Similarly, the root-mean-square distance, i.e., $E(x^2(t))^{1/2} = (Dt)^{1/2}$, increases sublinearly. This square root increase is the hallmark of a random walk (Berg 1993).

Enzyme kinetics and approximations   Consider the standard enzyme kinetic system in which a substrate S and enzyme E can reversibly bind to form a complex. The complex can undergo an irreversible reaction, generating a product P and releasing the enzyme.

Enzyme kinetics can be modeled via a system of nonlinear differential equations:

$$\frac{dS}{dt} = -k_+ SE + k_- C \tag{5.31}$$

$$\frac{dE}{dt} = -k_+ SE + (k_- + k_f)C \tag{5.32}$$

$$\frac{dC}{dt} = k_+ SE - (k_- + k_f)C \tag{5.33}$$

$$\frac{dP}{dt} = k_f C \tag{5.34}$$

where $S + C + P = N$, the total number of substrates in any particular form. This three-dimensional system of equations can be reduced further by envisioning that the complex reaches a quasi-stationary equilibrium. In that case, we can assume that $dC/dt \approx 0$, such that

$$C_q = \frac{k_+}{k_- + k_f} SE \tag{5.35}$$

where C_q denotes a quasi-stationary equilibrium. Note that the free enzyme concentration, $E = E_{tot} - C_q$, is equal to the total enzyme concentration minus that bound in complexes. Hence, for self-consistency, this should be

$$C_q = \frac{k_+}{k_- + k_f} S \left(E_{tot} - C_q \right) \tag{5.36}$$

such that

$$C_q \left(1 + \frac{k_+}{k_- + k_f} S \right) = \frac{k_+}{k_- + k_f} S E_{tot} \tag{5.37}$$

and further

$$C_q = \frac{k_+}{\left(k_- + k_f \right) + k_+ S} S E_{tot} \tag{5.38}$$

$$C_q = \frac{E_{tot}}{\frac{(k_- + k_f)}{k_+} + S} S \tag{5.39}$$

This quasi-steady state approximation can then be used to predict the product generation rate, $dP/dt = k_f C_q$, or

$$\frac{dP}{dt} = \frac{k_f E_{tot}}{\frac{(k_- + k_f)}{k_+} + S} S. \tag{5.40}$$

Usually, this form is reduced by identifying the maximum speed of the reaction as $v = k_f E_{tot}$ and the Michaelis constant as $K_M = \frac{(k_- + k_f)}{k_+}$. Note that because $k_- \gg k_f$ is typical, then $K_M \approx k_-/k_+$. This form can be expressed as

$$\frac{dP}{dt} = \frac{vS}{K_M + S}, \tag{5.41}$$

which is the Michaelis-Menten equation of enzyme kinetics. This equation is a saturating function of substrate (in the sense of being a Hill function) and is a reasonable approximation of dynamics (albeit in certain limits).

Beta distributions The distribution of tumbling angles can be approximated via a beta distribution. Beta distributions are a two-parameter family of probability distributions restricted to the $[0, 1]$ interval. The probability distribution function is

$$p(x) = \frac{x^{\alpha-1}(1-x)^{\beta-1}}{B(\alpha, \beta)} \tag{5.42}$$

where

$$B(\alpha, \beta) = \frac{\Gamma(\alpha)\Gamma(\beta)}{\Gamma(\alpha+\beta)} \tag{5.43}$$

and $\Gamma(z)$ denotes the Gamma function

$$\Gamma(z) = \int_0^\infty x^z e^{-x} \mathrm{d}x. \tag{5.44}$$

The beta distribution has a number of interesting limits and can be used to replicate other distributions. For example, when $\alpha = 1$ and $\beta = 1$, then it is equivalent to a uniform distribution. The mean of the distribution is $\alpha/(\alpha+\beta)$. The distribution can vary between being biased to one side or the other (or both sides) of the interval, as well as peaked in the middle of the interval. Its flexibility is useful when considering domain-limited data, like the angle of the next run for run-and-tumbling *E. coli*.

Chapter Six

Nonlinear Dynamics and Signal Processing in Neurons

6.1   WALKING IN THE PATH OF HODGKIN AND HUXLEY

How do living systems process, respond to, remember, and adapt to information from their environment? Chapter 5 described how bacteria, like *E. coli*, sense chemical gradients, thereby changing their behaviors to spend more time in good environments and less time in bad environments. In that case, a signal transduction pathway connects environmental signals to intracellular regulatory dynamics and, ultimately, to physical dynamics (the counterclockwise or clockwise rotations of the flagellar bundle). What *E. coli* does seems rather remarkable. The evolution of signaling cascades includes multiple components and nonlinear feedback operating across a wide range of input regimes such that a bacterium can sense/respond even without an actual brain, i.e., without white matter, gray matter, neurons, or glia cells. The chemotaxis circuit of *E. coli* illustrates—in an archetypal living organism—that the capability of sensing is embedded not in one particular "sensing" protein but rather in the system itself.

This substrate-independent approach to information processing is even more apparent when examining signal processing in the brain. In doing so, we will need to acquire a new vocabulary corresponding to the parts list, not just of neurons but of neural circuits. This chapter introduces new components; however, it does not begin and end with parts. Structure can often explain function, but it is the feedback and interactions among components that drive systemwide dynamics. Such a dynamical perspective is not universally adopted in introductory courses on neuroscience, but it is the perspective espoused by certain theoretical neuroscientists, including Eugene Izhikevich (2007), who offered this advice:

> Information processing depends not only on the electrophysiological properties of neurons but also on their dynamic properties.

> Even if two neurons in the same region of the nervous system possess similar electrophysiological features, they may respond to the same synaptic input in very different manners because of each cell's bifurcation dynamics.

This quote suggests that answers to deep problems in information processing will not be found by looking at cellular structure alone. Worse yet, we might even find the wrong answer by trying to leverage differences in cell physiology to explain observed differences

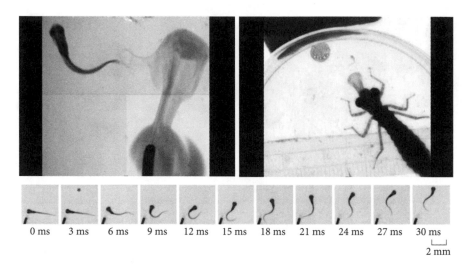

0 ms   3 ms   6 ms   9 ms   12 ms   15 ms   18 ms   21 ms   24 ms   27 ms   30 ms

2 mm

Figure 6.1: Electrical signaling in the central nervous system of the zebrafish larva underlies its rapid reaction to external stimuli. (Top left) Zoomed-in view of a zebrafish larva as it moves away from the water pulse. A zebrafish larva is typically ∼ 4 mm in length. (Top right) A dragonfly larva attempting to consume a zebrafish larva by rapidly extending part of its mouth (dragonfly larvae are typically 2–10 cm in length). (Bottom) The sequence of images illustrates the rapid reaction of a zebrafish larva to an experimental water pulse stimulus over a 30 ms period. Reproduced from Luo (2015).

in behavior. Instead, feedback in a neuronal system leads to dynamics that are not characteristic of lower scales of organization.

The gap between dynamics across scales is an example of what is called *emergence*, whereby living systems exhibit behavior that cannot be reduced or understood strictly in terms of the rules governing the underlying components (Anderson 1972). But, even if new features and principles emerge at different scales, that does not mean that we should forgo the biological details at smaller scales. Instead, this information is essential to understanding mechanisms governing different classes of dynamical behavior. Indeed, evolutionary adaptations have fixed and proliferated precisely because of the expanded dynamical features that could be accessed via distinct, physiological mechanisms. These physiological mechanisms extend from cells to whole organisms. For example, in the classic example of perturbations to a zebrafish larva, a stimulus experienced at the tail end is sensed within milliseconds by sensory neurons (Dill 1974a,b; Spence et al. 2008). Signals are then transmitted through the larva's nervous system to the central nervous system where the larva reacts through a combination of motor neurons and muscle neurons, rapidly changing its body conformation and swimming away from the stimulus, e.g., a water pulse initiated by the experimentalist. In this particular case, rapid sensing of changes in the fluid context provides protection from the actions of potential predators. Figure 6.1 depicts the movement of the zebrafish larva away from an experimental water pulse stimulus (top left) and away from a larval dragonfly (top right), with individual movements depicted over a 30 ms time series (bottom). This kind of integrative response has a core template: the reaction of individual neurons to stimuli.

This chapter focuses on neuronal dynamics as a first step toward understanding what a brain does. One of the key features of neuronal dynamics is excitability—i.e., a system filters out small perturbations while reacting to sufficiently large perturbations via marked

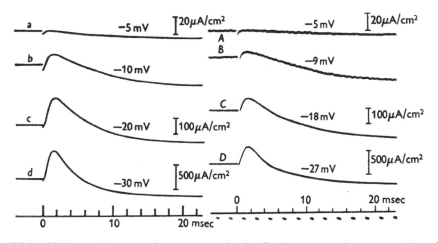

Figure 6.2: Rapid changes in transmembrane current when holding the transmembrane potential at a fixed voltage via the use of a voltage clamp. As sodium ions enter the cell during the initial depolarizaion, the release of potassium ions during the fall of the action potential. Note that the relaxation to rest state occurs much later (Hodgkin and Huxley 1952). The sequence of images include numerical calculations of model predictions (left) compared against experimental measurements (right) given variation in the externally fixed potential difference, a significant advance over prior (related) work (Cole and Curtis 1939).

departures from its rest state, only to relax back slowly to the initial prestimulus state. Such excitability is a template for information processing occurring at the level of single neurons. Figure 6.2, reproduced from Alan Hodgkin and Andrew Huxley's Nobel Prize–winning work, shows the change in associated conductance in a squid axon given different levels of a fixed voltage. The large size and accessibility of the axon made it ideal for the direct evaluation of changes in the action potential and comparison to model predictions. Critically, the response to changes in voltage are highly nonlinear, including a rapid spike in current followed by a slower relaxation. Figure 6.2 includes a comparison of model predictions (left) and experimental measurements (right); they are strikingly similar in terms of the nonlinear response and the relationship between the strength of the response and the magnitude of the fixed voltage difference. (Of some historical interest, a hand-operated mechanical calculator was used for the numerical calculations (Schwiening 2012).) As is already apparent, the spike response is not inevitable and varies with input—this feature is key to driving the excitability of neurons.

The concordance of model and experiment represents our challenge: trying to make sense of neurons by making sense of the Hodgkin-Huxley model. The Hodgkin-Huxley (HH) model and associated paper may seem like an impenetrable work from a different time. But it has stood the test of time precisely because the theory corresponds to the real world. As Jim Peebles (2019), the Nobel Prize–winning cosmologist, remarked:

> It is so easy for us theorists who build wonderful castles, beautiful ideas. Sometimes, it is remarkable, sometimes these beautiful ideas prove to be close to what the observations tell us. But often and also they turn out to be wrong. No great surprise, but time will tell, and it is the measurements that tell us.

The work of Hodgkin-Huxley is beautiful but complicated; it is beautiful because the idea has turned out to be right. The goal then of this chapter is to build up the HH model of a neuron and explain why its parts functioning together can give rise to information processing and the basis for learning, memory, and behavior. As we will find, the HH equations do not fit in the usual T-shirt-size aphorisms ($E = mc^2$ and the like). Hence, the first part of this chapter walks in the path of Hodgkin and Huxley to build up the model from (close to) first principles. Then the chapter focuses on how the HH model can be used to explain the dynamical behavior of single neurons. Finally, using the potential processing features of a single neuron, the last component of the chapter explains what happens when neurons are connected and how productive analogies to learning in vivo are the basis for learning (deep or otherwise) in silico.

6.2   THE BRAIN: MEMORY, LEARNING, AND BEHAVIOR

6.2.1   What is the brain for, really?

The nervous system is an information processing machine that serves as a gateway between the body and the environment. This gateway is important. Although what the brain does can be abstracted in terms of information and signal processing, the brain is not a disembodied entity. The processes of signaling, communication, memory, learning, and behavior are part of an extended physical and physiological system. Conventionally, this system is broken down into two categories: the central nervous system (CNS) and the peripheral nervous system (PNS). For humans, the central nervous system, aka the brain, is connected to the body via a system of nerves (e.g., cranial nerves and spinal nerves) as well as a spinal cord. This extended system also provides context for the flow of information. Interactions with the environment are mediated by sensory receptors, which initiate signals that are transmitted to the CNS via what is termed the *afferent path*. Then signals encoding the outcome of neuronal interactions in the CNS alter the state of the CNS and also trigger signals propagating from the CNS to effector cells and tissue via an *efferent path*. Such processes take time and are conducted in the presence of noise and uncertainty, but are also the basis for behavior, learning, and memory.

What does the brain do? In brief, the brain-body interactions lead to behavioral responses, but it is the brain that enables modifications to behavior via learning and memory. It would seem that terms so universally used as learning or memory require no definition. But definitions help center our goals. Here are a few:

Behavior: Actions taken by an organism, usually in the context of its environment

Learning: Adaptive change in behavior resulting from experience

Memory: The retention of learning, allowing learning (i.e., adaptive behavior) at a later time

These three intertwined concepts place the neuron, neuronal dynamics, and brain in context. In order for there to be adaptive behavior (i.e., learning), there must be some way for an organism to adapt. Adaptation requires that an organism recognizes and reacts to an environmental stimulus related to prior stimuli. This recognition implies an internal representation of experience, that is, a map (or transformation) of sensory input from the environment that can then be used as the basis for computations. That is what the nervous

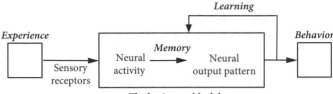

The brain as a black box

Figure 6.3: Schematic of behavior, memory, and learning in the brain. Experiences are detected by sensory receptors (and sensory neurons) that then transmit information to the central nervous system. Neural activity is then encoded as an output pattern, which is transmitted back to sensory, motor, and muscle neurons that can change behavior. The outcome of behavior then modifies the very neural activity and output patterns that drive behavior, connecting learning to long-term changes (i.e., memory).

system does. It takes in experiences via sensory receptors and then transmits electrical signals to the brain, where the brain generates neural output patterns that inform responses and changes in behavior. The unidirectional flow from environment to behaviors becomes a circuit when a set of the outputs is directed at the brain itself, thereby modifying behavior. This learning requires representations of the environments, at least in a statistical sense, that can be used as the basis for recognition of similar environments and/or contexts. Hence, learning requires a computation—first comparing a prior experience (via memory) to a present experience. The adaptive change in behavior requires that the outputs of the brain modify themselves (Figure 6.3).

A full description of the parts list of the nervous system lies outside the scope of this chapter but can be found in neuroscience textbooks (e.g., Luo 2015). In brief, the pathway from the environment (including both the internal and external components) involves the propagation of information and signals via sensory receptors at the surface of the body and internally via nerves and ganglia. This information is then transmitted to the central nervous system where it is processed, and then outputs are directed to the CNS as well as to motor components. The transmission of signals via the efferent path includes motor nerves as well as components of the visceral motor system, leading to changes in what are termed *effectors*, including muscles and glands. This repeated process links information and behavior, the physical and biological world, and yet at its core requires templates of computation. These templates are neurons.

6.2.2 Neurons: A synopsis

Neurons are cells essential to behavior, learning, and memory. There are many kinds of neurons, differing in their location (e.g., inside the brain or distributed in the body), structure (e.g., size and connectivity), and function (e.g., motor neurons, sensory neurons, or interneurons). Neurons are composed of a cell body, with electrical input coming from other neurons via axons and electrical output transmitted to other neurons via dendrites. The connections between axons and dendrites are called synapses. This electrical input encodes information transmitted between neurons. The dendrites collect this information and integrate it, and then the firing of the neuron transmits a new signal (and new information) to other neurons.

Figure 6.4 (Plate 4) provides two views of a neuron. On the left is a histological stain of components. This image depicts the physical positioning of dendrites conveying electrical

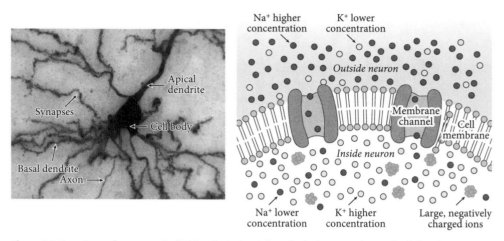

Figure 6.4: Two views of a neuron. (Left) Histological staining of a single neuron in a rat brain by Christiaan de Kock, including cell body, axon, apical dendrite, basal dendrite, and synapses. (Right) View of ion exchange across the membrane of neurons. Reproduced from Luo (2015).

signals into the cell body and that of the axon conveying electrical signals out of the cell body. On the right is a schematic of the localization and distribution of ions on either side of the neuron's cell membrane. As is apparent, there are different concentrations of ions on either side of the cell membrane. A difference in the concentration of charged particles implies there is a difference in the *electrical potential* (Benedek and Villars 2000a). The convention in neuroscience is to denote this potential difference as the *transmembrane potential* or *transmembrane voltage*. Differences in electrical potential lead to forces, and forces lead to changes in the mobility and positions of particles, in this case ions. The propagation of electrical signals between neurons is the basis for behavior, learning, and memory.

In vertebrates, information generally flows from dendrites to cell bodies to axons as part of networks. A few examples of these networks can be seen in Figure 6.5 (Plate 5). As is apparent, the input can involve integration from multiple dendrites, and outgoing signals can then split, enabling distribution of signals to multiple neurons. For example, sensory information transmitted from a sensory neuron through nerves in the spinal cord would be transmitted via the brain stem and thalamus to the primary somatosensory cortex. After processing, a signal would be relayed via the primary motor cortex through nerves in the spinal cord to a motor neuron, which would trigger the movement of muscles as part of voluntary movement. Voluntary triggering of muscle neurons in response to a stimulus is examined in detail by Ron Milo and colleagues as part of their BioNumbers project (Milo et al. 2010). In this example, they ask the question, "How fast can Olympic athletes respond to the starter's pistol?" Here is their answer (Shamir et al. 2016):

> Upon hearing the shot, athletes process and propagate an electric impulse from the brain all the way to their feet (1 m). Considering the speed of the action potential (10–100 m/s), this implies a latency of 10–100 ms regardless of other processes, such as the speed of sound and signal processing in the brain. The best athletes respond after 120 ms, and a reaction time below 100 ms is immediately disqualified as a false start.

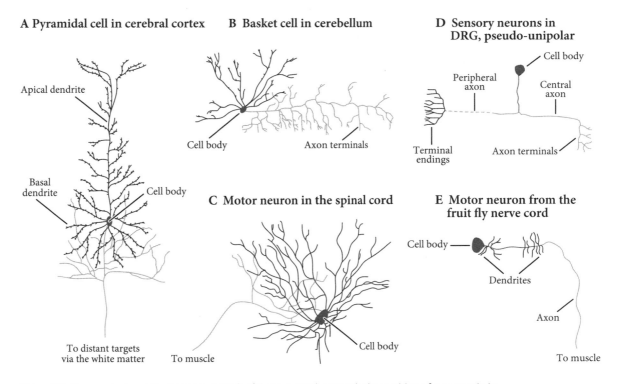

A Pyramidal cell in cerebral cortex   **B Basket cell in cerebellum**   **D Sensory neurons in DRG, pseudo-unipolar**

Apical dendrite

Basal dendrite

Cell body

To distant targets via the white matter

Cell body

Axon terminals

To muscle

C Motor neuron in the spinal cord

Cell body

Cell body

Peripheral axon

Central axon

Terminal endings

Axon terminals

E Motor neuron from the fruit fly nerve cord

Cell body

Dendrites

Axon

To muscle

Figure 6.5: Neurons in a vertebrate context. Each of these examples reveals the position of a neuron in its context, spanning neurons in the brain to sensory neurons to motor neurons. Reproduced from Luo (2015).

This insight is telling. It connects action potential propagation at one scale to a familiar behavior, while placing limits on the behavior precisely because of the biophysical nature of the nervous system. This example reinforces the idea that the nervous system is not just an information processing system; it is part of a living organism. Yet the insight has limitations. Why is the speed of the action potential 10–100 m/s? What allows the action potential to propagate? And how does a signal turn into a response in the first place? These are deeper problems, and taking a distance divided by a velocity to get a time provides an answer but an incomplete one. A deeper answer requires delving into the neuron itself.

In walking the path of Hodgkin and Huxley, we will focus on the dynamics associated with the movement of ions across a neuron's membrane in response to an electrical signal that arrives via the dendrites to the neuronal cell body. This process provides the basis for an answer not just to the question of how fast is too fast to start running in response to the sound of a starter pistol but to a whole class of problems related to information processing and excitability.

6.3   OF IONS AND NEURONS

6.3.1   The HH equations: A far-off destination

The 1952 paper by Hodgkin and Huxley is remarkable on many levels. It combines theory and experiment. It is quantitative. It reveals basic principles of how excitation works in neurons. It remains deeply influential. And yet, it is also difficult to read and, at some level, intimidating. The following equations represent what is known as the Hodgkin-Huxley (HH) model:

$$C_M \frac{dV}{dt} = I - \bar{g}_K n^4 \left(V - V_K \right) \ldots$$

$$- \bar{g}_{Na} m^3 h \left(V - V_{Na} \right) - \bar{g}_l \left(V - V_l \right) \tag{6.1}$$

$$\frac{dn}{dt} = \alpha_n(V)(1 - n) - \beta_n(V)n \tag{6.2}$$

$$\frac{dm}{dt} = \alpha_m(V)(1 - m) - \beta_m(V)m \tag{6.3}$$

$$\frac{dh}{dt} = \alpha_h(V)(1 - h) - \beta_h(V)h \tag{6.4}$$

These four equations describe, in turn, the dynamics of a voltage V (i.e., the transmembrane potential) and phenomenological representations of the extent to which the membrane is permeable to the exchange of potassium, sodium, and chloride ions as represented by n, m, and h. In this case, voltage changes are driven by the balance of internal currents (with associated conductances $\bar{g}_K$, $\bar{g}_{Na}$, and $\bar{g}_l$), voltage constants, V_k, V_{Na}, and V_l, capacitive current (with associated capacitance C_M), and an applied current I. The permeability to ion exchange is characterized by kinetic rate constants (denoted by α and β) that are nonlinear functions of the voltage V. The purpose of the next series of sections is to break down this model, part by part, and reconstruct the meaning of each term. Then, with the model in hand, the chapter explores how a neuron operating given these principles can exhibit an astonishing array of dynamical behaviors, including nonlinear response to stimuli and excitability.

6.3.2   Where do the ions go?

The information processing activity of neurons is mediated by the exchange of ions across the cell membrane. The three principal ions are potassium (K^+), sodium (Na^+), and chloride (Cl^-). The state of a neuron may be described in terms of the flow of ions across the membrane. Because ions are charged particles, it is possible to break down the flow of ions into two components: (i) flow mediated by concentration differences; (ii) flow mediated by electropotential differences.

To begin, first consider what would happen if there were more uncharged particles inside the cell relative to those outside of a cell. A cellular membrane is permeable to the passage of certain molecules (like water), while many molecules include sugars like glucose or charged ions that may only pass through via specialized transport mechanisms, including facilitated diffusion, transporters, and/or channels. In the event that passage is equally likely in either direction for a given molecule, one would expect that more molecules would leave the cell than enter (when the concentration is higher inside than outside). Over time, even if the movement of individual molecules were stochastic, there would be more molecules leaving than entering the cell. As a result, the concentration difference would decrease until there was the same concentration of particles inside as outside of the cell. The same logic applies if there were initially fewer particles inside the cell than outside. In effect, a concentration difference drives a flux that moves against the gradient (from high to

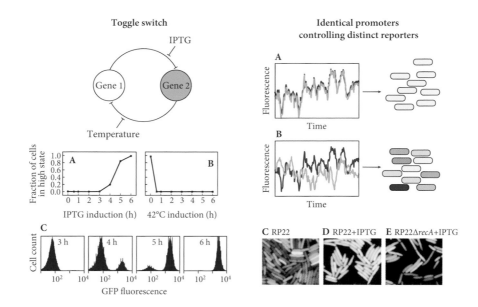

Plate 1: Stochasticity leads to variation in cellular-level outcomes, whether in bistable switches or in identically controlled reporters. (Left) A genetic toggle switch is a synthetically engineered bistable feedback loop where two genes mutually inhibit each other. Activities of the two genes are controlled by external factors—IPTG and temperature. When IPTG is high, the system turns on (as evidenced by the expression of GFP (green fluorescent protein); when temperatures are increased, the system turns off. However, as shown in panel c, individual cells do not move uniformly between states; instead, there are marked differences in cellular state associated with the bistability of the circuity. (Right) Gene expression variability induced by intrinsic noise. In the event that the same promoter is used to control distinct reporters that express GFP and RFP, it is possible to evaluate the extent to which the expression of a promoter is identical (as in A, in the absence of intrinsic noise) or variable (as in B, in the presence of extrinsic noise). The correlation between identical promoters is one way to measure intrinsic noise. Panels C–E show microscopy images of synthetically engineered *E. coli* with the two-color reporter system, given the absence and presence of IPTG (C and D, respectively). As is apparent, constitutive expression leads to differences that are then decreased in the presence of strong expression induced by IPTG. Images reproduced from Gardner et al. (2000) and Elowitz et al. (2002).

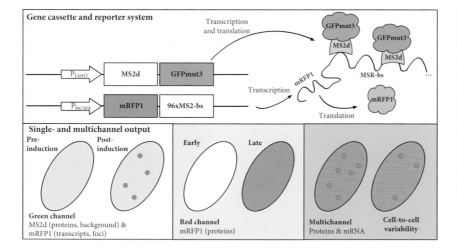

Plate 2: Gene reporter system for measuring transcripts per cell via an mRNA detection system in *E. coli*. (Top) The two-promoter reporter system, including a binding site for a fusion protein on RNA. (Bottom) Expected output via the green channel, red channel, and multichannel, including cell-to-cell variation. For more details, see the description in the text and refer to Golding and Cox (2004) and Golding et al. (2005).

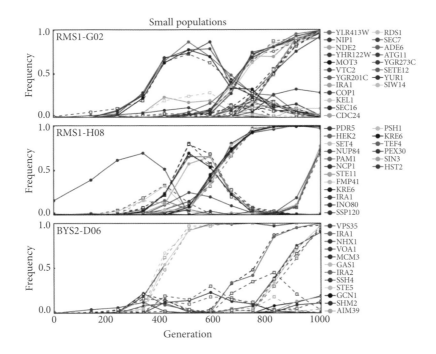

Plate 3: Trajectories of spontaneous mutations in evolving yeast populations. Solid lines denote synonymous mutations and dashed lines denote nonsynonymous or intergenic mutations. The labels on the right-hand side denote the mutation identity. Reproduced from Lang et al. (2013).

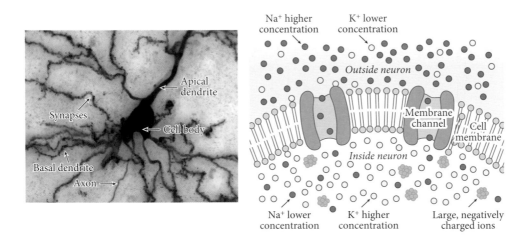

Plate 4: Two views of a neuron. (Left) Histological staining of a single neuron in a rat brain by Christiaan de Kock, including cell body, axon, apical dendrite, basal dendrite, and synapses. (Right) View of ion exchange across the membrane of neurons (Reproduced from Luo 2015).

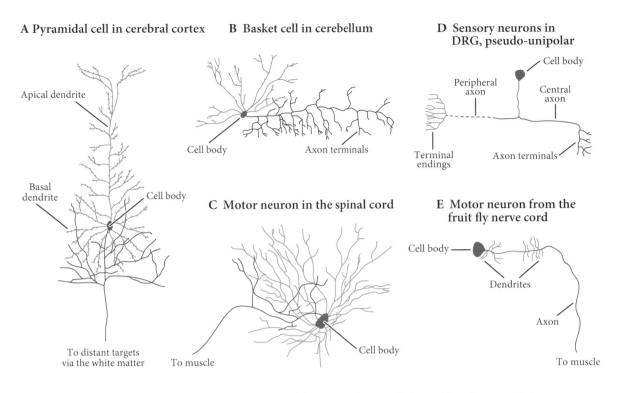

A Pyramidal cell in cerebral cortex

Apical dendrite

Basal dendrite

Cell body

To distant targets via the white matter

B Basket cell in cerebellum

Cell body

Axon terminals

C Motor neuron in the spinal cord

To muscle

Cell body

D Sensory neurons in DRG, pseudo-unipolar

Cell body

Peripheral axon

Central axon

Terminal endings

Axon terminals

E Motor neuron from the fruit fly nerve cord

Cell body

Dendrites

Axon

To muscle

Plate 5: Neurons in a vertebrate context. Each of these examples reveals the position of a neuron in its context, spanning neurons in the brain to sensory neurons to motor neurons. Reproduced from Luo (2015).

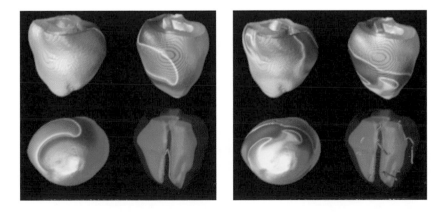

Plate 6: Spatially extended dynamics simulated in the rabbit ventricular geometry. (Left) A single spiral wave. (Right) Multiple spiral waves. Images reproduced from Figure 31 of Cherry and Fenton (2008).

Flapping mechanism

Corkscrew mechanism

$t = 5$ sec

$t = 25$ sec

$t = 45$ sec

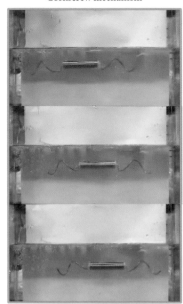

Plate 7: Examples of mechanical swimming at a low Reynold's number using rubber band–powered swimmers that include flapping (Left) and corkscrew (Right) mechanisms. The flapping swimmer does not move forward over time, whereas the corkscrew swimmer does. For videos see https://uwmadison.app.box.com/s/384lkhmyaj6jel-nagl74vbbeusd83r7r. Videos and images courtesy of Michael Graham and Patrick Underhill, used with permission. Videos were taken by Patrick Underhill, Diego Saenz, and Mike Graham at the University of Wisconsin–Madison, with funding from the National Science Foundation. Swimmer designs were inspired by the work of G. I. Taylor (1951); additional context in Graham (2018).

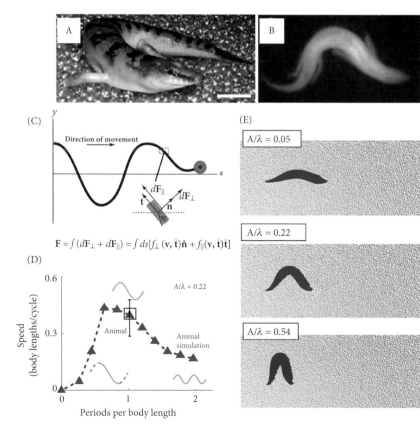

(A)

(B)

(C)

y

Direction of movement

$dF_\parallel$

$dF_\perp$

$\mathbf{t}$ $\mathbf{n}$

x

$$\mathbf{F} = \int (d\mathbf{F}_\perp + d\mathbf{F}_\parallel) = \int ds[f_\perp(\mathbf{v}, \hat{\mathbf{t}})\hat{\mathbf{n}} + f_\parallel(\mathbf{v}, \hat{\mathbf{t}})\hat{\mathbf{t}}]$$

(D)

Speed (body lengths/cycle)

0.6

0.3

0

0   1   2

Periods per body length

$A/\lambda = 0.22$

Animal

Animal simulation

(E)

$A/\lambda = 0.05$

$A/\lambda = 0.22$

$A/\lambda = 0.54$

Plate 8: Lessons from a sandfish lizard swimming through sand. The organism *S. sincus* is shown above the surface (A) and in the midst of subsurface locomotion (B) via X-ray tracking (reprinted from Maladen et al. 2011). (C) Schematic of resistive force theory; for origins, see Gray and Hancock (1955). (D) Comparison of propulsion speed in terms of body lengths per cycle versus the wave periods per body length (adapted from Maladen et al. 2011, with data from Maladen et al. 2009). (E) Discrete element method simulations of sandfish swimming with different aspect ratios for the amplitude of body oscillations relative to body length (courtesy of Daniel Goldman; images by Yang Ding).

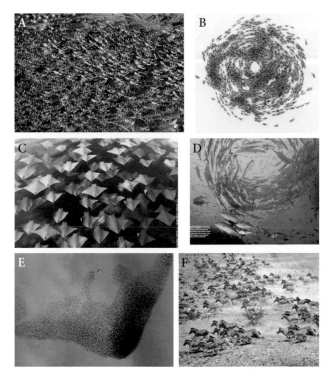

Plate 9: Gallery of collective behavior, including snapshots in time of (A) wingless locusts, (B) army ants, (C) golden rays, (D) fish, (E) starling flocks, and (F) zebra herds. Reproduced from Vicsek and Zafeiris (2012).

Plate 10: Flock of European starlings and emergent order. A snapshot of $N = 1246$ birds from a flocking event, which has a linear effective size of 36.5 m. Reproduced from Bialek et al. (2012).

Plate 11: Banded patterns of mussels (*Mytilus californianus*), starfish (*Pisaster ochraceus*), and seaweed/kelp in the rocky intertidal. Photo by Dave Cowles, Goodman Creek, WA, July 2002, used with permission.

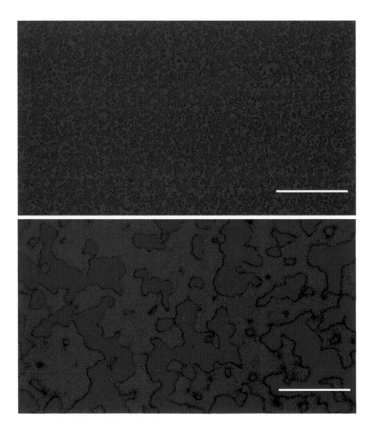

Plate 12: Dynamics of *V. cholerae* in T6SS– mutants (top) and in interactions with a functional T6SS system (bottom). In the top panel cells cannot kill each other locally, while in the bottom panel cells can kill each other locally via the T6SS mechanism. The phase separation in the T6SS case in the bottom panel is a hallmark of coarsening and a direct result of the coordination nature of this bacterial game. Scale bars denote 100 μm. Reproduced from McNally et al. (2017).

Plate 13: Spatiotemporal dynamics of a T6SS game in space, from random initial conditions to labyrinthine patterns. This game uses G = 10 and C = 18. The times shown here are after 0.2 time units (the equivalent of 20 iterations, each with a dt = 0.01). The time progresses across each row from left to right and then down from top to bottom. More details on how to develop a stochastic simulation of T6SS-like games are described in the computational laboratory associated with this chapter.

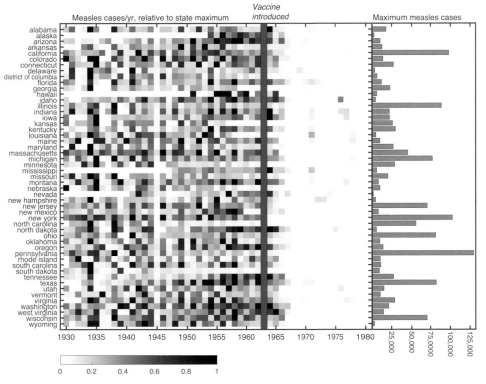

Plate 14: Measles incidence before and after the introduction of vaccines in 1963. (Left) Measles incidence per year relative to yearly state maximum (including the District of Columbia). (Right) The maximum number of annual measles cases per state in the time period shown in the left panel. Adapted from a 2015 Project Tycho data visualization. Project Tycho is a global health data initiative based at the University of Pittsburgh: https://www.tycho.pitt.edu.

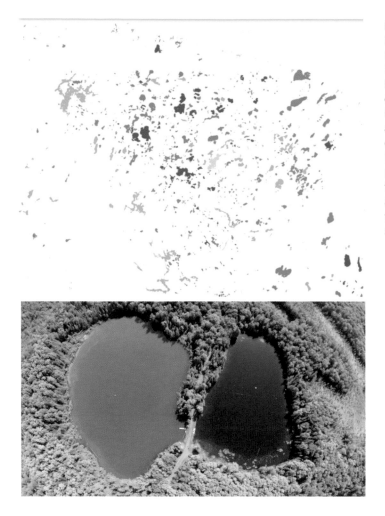

Plate 15: Alternative stable states in shallow lakes in Wisconsin (Top) and Michigan (bottom). (Top) Lake water clarity derived from Landsat images of Iron County, Wisconsin from July 15, 2004. Source: https://lakesat.org/LakesTSI.php at the University of Wisconsin-Madison (credit: Sam Batzli). Darker blue corresponds to more transparent conditions and lighter green corresponds to more turbid conditions. (Bottom) Photograph of Peter (Left) and Paul (Right) Lakes. Peter Lake was subject to an experimental manipulation resulting in a eutrophic, turbid state, while Paul Lake was left undisturbed and served as the control of an oligotrophic, clear water state. Photo credit: Steve Carpenter/University of Wisconsin–Madison Center for Limnology.

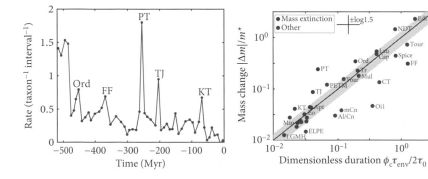

Plate 16: Carbon cycle disruptions and mass extinction. (Left) The loss of taxa over time (in millions of years) where five mass extinctions are labeled, including the end-Cretacious (KT), end-Triassic (TJ), end-Permian (PT), end-Ordovician (Ord), and Frasnian-Fammenian (FF). Reproduced from Arnscheidt and Rothman (2022) using data from Alroy (2014). (Right) Influx of carbon into the ocean as a function of a dimensionless duration for 31 different disruptions of the global carbon cycle over the past 540 million years. Reproduced from Rothman (2017).

low concentrations). Formally, this is termed Fick's law, such that the net flux of uncharged particles is proportional to the concentration gradient:

$$J = -D\frac{\partial n}{\partial x} \tag{6.5}$$

where D is the diffusivity (e.g., cm^2/sec), $n(x, t)$ is the concentration (e.g., cm^{-3}), and J is the particle flux (e.g., cm^{-2}sec^{-1}). The technical appendix show how to derive this law from first principles.

Fick's law can be applied to the movement of each ionic type across the cell membrane. Denoting the radial distance from the center of the cell as r and assuming a symmetric membrane, the diffusion should be

$$J = -D\nabla_r n(r, t) \tag{6.6}$$

where D is the diffusivity of the molecules and ∇_r denotes the derivative of the concentration with respect to the radial direction. When there is a higher concentration outside the cell than inside, the net flux is negative (molecules move into the cell). In contrast, when there is a higher concentration inside the cell than outside, the net flux is positive (molecules move out of the cell). Here D has units of area per time, and $n(r, t)$ has units of molecules per volume such that J has units of molecules per unit area per unit time. Now, if the molecules were uncharged and were not differentially produced inside or outside, we would expect that over time any imbalance in concentration would disappear. That is to say, diffusion of molecules leads to balanced concentrations. However, ions are charged molecules, and that makes a difference.

Consider a case where there is an excess of potassium ions inside the cellular cytoplasm compared to the concentrations in the extracellular medium and where the system is otherwise charge balanced. Both the inside and outside of the cell have zero net charge, and channels are closed such that ions cannot mobilize across the membrane (Figure 6.6A). What would happen if channels for the positively charged potassium ions were to open while channels for the negative ions remained closed? Fick's law states that potassium ions should diffuse out against the concentration gradient. This diffusion is in the opposite direction of the concentration gradient, which increases inward. Acting alone, diffusive flux would eventually lead to the equilibration of concentrations inside and outside of cells. But the diffusing molecules are charged ions. Hence, the diffusive flux leads to the net accumulation of positive charges outside the cell and a net negative charge inside the cell (Figure 6.6B). As potassium ions diffuse out, there will be a slight excess of net positive charge just outside the cell membrane and, by extension, a slight excess of negative charge just inside the cell membrane (Figure 6.6C). This charge difference generates an electric field. The electric field points in the direction of the force on a positive test charge. In this case, the force points inward, driving ions back inside the cell (Figure 6.6D). These two fluxes can lead to an equilibrium where the concentrations of ions inside and outside the cell differ.

To explore this balance in detail, denote the electric potential across the cell membrane as ϕ such that the electric field is $E = -\nabla\phi$. The flux of charged ions is proportional to the strength of the electric field (Zangwill 2013):

$$J^{(e)} = -qn\hat{\mu}\nabla\phi \tag{6.7}$$

where the charge is $q = 1.60 \times 10^{-19}$ coulombs, mobility $\hat{\mu} = D/kT$ where $k = 1.38 \times 10^{-23}$ m^2 kg sec^{-2} K^{-1} is the Boltzmann constant, and temperature T is in kelvin (K). Rewriting the mobility allows us to rewrite the voltage-driven flux as

$$J^{(e)} = -zqn\frac{D}{kT}\nabla\phi \tag{6.8}$$

where z is the charge valence (e.g., -1 for chloride, $+1$ for potassium and sodium, and $+2$ for calcium), and $kT/q \approx 27$ mV at a body temperature $T = 310$ K. In summary, imbalances in

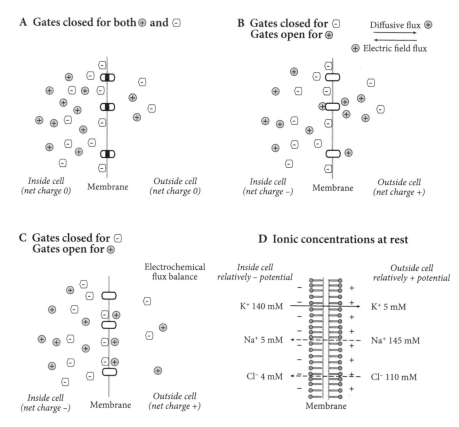

Figure 6.6: Depiction of the link between charges, diffusive flux, and electric field–induced flux. (A) Initially, there is an equal concentration of positive and negative ions both inside the cell and outside the cell. Channels permit the flow of positive ions but are initially closed. (B) Channels are open, enabling flux of positively charged ions across the cell membrane. Because of the concentration imbalance, there is a diffusive flux outward. However, the diffusive flux causes a charge imbalance, leading to an electric field–induced flux inward. (C) This balance of fluxes can lead to a sustained potential difference across the membrane, where there is an accumulation of negative charge inside the membrane and an accumulation of positive charge outside the membrane. (D) At equilibrium, the concentration of ions differs inside versus outside cells. Typical ranges illustrated here for each ion are as follows: potassium differs from 140 mM to 5 mM, sodium differs from 5 mM to 145 mM, and chloride differs from 4 mM to 110 mM, comparing inside to outside, respectively. The interior of the cell is at a relatively lower potential than the outside (noted by the − and + signs inside and outside the membrane), such that the chemical gradient drives potassium outward and the electrical gradient drives potassium inward. In its rest state, the cell membrane is relatively more permeable to K$^+$ ions (solid line) than to Na$^+$ and Cl$^-$ (dashed lines), respectively. Extended details in (Luo 2015).

ionic density lead to two fluxes working in opposite directions: (i) diffusive fluxes against the concentration gradient; (ii) electrical field–induced fluxes that can sustain a concentration gradient. So, rather than just counting the diffusivity of neutral particles, we must account for both contributions to the flux of charged ions across the membrane:

$$J_{tot} = - \overbrace{zqn\frac{D}{kT}\nabla\phi}^{\text{electrostatic flow}} - \overbrace{D\nabla n}^{\text{diffusive flow}} , \tag{6.9}$$

using the convention that the positive direction points outward.

Hence, if there are more positive ions initially inside the cell than outside the cell, then there should be diffusive flux outward (meaning the second term in Eq. (6.9) is positive). The diffusive flux of positively charged ions induces an electrical field that points inward (meaning the first term is negative). At equilibrium, the total flux of charged ions should be zero:

$$\frac{zq}{kT}n\frac{d\phi}{dr} = \frac{dn}{dr} \tag{6.10}$$

where the derivative is in the radial direction perpendicular to the membrane. This is known as the Nernst-Planck equation, which can be rewritten as

$$d\phi = \frac{kT}{zq}\frac{dn}{n} \tag{6.11}$$

such that integrating across the membrane from outside to inside in the case of a positively charged ion yields

$$\phi_{out} - \phi_{in} = \frac{kT}{zq}\log\frac{n_{out}}{n_{in}} \tag{6.12}$$

where we identify $E = \phi_{out} - \phi_{in}$ as the Nernst potential specific to each ion. This relationship between concentration differences and a potential difference can be rewritten as

$$\frac{n_{out}}{n_{in}} = e^{\frac{-\Delta U}{kT}} \tag{6.13}$$

where ΔU is the energy difference across the membrane equal to $zq\Delta\phi$. In other words, the Nernst potential arises naturally as the potential difference in a Boltzmann distribution, such that the occupancy of ions in two different states (inside or outside of cells) is exponentially distributed with respect to the energy differences in those two states (Benedek and Villars 2000a,c).

The ion-specific Nernst potentials can be solved for typical ionic concentrations inside and outside neurons. Given typical ranges of intracellular and extracellular ionic concentrations (Figure 6.6D) yields the following typically potential values:

$$E_{K^+} = 27\log\frac{5}{140} = -89\text{mV} \tag{6.14}$$

$$E_{Na^+} = 27\log\frac{145}{5} = 90\text{mV} \tag{6.15}$$

$$E_{Na^+} = 27 \log \frac{145}{15} = 61\text{mV} \tag{6.16}$$

$$E_{Cl^-} = -27 \log \frac{110}{4} = -90\text{mV} \tag{6.17}$$

where the potential values differ by cell type and condition (Borisyuk et al. 2005). These Nernst potentials correspond to the resting state of a system with only one kind of diffusing, charged molecule. The Nernst potential can also be thought of as a reverse potential given that the current I changes sign as V crosses the value E—given a single ion. But a neuron has more than one kind of ion. In order to understand how a system of ions interacts with signals (in the form of currents incoming from axons), one has to begin to connect the links in the entire neuronal "circuit."

6.3.3   Feedback between voltage, capacitance, and conductance

The total current flowing across the membrane is equal to the sum of ionic currents (associated with each of the different charged ions), the applied current, and the "capacitive current." The capacitance, C, quantifies the ability of the membrane to store charge, and the associated current, $C\dot{V}$, quantifies the current arising from changes in the voltage across the membrane, or what is termed the *transmembrane potential*. According to Kirchhoff's first law, the currents flowing into a junction must equal those flowing out of a junction. Consider the input of an applied current into a cell I and consider the junction as corresponding to the cell membrane. As a result, the input (or applied) current must equal the currents flowing across the membrane:

$$I = C\dot{V} + I_{Na} + I_{Ca} + I_K + I_{Cl}. \tag{6.18}$$

This law can be rewritten switching the location of the terms, such that the dynamics of V are on the left-hand side:

$$C\dot{V} = I - I_{Na} - I_{Ca} - I_K - I_{Cl}. \tag{6.19}$$

This appears to be the start of a *linear* differential equation insofar as the voltage changes linearly with respect to variation in the currents. But the ionic currents are not fixed—they depend on the potential difference across the membrane. Ohm's law establishes the link between current and potential difference: $V/I = R$. This law says that the ratio of voltage to current is the resistance, R (a feature of the material). This law can be rewritten as $I = V/R$, such that for any given voltage difference the current will increase (decrease) with decreasing (increasing) resistance (Benedek and Villars 2000a). The inverse of resistance is termed the *conductance*, $g = 1/R$. Hence, in this example, the current associated with the movement of potassium ions is

$$I_K = g_K(V - E_K) \tag{6.20}$$

where E_K is the Nernst potential identified above. In essence, if potassium were the only ion and there were no applied current, then the dynamics would have an equilibrium when $\dot{V} = 0$ such that the transmembrane potential, V, would converge to the Nernst potential, E_K. But there are other ions. If voltage is held fixed, then the flow of ions depends on the relative value of the current potential vis á vis the Nernst potential. For multiple ions, the dynamics of the current can be rewritten as

$$C\dot{V} = I - g_K(V - E_K) - g_{Na}(V - E_{Na}) - g_{Cl}(V - E_{Cl}). \tag{6.21}$$

At equilibrium, $\dot{V} = 0$. This is satisfied when the resting potential of the neuron is

$$V^* = \frac{g_K E_K + g_{Na} E_{Na} + g_{Cl} E_{Cl}}{g_K + g_{Na} + g_{Cl}}. \tag{6.22}$$

This can be interpreted as a weighted average of the Nernst potentials, weighted by the conductance associated with each ionic channel. Notably, the conductances, g, are not fixed values. Instead, g_K, g_{Na}, and g_{Cl} are voltage dependent. Voltage dependent gating is essential to understanding why neurons have nonlinear feedback properties, including the dynamical feature of excitability.

How can ionic conductance depend on voltage? The answer lies in the relationship between the configuration of channels and the mobility of ions. Ion-specific channels are not passive—they can be thought of as having different configurations that, in the parlance of HH, correspond to opening and closing gates. Ions can move across the channel when the opening gates are open and the closing gates are not closed (Hodgkin and Huxley 1952). The configuration of these gates can change with time. Phenomenologically, the gate variables represent the effective number of molecular gates that enable effective flux—higher coordination numbers imply higher nonlinearities in response. The open or closed configuration status of these gates changes with time. For example, if x is the fraction of open gates of a certain type, then the dynamics of $x(t)$ represent the result of a two-state transition process, between off to on and between on to off. "Off" and "on" correspond to whether or not the gate is in the wrong or correct position for ion diffusion. Given transition rates of k_{on} (from off to on) and k_{off} (from on to off), then the dynamics of the fraction of open gates are

$$\dot{x} = k_{on}(1 - x) - k_{off} x \tag{6.23}$$

$$= k_{on} - (k_{on} + k_{off}) x \tag{6.24}$$

$$= \left(\frac{k_{on}}{k_{on} + k_{off}} - x\right) / \tau \tag{6.25}$$

$$= (x^* - x) / \tau \tag{6.26}$$

where the time constant for relaxation is $\tau \equiv 1/(k_{on} + k_{off})$ and the equilibrium is $x^* = k_{on}/(k_{on} + k_{off})$. This relaxation implies that the frequency of On gates would exponentially relax from an initial configuration back to the equilibrium:

$$x(t) = x_0 e^{-t/\tau} + x^* \left(1 - e^{-t/\tau}\right). \tag{6.27}$$

The technical appendix presents a brief derivation of the relaxation dynamics in this class of kinetic problem. In this example, the kinetic constants are fixed. For neurons, the kinetic constants are voltage dependent, i.e., $k_{on} = k_{on}(V)$ and $k_{off} = k_{off}(V)$. This nonlinear dependency of kinetic rates on voltage has significant consequences.

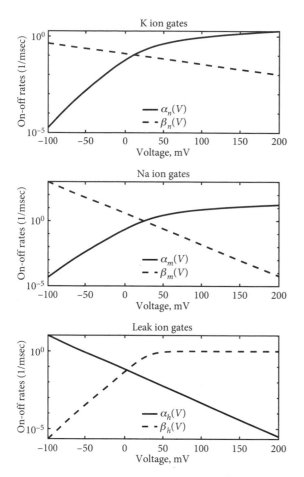

Figure 6.7: Transmembrane ion channels can be in an open and closed configuration such that switches between states (e.g., between open and closed states of the channel) are voltage dependent. The switch rates for K, Na, and leak ions are α_n, α_m, and α_h, respectively, for On rates and β_n, α_m, and β_h, respectively, for Off rates.

Returning to the original work of HH, if a voltage is held fixed (i.e., clamped), it is possible to isolate the interaction between a fixed voltage and the change in the frequency of open gates. The frequency of open gates changes the flow of currents. Ionic currents induce a change in voltage that shapes the functional dependence of the $\alpha(V)$ and $\beta(V)$ terms. This dependence can potentially intensify system dynamics, including leading to excitability. The voltage dependence is, in essence, an empirical problem, linking evolved cellular physiology to system dynamics. The voltage-dependent On rates and Off rates, denoted using activation variables $\alpha(V)$ and inactivation variables $\beta(V)$, respectively, are

$$\alpha_n(V) = 0.01 \frac{10-V}{e^{\frac{10-V}{10}}-1} \tag{6.28}$$

$$\beta_n(V) = 0.125 e^{-V/80} \tag{6.29}$$

$$\alpha_m(V) = 0.1 \frac{25-V}{e^{\frac{25-V}{10}}-1} \tag{6.30}$$

$$\beta_m(V) = 4 e^{-V/18} \tag{6.31}$$

$$\alpha_h(V) = 0.07 e^{-V/20} \tag{6.32}$$

$$\beta_h(V) = \frac{1}{e^{\frac{30-V}{10}}+1} \tag{6.33}$$

The quantitative dependencies of these rates are shown in Figure 6.7. As is apparent, at negative voltages, Off rates for both K and Na are high, whereas the Off rate for the leak current is low. In contrast, at high voltages On rates for both K and Na are high, whereas the On rate for the leak current is low. These specific functional forms were inferred by Hodgkin and Huxley via numerical fitting to measurements on the giant squid axon.

Finally, we can return to the original HH equations, but hopefully with a newfound sense of understanding. The equations below include overbraces to denote the meaning of each component, albeit here shifting the voltage dynamics to the left-hand side to clearly delineate the nonlinear nature of the dynamics:

$$\overbrace{C_M \frac{dV}{dt}}^{\text{capacitive current}} = \overbrace{I}^{\text{applied current}} \overbrace{-\bar{g}_K n^4 (V - V_K)}^{\text{potassium current}} \dots$$

$$\overbrace{-\bar{g}_{Na} m^3 h (V - V_{Na})}^{\text{sodium current}} \overbrace{-\bar{g}_l (V - V_l)}^{\text{leak current}} \tag{6.34}$$

$$\overbrace{\frac{dn}{dt}}^{\text{potassium channels}} = \overbrace{\alpha_n(V)(1-n)}^{\text{on}} - \overbrace{\beta_n(V)n}^{\text{off}} \tag{6.35}$$

$$\overbrace{\frac{dm}{dt}}^{\text{sodium channels}} = \overbrace{\alpha_m(V)(1-m)}^{\text{on}} - \overbrace{\beta_m(V)m}^{\text{off}} \tag{6.36}$$

$$\overbrace{\frac{dh}{dt}}^{\text{leak channels}} = \overbrace{\alpha_h(V)(1-h)}^{\text{on}} - \overbrace{\beta_h(V)h}^{\text{off}} \tag{6.37}$$

In essence, the HH model arises via the following biophysical feedback: (i) Ohm's law, i.e., that a voltage is equal to the current multiplied by the resistance (or the inverse of conductance); (ii) Kirchhoff's law, i.e., that input and output currents must balance at a junction; (iii) that the opening and closing of ionic channels are voltage dependent; (iv) that ionic channels modify current flow in a nonlinear fashion. This is a beautiful theory that, sensu Peebles, is beautiful because it accords with measurements. The next section begins to explore the properties of neurons arising from this combination of physiological mechanisms.

6.4 DYNAMICAL PROPERTIES OF EXCITABLE NEURONAL SYSTEMS

6.4.1 Overview

The HH model exhibits excitability and other features as a direct result of the feedback between gate status and voltage dynamics. Some of these features include (i) filtering of small perturbations; (ii) excitability given sufficiently large perturbations; (iii) refractory periods; (iv) beating. In addition to discussing them here, each of these features is also explored at length in the accompanying computational lab guide and in the problems at the end of this chapter. In all of the examples below, the constituent HH equations are simulated using standard parameters (Hodgkin and Huxley 1952) such that the resting voltage is 0. These parameters are $\bar{g}_K = 36$ mS/cm^2, $\bar{g}_{Na} = 120$ mS/cm^2, $\bar{g}_L = 0.3$ mS/cm^2, $E_K = -12$ mV, $E_{Na} = 120$ mV, $E_L = 10.6$ mV, and $C = 1$ muF/cm^2. The initial voltage and gating states varies in the following series of examples. In the event that the neuron is initially at rest, then $V(0) = 0$ and $n(0) = n^*$, $m(0) = m^*$, and $h(0) = h^*$. Note that at equilibrium the activation and inactivation processes must be in balance—and they themselves depend on the voltage V^*. For this to hold, then

$$n^* = \alpha_n(V^*)/(\alpha_n(V^*) + \beta_n(V^*))$$

$$m^* = \alpha_m(V^*)/(\alpha_m(V^*) + \beta_m(V^*))$$

$$h^* = \alpha_h(V^*)/(\alpha_h(V^*) + \beta_h(V^*))$$

At these particular values, the system is at rest. Gating values that diverge from these equilibria can lead to an increase in conductances, including the possibility of membrane *depolarization*, which is marked by the rapid increase of the membrane potential and the inflow of sodium ions into the cell such that the system appears to "fire" even if no stimulus or a weak stimulus is applied (Figure 6.8). Eventually, the membrane will repolarize, gates will close, and the system will return to rest. This baseline dynamic—and more—is shown below.

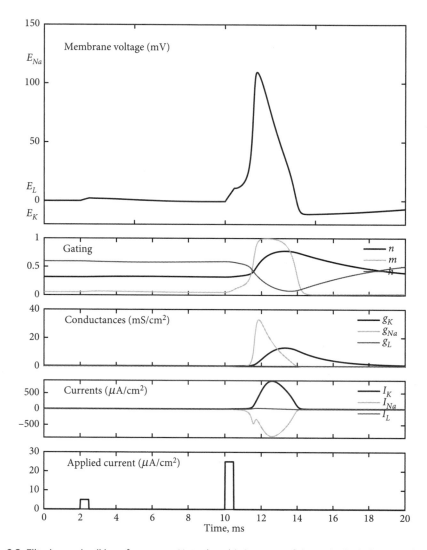

Figure 6.8: Filtering and spiking of a neuron. Note that this instance of dynamics includes a weak applied current (from 2 to 2.5 ms) followed by a stronger applied current (from 10 to 10.5 ms). The first current is filtered, whereas the second current leads to a spike of the transmembrane voltage associated with rapid changes in conductance. The units are mV for voltage, mS/cm² for conductances, and $\mu A/cm^2$ for currents. The format of this and subsequent figures depicting the reaction of neurons to perturbations follows the layout style utilized in Izhikevich (2007). In this example, the HH model is simulated using $\bar{g}_K = 36$ mS/cm², $\bar{g}_{Na} = 120$ mS/cm², $\bar{g}_L = 0.3$ mS/cm², $E_K = -12$ mV, $E_{Na} = 120$ mV, $E_L = 10.6$ mV, and $C = 1$ muF/cm², and using the equilibrium conditions for gating variables, n^*, m^*, and h^*, as described in the text.

6.4.2   Filtering and excitability—the system and its input

Figure 6.9 presents a time-dependent example of the response of a neuron to the application of a series of external currents. Initially at rest (V^*, n^*, m^*, h^*), the application of a small current perturbs the membrane (via depolarization—described above). The application of the current leads to a change in voltage and then a subsequent change in gate states. However, the application of a small current is insufficient to trigger a response and the membrane repolarizes—in essence, the rest state of the neuron is locally stable. But local stability only describes the dynamics of a system with respect to small perturbations. Indeed, the application of a larger current leads to larger changes in the voltage that depolarize the membrane, changing conductances. This feedback further depolarizes the membrane. Hence, there is inward Na+ current, followed by outward K+ current, and small leak current. The neuron is "excited" and fires. These two contrasting examples in Figure 6.9 reveal a holistic view of neuronal dynamics. The panels represent (top to bottom) voltage dynamics, gating variables, conductances, currents, and applied currents. As is apparent in the first example, there are two distinct perturbations—a small current at 2 ms and a larger current at 10 ms. Only in the latter case does the neuron spike. The smaller current does not sufficiently depolarize the membrane to initiate an excitatory response. In contrast, the larger current sufficiently depolarizes the membrane to initiate a response in which the neuron is "excited" and fires. This firing of the neuron includes a large voltage spike. This spike is the basis for signal propagation, information processing, and the interaction between neurons

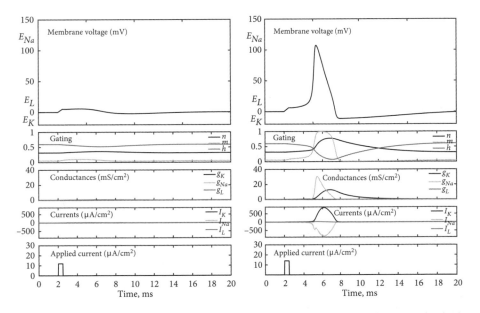

Figure 6.9: Excitability implies that small differences in perturbations near the critical point can lead to large changes in response. In both cases, a current is applied from 2 to 2.5 ms, albeit of 12 versus 14 $\mu A/\text{cm}^2$. This slight difference leads to a dramatic change in output. The first current is filtered, whereas the second current leads to a spike of the transmembrane voltage associated with rapid changes in conductance. Note that the difference can be made even smaller, resulting in a discontinuity (which the homework problems explore in detail). The units are mV for voltage, mS/cm^2 for conductances, and $\mu A/\text{cm}^2$ for currents.

as part of a connected system. (For more on connected neurons and excitable cells, see Chapter 7).

The excitability of the system encapsulates the differing responses of the neuron to small and large perturbations. As noted above, the system is stable, insofar as small perturbations lead to the relaxation of the system back to the fixed point without long excursions. However, when perturbed with a sufficiently large applied current, then the state variables $V(t), n(t), m(t)$, and $h(t)$ undergo a long trajectory away from then initial state before returning to the equilibrium or rest state. This difference between small and large perturbations is precisely the mechanism underlying the ability of neurons to filter sufficiently small input and react nonlinearly with an impulse-like output to sufficiently large output. The excitability also implies that there is a critical transition between large and small outputs rather than a smooth change in dynamics with increasing input size. Figure 6.9 shows this discontinuity in response to a 16% increase in applied current over a 0.5 ms period. Figure 6.9 (left) shows the filtering of an applied current of magnitude 12 μA/cm^2, while Figure 6.9 (right) shows the characteristic firing of an action potential given an applied current of magnitude 14 μA/cm^2. This small difference leads in the first case to filtering of a signal and in the second case to an excitatory response. The full dynamics take place in a four-dimensional state space. Viewing the dynamics reveals that the voltage changes quickly even as the gating variables relax more slowly. This difference between fast and slow variables leads to simplifications of the HH dynamics.

6.4.3   Refractory period

The concept of excitability is usually accompanied by a period in which further application of current does not lead to another large deviation in system dynamics. This period in which the system is relatively nonresponsive to perturbations is termed the refractory period. The *refractory period* arises because an excitation involves a long excursion in state space from a resting condition, given a sufficiently large initial perturbation. These long excursions include a period in which voltages spike, leading to systematic shifts in the relative conductivities in the ion channels. The ion channel states relax over longer time scales. Hence, the subsequent addition of an external current after an initial excitation does not necessarily alter the state of gating and therefore has little impact on conductance and currents. Instead, the system continues to relax in the four-dimensional state space of $V(t)$, $n(t)$, $m(t)$, and $h(t)$ until the membrane and gates have returned closer to their rest state. Figure 6.10 shows an example of such a refractory period by examining the neuronal dynamics as the perturbation interval increases from 10 ms to 30 ms. For short periods, the system is relatively insensitive to the second application of applied current. In contrast, the neuron spikes a second time when there is a large enough gap between the first and second perturbations. In this model, a critical, minimum timing between the application of applied currents is required for a second voltage spike (see homework problems).

6.4.4   Beating

The HH model can recapitulate neuronal filtering of relatively small applied currents, the firing of an action potential in response to large applied currents, and the emergence of a refractory period after a spike in which further applications of large currents do not lead

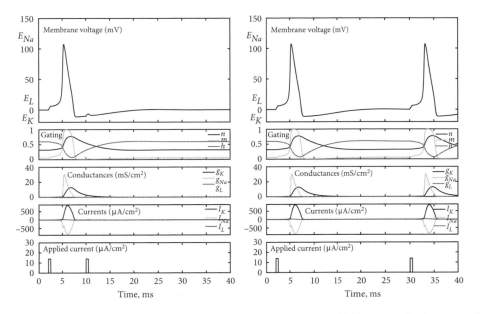

Figure 6.10: Refractory periods after an initial excitation. In both cases, an initial large-amplitude current of $14\ \mu A/\text{cm}^2$ is applied between 2 and 2.5 ms. The left and right panels depict a second pulse of equal amplitude applied between 10 and 10.5 ms (left) and 30 and 30.5 ms (right). Note that the second stimulus only leads to an action potential spike in the case of the larger interval (which the homework problems explore in detail). The units are mV for voltage, mS/cm^2 for conductances, and $\mu A/\text{cm}^2$ for currents.

to more spike-like responses. Together, these dynamical phenomena suggest how neurons might exhibit persistent oscillatory dynamics given a constant input, i.e., beating. Beating is characterized by oscillatory spiking behavior followed by the slow relaxation toward a rest state before the next spike. Figure 6.11 shows an example of beating dynamics using the same initial conditions and physiological parameters as in the prior examples with a constant applied current of $25\ \mu A/\text{cm}^2$. This constant input current leads to an oscillatory output. To understand why, consider the initial portion of the input current. The initial application of a constant current is sufficiently large to initiate an excitatory response—via the same mechanisms as shown in Figure 6.9 (right). But after the spike, the system enters into a refractory period. As a result, the continued application of current does not necessarily lead to another excitation—via the same mechanisms as shown in Figure 6.10. Eventually, the gating variables will relax sufficiently close to their rest states, at which point the constant applied current will lead, yet again, to a spike. The repetition of this sequence leads to the phenomenon of beating: the concatenation of a sequence of excitations and refractory periods given a constant applied current input.

6.5   FROM NEURONS TO NEURAL NETWORKS AND INFORMATION PROCESSING

Neurons have a remarkable range of dynamic properties. Coupled together, neurons can do even more. The excitable dynamics of individual nerve cells are coupled to each other in the central nervous system. A full treatment of neural networks and collective information processing can be found elsewhere (Rolls and Treves 1997; Izhikevich 2007). Yet it is worth taking a step in this direction by examining some of the potential outcomes in multiple groups of neurons—by connecting highly nonlinear input-output relationships together. In doing so, it is useful to make certain simplifications; i.e., rather than considering the full nonlinear dynamical profile of an excitatory neuron, we assume that such a profile represents a type of integrate-and-fire mechanism. The integration is that of accumulating the application of an external current, and the firing is that of an action potential. Such action potentials, if connected to other neurons, can lead to information processing in a generalized sense.

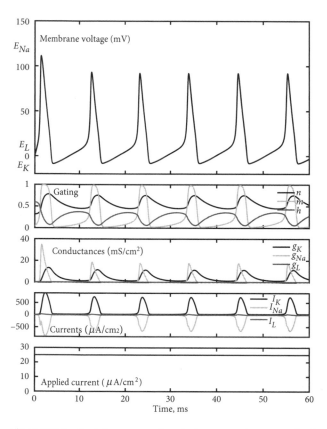

Figure 6.11: Beating phenomenon in a single neuron given a constant applied current of 25 μA/cm^2. The units are mV for voltage, mS/cm^2 for conductances, and μA/cm^2 for currents.

6.5.1 Integrate and Fire

Consistent with the full HH model of a neuron, the dynamics of the transmembrane potential, i.e., the voltage V, can be written in the generic form

$$C\frac{dV}{dt} = I - \sum_i g_i(V - E_i) \qquad (6.38)$$

where i denotes the ion type that is exchanged across the membrane and E_i are the associated ion-specific Nernst potentials. Although the full set of equations includes multiple ions, it is instructive to focus on one ion exclusively as a means to reduce the complexity of the HH equations into what is tantamount to an integrate-and-fire model.

First, note that the resistance is defined as the inverse of the conductance, $R_i = 1/g_i$. Hence, assuming there is one particular ion and dividing both sides by g_i yields

$$\tau\frac{dV}{dt} = E_i - V + R_iI \qquad (6.39)$$

where τ is a characteristic time scale C/g_i. Hence, when $V > E_i + R_iI$, then the voltage should increase, and when $V < E_i + R_iI$, then the voltage should decrease. In this light, the gating variables g_i mediate feedback control, changing the passage of ions as part of a rapid spike—including an increase and then decrease, with overshoot, of the voltage—followed by a slow relaxation of the gating variables. A simplified view of these dynamics would be to replace the fully nonlinear gating responses with a rule: $V(t_c) = V(t_c) + V_s$ where V_s is a short-duration spike (akin to the rapid increase in the action potential). The critical point would correspond to the moment when V exceeds a critical value V_c. Following the spike, the system is reset at $V(t_c + \delta t) = 0$. Figure 6.12 shows such a simplified integrate-and-fire model in action. In practice, there is a small-duration spike, followed by a reset in which we arbitrarily reset the voltage to the Nernst potential. In the limit of a strong applied current, the time between spikes should be approximately equal to the time it takes for the voltage to increase from E_i to V_c. The speed of voltage increase is $R_i I / \tau$ or $R_i I g_i / C$ such that the interspike interval will be $\frac{CV_c}{R_i I_c g_i}$, assuming that the spike period δt is much less than the time to integrate and fire. Figure 6.12 shows that in each cycle there is a buildup of voltage before the system reaches a critical point that is followed by a rapid spike and reset. This approximation of the full HH model via an integrate-and-fire model provides a simplified view of neuronal dynamics. Via an integrate-and-fire model, the action of a single neuron reduces to the question of whether or not the incoming current exceeds a threshold. If it does, the neuron fires; if not, the neuron remains quiescent. Associating a firing event with the value of a single bit (either 1 or 0), the integrate-and-fire model provides a means to map physiological signals to information, computation, and perhaps even cognition.

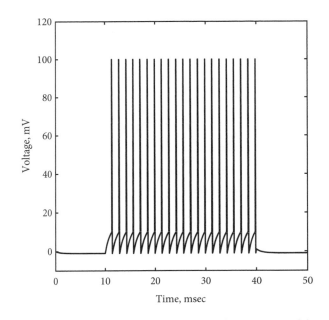

Figure 6.12: Dynamics in an integrate- and -fire neuron model. A constant current of 10 μA/cm^2 is applied from 10 to 40 ms, otherwise there is 0 applied current. Here we use an illustrative model with $g_i = 1$ mS/cm^2 and $E_i = -1$ mV, where the simulation includes a spike of 100 mV in a 0.5 ms pulse. As is apparent, the dynamics build up and then fire during the period of applied current and not otherwise.

6.5.2   The basis for neural logic to computation

The integrate-and-fire mechanism is a simplification of neuronal dynamics. Nonetheless, it provides a route to explore the basis for information processing. To begin, consider a circuit in which two neurons are connected to a third, as in Figure 6.13. In this case, the total input into the focal neuron is $z = \sum_i w_i y_i$. This input is then processed such that the output is $f(z)$. The computational features of neurons arise as a consequence of nonlinear responses, as shown in both the full HH model and in the integrate-and-fire simplification in the prior section. A few examples of input-output relationships are shown in the bottom panels of Figure 6.13, including a Boolean response, rectified linear unit (ReLU), and the sigmoid response. Each of these three functions is used in the neural network processing field as a simplified representation of potential input-output relationships. Here we use them as the basis to explore logical computations via the judicious choice of input weights.

Figure 6.14 depicts three potential logic gates where the two inputs (x_1 and x_2) are denoted in the row and column, respectively, and the four potential outputs are shown in the entries. The outputs represent the possible outcomes given a Boolean logic circuit and given the potential Boolean state of the inputs, either 0 or 1. It is not evident that combining the output of two neurons can yield simple logical calculations, without appropriate adjustment of weights. First, consider an AND gate. Recall that the input to $f(z)$ can take in an arbitrary constant, which we denote as the bias b. In the case of an AND

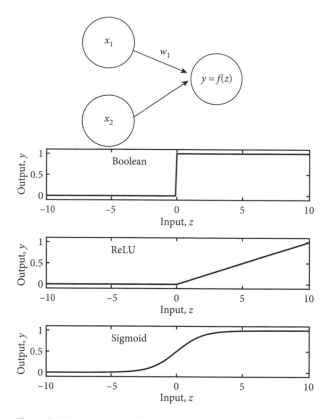

Figure 6.13: Neuron connections in a simple network, with two input and one output neuron. (Top) Schematic of connections. (Bottom) Input-output relationships for different computational logical functions: Boolean, ReLU, and sigmoid. Here $z = \sum_i w_i x_i$, where w_i are the weights and x_i are the states of the neurons. Actual connectivity would require that signals from x_i propagate along axons via synapses and dendrites into the cell body of neuron y. The functions are (i) Boolean: $f(z) = \Theta(z)$; (ii) ReLU $f(z) = z\Theta(z)$; (iii) sigmoid $f(z) = \frac{1}{1+e^{-z}}$, where $\Theta(z)$ is the Heaviside function, such that it is 1 if $z \geq 0$ and 0 otherwise.

gate, the input must exceed 0, but given that the output is either 0 or 1, this suggests the need to have a *negative* bias, i.e., $z = b + w_1 x_1 + w_2 x_1$. Consider $b = -1$, in which case the output $f(z)$ should only be turned on (i.e., $f(z) > 0$) when both inputs are on. This holds for any choice of $w_1 > 0.5$ and $w_2 > 0.5$. Likewise for an OR gate, the use of a negative bias, e.g., $b = -1$, and a sufficiently large weight implies that either input signal can be on to generate an output fire, i.e., $w_1 > 1$ and $w_2 > 1$ (obviously, to avoid marginal cases, a value outside of the margin of system variability would be ideal). A third example is perhaps even more telling: that of the NAND gate. The NAND gate is a "universal gate," as it can be used in combination to generate AND, OR, and even NOR gates. As a result, the ability of two inputs to be combined to generate a NAND logical output is a gateway to information processing and arbitrarily complex computation. For the NAND gate, then $z = b + w_1 x_1 + w_2 x_2$ should turn on insofar as the bias is positive and the weights are negative, e.g., $b = 1$ and $w_1 = 0.75$ and $w_2 = 0.75$. In that event, only when $x_1 = 1 = x_2$ is $z < 0$. Note that evaluating the logical output can be thought of as a line defined by $z = 0$, which given the linear combination of inputs defines a line in the $x_1 - x_2$ space. Hence, for any combination of $(0, 1)$, $(1, 0)$ and $(0, 0)$, the output will be on, and only in the case of $(1, 1)$ will the output be off. The lower panels of Figure 6.14 depict these relationships. The XOR gate brings on new challenges and is the topic for one of this chapter's homework problems.

Combining two inputs is not necessarily sufficient for more complex logic operations despite the fact that the combination of two inputs is sufficient to generate the AND, OR, and NAND logic gates. As an example, consider the addition of two signals defined by the following table:

A	B	Output	Carry	Sum
0	0	0	0	0
0	1	1	0	1
1	0	1	0	1
1	1	2	1	0

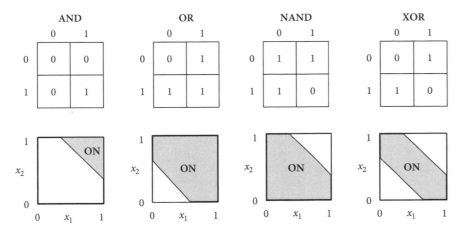

Figure 6.14: Logic gates—AND, OR, NAND, and XOR. The top panels denote the Boolean logic equivalents. The bottom panels denote the expected output given the condition $z > 0$ if $z = b + w_1 x_1 + w_2 x_2$. The solutions for potential biases b and weights w_1 and w_2 are described in the text and in the homework problems.

In this case, the sum of two inputs can be expressed in terms of a carry and sum bit. Note that the logic of the carry bit is an AND gate; however, the logic of the sum bit is an XOR gate. Notice that in Figure 6.14 the last column includes XOR logic. However, the associated diagram in Figure 6.14 suggests that it is impossible to combine two Boolean inputs in a single output to generate an XOR output. The reason is that the On cases lie between two lines in the plane. Instead, the answer requires specifying two lines and suggests that a two-layer network is required to recapitulate XOR logic. As the output of individual neurons is combined, more and more inputs can influence an output. This first step toward complex computation moves us beyond nerve cells and single input-output relationships to networks. The homework problems provide an opportunity to try to figure out the structure of this two-layer network on your own.

6.6 TAKE-HOME MESSAGES

- Information processing underlies organismal behavior, learning, and memory.
- Responses of organisms to stimuli are mediated by neurons in the central nervous system (i.e., the brain) as well as sensory neurons, motor neurons, and muscle neurons that are part of the peripheral nervous system.
- The template for nonlinear processing of information is the neuron.
- The neuron (or nerve cell) is an excitable system that responds to electrical stimuli through a nonlinear mechanism of electrochemical flux.
- Hodgkin and Huxley's foundational work in the late 1940s and early 1950s led to a model of the joint dynamics of voltage changes and transmembrane currents that matched experimental data.
- The model of a neuron has the feature of excitability: a neuron's dynamical system state can filter out small perturbations (e.g., in the form of an applied current) while reacting with large departures from the rest state (e.g., via an action potential or "spike").

- Excitable dynamical systems have a number of key features, including filtering and refractory dynamics, and can be driven into beating modes through application of constant currents.
- The study of excitable systems also provides a template for understanding information processing in connected neuronal systems.
- Combining output from neurons is a gateway to information processing and universal computation.

6.7 HOMEWORK PROBLEMS

These problems leverage lessons learned in the computational laboratory guide that accompanies this chapter to deepen your understanding of neuronal dynamics. The guide covers the following techniques:

- Developing a computationally robust HH simulation
- Including time-dependent impulses in a coupled system of differential equations
- Improving visualization methods, including overlaying multiple outputs on the same and/or stacked axes
- Improving system plots aesthetically

As a whole, the computational lab also includes steps toward solving each of the homework problems. In working through the accompanying code available on the book's website, it is strongly recommended to enter in these extended codes or, at minimum, to read through them carefully line by line. Given the complexity of the HH model, it is important to understand both the dynamics and the visualization components of this analysis before beginning.

PROBLEM 1. Excitability and Refractory Dynamics in the Hodgkin-Huxley Model

The Hodgkin-Huxley model can be written as

$$C\dot{V} = I - \bar{g}_K n^4 (V - E_K) - \bar{g}_{Na} m^3 h (V - E_{Na}) - g_{Cl}(V - E_{Cl})$$
$$\dot{n} = \alpha_n(V)(1-n) - \beta_n(V)n$$
$$\dot{m} = \alpha_m(V)(1-m) - \beta_m(V)m$$
$$\dot{h} = \alpha_h(V)(1-h) - \beta_h(V)h$$

The governing parameters are $\bar{g}_K = 36$ mS/cm^2, $\bar{g}_{Na} = 120$ mS/cm^2, $\bar{g}_L = 0.3$ mS/cm^2, $E_K = -12$ mV, $E_{Na} = 120$ mV, $E_L = 10.6$ mV, and $C = 1$ muF/cm^2, which correspond to a resting state with $V^* = 0$ mV. This model and scripts with associated parameters are available on the book's website. As described in the technical appendix, a naive integration of this model can lead to a singularity issue. Hence, the files on the website explicitly resolve the apparent singularity $\alpha_n(V)$ and $\alpha_m(V)$ through an interpolation scheme valid for all currents between -100 mV and 200 mV. It is recommended that you develop a scheme to resolve this issue in advance if you decide to write your own version.

    a. Given the parameters above, identify the minimum input current, I, over a duration of 0.5 msec necessary to stimulate an excitatory response in a neuron initially at rest. Explain and interpret your findings.

    b. Using an input pulse $I = 25$ over 0.5 msec, first stimulate an excitation. Then try to stimulate a second excitation with the same pulse structure. In doing so, identify the minimum time delay, i.e., the "refractory period" necessary to excite a second action potential pulse. Explain and interpret your findings.

PROBLEM 2. Physiological Differences and Excitability

The response of the gating variables influences the excitatory features of the model. Modify the gating dynamics by slowing down the rate of the opening and closing of n, m, and h gating variables by a factor of ϵ. Specify how you did so. Next, by setting $\epsilon = 0.1$ (10-fold reduction in response speed), calculate the duration of an action potential pulse given the input current of $I_a = 25$ over 0.5 msec. Try to find a critical value of ϵ such that pulses disappear altogether given that input current. Interpret your findings on how excitability might be related to the speed of feedback to the opening and closing of ionic channels.

PROBLEM 3. Controlling the Beating of Neurons

Using the parameters and system from Problem 1, modulate the constant input current I from 2 to 50 $\mu A/cm^2$. Evaluate how the neuron responds to this systematic change in input. Specifically, evaluate if there is a critical point that leads to beating. After identifying a beating regime, estimate the frequency of beats and explain if (and how) the frequency changes with the strength of the driving current.

PROBLEM 4. Interacting Neurons

Consider two neurons, A and B, that interact with each other. Each are excitable dynamical systems, but their voltages can transfer from one to the next (e.g., via transmission of an action potential/nerve impulse along an axon), as in the following schematic:

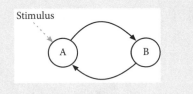

As a simplification, consider the following dynamical system representing coupled neurons:

$$C\dot{V}_i = I - \bar{g}_K n_i^4 (V_i - E_K) - \bar{g}_{Na} m_i^3 h_i (V_i - E_{Na}) - g_{Cl}(V_i - E_{Cl}) - \epsilon(V_i - V_j)$$

$$\dot{n}_i = \alpha_n(V_i)(1 - n_i) - \beta_n(V_i)n_i$$

$$\dot{m}_i = \alpha_m(V_i)(1 - m_i) - \beta_m(V_i)m_i$$

$$\dot{h}_i = \alpha_h(V_i)(1 - h_i) - \beta_h(V_i)h_i$$

where ϵ denotes a coupling strength and the last term should be interpreted as follows: if $i = 1$, then $V_1 - V_2$, or if $i = 2$, then $V_2 - V_1$. The last term in the voltage dynamics denotes the transmission from node A to B and from node B to A. In this problem, use the default parameters for Hodgkin-Huxley and assume that the applied current only affects the state of neuron A, as you answer the following:

a. Extend your code to include a pair of coupled neurons. Describe how you modeled the coupled systems and include a code snippet of your updated model.
b. Using the value $\epsilon = 0.1$, does neuron B fire given an impulse schedule of 2 between $t = 2$ and $t = 2.5$ and a second impulse of 25 between $t = 10$ and $t = 10.5$? If so, does neuron A fire a second time, or only once? Visualize your findings and interpret them.
c. Find a critical value of ϵ that enables activation of neuron A to also turn on neuron B. Visualize/justify your answer.
d. Now, modify the input to remain constant and identify a critical level that can trigger beating in both A and B. Is every A action potential followed by a B? Why or why not?

PROBLEM 5. Neural Addition

Develop a neural network that adds together two neuronal states, in terms of a carry and sum bit. In each case, use the logical output $z = b + w_1 x_1 + w_2 x_2$. As explained in the chapter, it is likely that you will need to include a two-layer neural network that generates XOR logic to produce the sum bit. Verify your answer for each of the four possible input values.

6.8   TECHNICAL APPENDIX

Fick's law   Consider a molecule whose local concentration is denoted as $n(x, t)$ given position x at time t, where the units of concentration are per volume (e.g., cm^{-3}, equivalent to ml^{-1}). It is possible to subdivide the domain so as to focus on a volume $(\Delta x)^3$ so that the concentration with x is $n(x - \Delta x, t)$, $n(x)$, and $n(x + \Delta x, t)$:

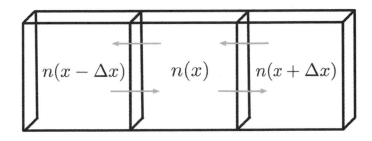

In a suitably small interval Δt, assume there is a probability v that a molecule in the volume centered at position x will move out of its location and into the small volume to its right at position $x + \Delta x$. As a result, there will be a flux of particles to the right of $J_+(x) = \frac{v}{\Delta t} n(x) \frac{(\Delta x)^3}{(\Delta x)^2}$, which can be reduced to $J_+(x) = \frac{v}{\Delta t} \Delta x n(x)$ with units of $\text{cm}^{-2}.\text{sec}^{-1}$. The same logic applies to the adjacent volume, such that the leftward flux of particles from the volume at $x + \Delta x$ is $J_-(x) = \frac{v}{\Delta t} \Delta x n(x + \Delta x)$. The net flux in the positive x direction at the right interface of the central volume is therefore the difference between these two fluxes, or

$$
\begin{aligned}
J &= \frac{v}{\Delta t} \Delta x n(x) - \frac{v}{\Delta t} \Delta x n(x + \Delta x) \\
&= \frac{v}{\Delta t} \Delta x n(x) - \frac{v}{\Delta t} \Delta x n(x) - \frac{v(\Delta x)^2}{\Delta t} \frac{\partial n(x)}{\partial x} \\
&= -\frac{v(\Delta x)^2}{\Delta t} \frac{\partial n}{\partial x}
\end{aligned}
\tag{6.40}
$$

or in the limit of infinitesimal times and volumes

$$
J = -D \frac{\partial n}{\partial x}. \tag{6.41}
$$

Here D is the diffusivity with units of cm^2/sec, and is defined as the appropriate limit of $v(\Delta x)^2/\Delta t$ when both Δx and Δt approach zero. The random movement of particles leads to a net flux from high to low concentrations, thereby homogenizing a system.

Relaxation given on-off kinetics Consider a system with two states, on and off, and transition rates of k_{on} (from off to on) and k_{off} (from on to off), where x denotes the frequency of the On state. This defines a two-state Markov chain problem, which can be described using a single dimension: $x(t)$. The dynamics of $x(t)$ obey

$$
\dot{x} = k_{on}(1 - x) - k_{off} x, \tag{6.42}
$$

which can be reduced (as described in the chapter) to

$$
\dot{x} = (x^* - x)/\tau \tag{6.43}
$$

where the time constant is $\tau \equiv 1/(k_{on} + k_{off})$ and the equilibrium is $x^* = k_{on}/(k_{on} + k_{off})$. This system can be solved exactly given some initial value $x(t=0) = x_0$. Note that one can transform the system to one in which $u(t) = x(t) - x^*$ denotes the deviation from the expected equilibrium. Because $\dot{u} = \dot{x}$, then

$$\dot{u} = -u/\tau \tag{6.44}$$

or

$$u(t) = u_0 e^{-t/\tau}. \tag{6.45}$$

Restoring the transformed variable leads to an equation for the relaxation for the frequency of On gates:

$$x(t) = x_0 e^{-t/\tau} + x^* \left(1 - e^{-t/\tau}\right). \tag{6.46}$$

Chapter Seven

Excitations and Signaling, from Cells to Tissue

7.1   FROM EXCITABLE CELLS TO EXCITABLE SYSTEMS

Neurons are excitable cells whose behavior is induced by electrical stimuli. As was shown in chapter 6, the excitability of neurons implies that cells can filter out small perturbations while responding dramatically to large perturbations. The large perturbations include a voltage spike over a short, millisecond-scale period. This spike is followed by a longer, refractory period, typically on the order of 10 ms. During this refractory period, the neuron filters out additional stimuli—even ones that would normally excite the neuron if at rest. This type of behavior is enabled by feedback in neurons that induces large deviations from a rest (or steady) state: the hallmark of excitability. In the case of neurons, the excitability is made possible by voltage-dependent gates that differentially enable the passage of ions, including Na (+), K (+), and Cl(−), through specific ion transporters. Yet, despite demonstrating that a highly resolved biophysical model of neurons can, in fact, lead to excitability, the reality is that the very same biophysical detail represents a barrier to intuition. Are all the components of the Hodgkin-Huxley model of neuronal excitation necessary to understand the principle? If not, then perhaps a simpler model can help identify common principles of excitable, electrical systems—from nerve cells to cardiac cells to tissue dynamics as a whole.

Figure 7.1 provides context for the rationale that common principles may be found across distinct systems; here nerve cells, muscle neurons, and the electrically induced beating of a heart. In all of these cases, the external application of a current leads to excitation of a cell, which then relaxes over time. Continuous application of current, whether because of external stimulus or internal propagation of signals, can lead to regular patterns. In the case of spatially extended excitable systems—like the heart—the excitation of one cell can induce adjacent cells, which can induce others, such that a signal soon propagates over time and space. This spatiotemporal excitation is the basis for the homeostatic maintenance of cardiac rhythms. Because excitable systems are dynamical systems, perturbations or variations in the status of cells (or connectivity between cells) can also lead to dysregulation and disease.

In moving from cells to organisms, it is critical to be cognizant that organisms are not a linear aggregation of nonlinear components. Instead, the connectivity and interactions between components implies that collective properties need not scale as might be expected

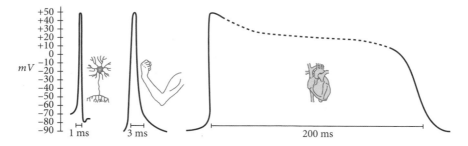

Figure 7.1: Action potential spikes over time for nerve cells, muscle neurons, and cardiac tissue. Reproduced via an image from Flavio Fenton.

from the summation of cellular dynamics. Heart rate is an ideal example to illustrate this point. The interbeat interval increases with the size of organisms, just as heart rate (or cardiac output frequency) decreases with the size of organisms. The study of biological scaling goes back a century to early work by Max Kleiber, who strove to measure the relationship between size and metabolic rate (Kleiber 1932, 1947, 1961). For decades, the key insight of metabolic scaling was to highlight that, despite the fact the surfaces of organisms with body mass M should seemingly scale like $M^{2/3}$, physiological measurements suggested that metabolic rate scaled like $M^{3/4}$ (McMahon 1973; Peters 1983; Schmidt-Nielsen 1984)—with similar ideas extending to principles of plant growth (Niklas 1994). The study of scaling in biology and ecology was rejuvenated by a proposal that "quarter-law scaling" was driven by the propagation of blood through complex networks (West et al. 1997; Gillooly et al. 2001; Brown et al. 2004). The reality is more complicated. Reexamination of the underlying empirical observations suggested that metabolic rates do not obey a single power law, including break points in scaling, and that the measurements are compatible with both dimensional analysis arguments and quarter-power scaling (Dodds et al. 2001; White et al. 2007; Price et al. 2012). Yet, for our purposes, it is worthwhile to highlight the fact that heart-rate is not universally the same even if the elements—cardiac cells—are relatively invariant across mammals. Instead, heartbeat frequency decreases with size, very roughly like $f_H \sim M^{-0.25}$ (Savage et al. 2004). Although this scaling analysis comes with uncertainty—and could be compatible with a range of similar exponents—the implications are striking.

Note that the life span of mammals and birds increases with body size, and despite significant variation, the increases are close to $M^{0.25}$, albeit there are strong exceptions to this rule. Nonetheless, the decreasing heart rate and increasing life span of mammals and of birds suggest interesting invariants. Consider the example of humans. Individuals can typically live to 80 years, such that the total number of heartbeats in a lifetime is

$$60 \text{ bpm} \times 60 \text{ min/hr} \times 24 \text{ hr/day} \times 365 \text{ days/yr} \times 80 \text{ yr} = 2.5 \times 10^9 \text{ bpm}.$$

In contrast, hummingbirds have a much faster heartbeat but also a far shorter life span. For hummingbirds, the total number of heartbeats can be estimated as

$$1000 \text{ bpm} \times 60 \text{ min/hr} \times 24 \text{ hr/day} \times 365 \text{ days/yr} \times 5 \text{ yr} = 2.6 \times 10^9 \text{ bpm}.$$

A direction for a different type of book (or TED Talk) might begin with this question: what are you going to do with your 2.5 billion beats? Here the question is somewhat more mundane: *How can we model the regular and potentially irregular nature of heart rhythms?* Note that, for some scientists, striving to answer this second question also becomes the de facto solution to the first question.

In practice, despite the finding of a characteristic shape for a single nerve cell excitation— a similar process of which shapes cardiac cell excitation—the dynamics of heartbeats at the organismal cell level need not be regular. Instead, the dynamical system as a whole can malfunction, leading to heart rates that vary over minute time scales or that are entrained in a particular diseased rhythm. The link between cardiac rhythms and disease is illustrated in Figure 7.2, which includes one example of a healthy heart and three examples of cardiovascular dysfunction (Goldberger et al. 2002). The striking piece of these time series rhythms is the extent to which what you might think is healthy is not; indeed, when including these images in the classroom, the bulk of students do not in fact choose the healthy heart out of these options. Can you? The answers are provided in the technical appendix, but it is worthwhile to try to reason through the options—before looking.

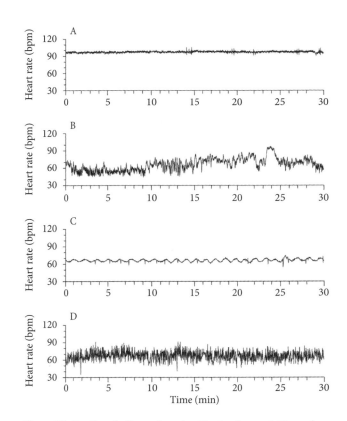

Figure 7.2: Cardiac rhythms—from healthy to diseased. These four examples include one example of a "healthy" heart and three examples of disease, including severe congestive heart failure as well as cardiac arrhythmia (a form of atrial fibrillation) (Goldberger et al. 2002). The answers to the characterization are available in the technical appendix—before looking, it is worthwhile to hazard a guess which rhythm is healthy and which are not.

Hence, the goal of this chapter is to begin to take a step toward understanding organismal-scale dynamics. But to do so requires taking a step back from the complex representation of nerve cells. Instead, this chapter strives to develop a model of a model: translating the Hodgkin-Huxley model into a simplified form. As described in Chapter 6, the Hodgkin-Huxley model can be written as

$$
\overbrace{C_M \frac{dV}{dt}}^{\text{capacitive current}} = \overbrace{I}^{\text{applied current}} \overbrace{- \bar{g}_K n^4 (V - V_K)}^{\text{potassium current}} \dots
$$
$$
\underbrace{- \bar{g}_{Na} m^3 h (V - V_{Na})}_{\text{sodium current}} \underbrace{- \bar{g}_l (V - V_l)}_{\text{leak current}} \tag{7.1}
$$

$$
\underbrace{\frac{dn}{dt}}_{\text{potassium channels}} = \overbrace{\alpha_n(1-n)}^{\text{on}} - \overbrace{\beta_n n}^{\text{off}} \tag{7.2}
$$

$$\underset{\text{sodium channels}}{\underbrace{\frac{dm}{dt}}} = \overset{\text{on}}{\overbrace{\alpha_m(1-m)}} - \overset{\text{off}}{\overbrace{\beta_m m}} \tag{7.3}$$

$$\underset{\text{leak channels}}{\underbrace{\frac{dh}{dt}}} = \overset{\text{on}}{\overbrace{\alpha_h(1-h)}} - \overset{\text{off}}{\overbrace{\beta_h h}} \tag{7.4}$$

The goal here is to explain how the essential excitability dynamics of the HH model can be embedded in a two-dimensional system known as the FitzHugh-Nagumo (FN) model (FitzHugh 1961; Nagumo et al. 1962), including a voltage v and gate variable w:

$$\underset{\text{capacitive current}}{\underbrace{\frac{dv}{dt}}} = \overset{\text{current}}{\overbrace{v(a-v)(v-1)}} - w + \overset{\text{applied current}}{\overbrace{I_a}} \tag{7.5}$$

$$\underset{\text{channel dynamics}}{\underbrace{\frac{dw}{dt}}} = \overset{\text{relaxation dynamics}}{\overbrace{bv - \gamma w}} \tag{7.6}$$

As in Chapter 6, rather than explaining each term at the outset, this system of equations and its interpretation will be derived. The reduction of dimensions implies that it is possible to understand key principles through detailed mathematical investigation of the feedback between voltage and gating variables. With this simplified model in place, it is then possible to envision connecting cells together—whether as part of neural networks or as part of cardiac tissue—that could lead to signal propagation underlying healthy function as well as oscillations and even spatiotemporal chaos underlying disease and dysfunction. To get there requires starting at the beginning, with the basics of oscillatory systems and how modulating certain types of feedback can connect principles of a swinging pendulum to that of an excitable neuron cell.

7.2   PRINCIPLES OF OSCILLATORY DYNAMICS

7.2.1   Springs, pendulums, and other fundamentals of oscillatory dynamics

Early physics classes introduce the concept of simple oscillatory models via mechanical examples (Figure 7.3): (i) the bead or ball on a spring; (ii) the swinging pendulum. In both of these cases, the spring or pendulum arm is considered to be massless. Instead, the dynamics characterize the forces acting on the object at the end of the spring or pendulum of mass m. In the case of the spring, Hooke's law dictates that the forces scale linearly with the displacement and act to restore the spring relative to some rest length, $F = -kx$ (the displacement should be thought of as relative to that rest length). In the case of the pendulum, gravity acts downward on the mass with magnitude $-mg$ where $g \approx 9.8$ m/sec^2. However, because the pendulum arm is rigid, the force acting perpendicular to the pendulum arm is $-mg\sin\theta$. These external forces act on masses via Newton's second law of motion,

$$F = ma, \tag{7.7}$$

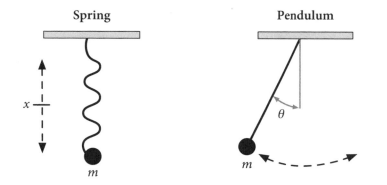

Figure 7.3: Canonical examples of oscillators: the spring and pendulum with a mass attached on one end. In the case of the spring, the extension, $x(t)$, changes with time under the action of a restoring spring force. In the case of the pendulum, the angle changes as a result of the action of gravity (acting downward).

where the acceleration is the derivative of velocity, or $a = \frac{dv}{dt} = \ddot{x}$. This force balance dictates the resulting dynamics.

First, consider a spring with a mass m attached on one end, displaced a length x from its intrinsic or resting length. The equations of motions are

$$-kx = m\ddot{x}. \tag{7.8}$$

This is a second-order, *linear* ordinary differential equation. The term second-order denotes the fact that there is a second derivative of a variable—the position x. Although this may appear to cause new complications, there is an advantage, thus far: the system only depends linearly on x. In general, oscillatory solutions are required to solve problems where the second derivative of a time-dependent function $x(t)$ is proportional to the original function. To understand why, note that the second derivatives of an oscillatory function—like sine and cosine—are just the original function multiplied by a constant (see the technical appendix for a refresher if needed).

Hence, consider the case $x(t) = A \sin \omega t$. If this were the correct solution, then we must find both A and ω, the amplitude and frequency, respectively. Note that

$$x(t) = A \sin \omega t$$
$$\dot{x}(t) = \omega A \cos \omega t$$
$$\ddot{x}(t) = -\omega^2 A \sin \omega t$$
$$\ddot{x}(t) = -\omega^2 x(t) \tag{7.9}$$

Substituting this candidate solution into the equation of motion yields

$$-kx = -m\omega^2 x \tag{7.10}$$

such that $w = \sqrt{k/m}$, where the prefactor A (and indeed the phase) depends on the initial conditions. The solution here goes on indefinitely as some initial energy is injected into the system at first by either compressing or extending the spring. This sounds nice, in theory. But oscillations are not forever, neither for physical systems nor biophysical systems.

7.2.2 Oscillatory dynamics with state-dependent drag

In practice, actual systems are subject to drag, dissipation, entropy, and all sorts of feedback that means there is no free lunch. One way to modify this model is to add an explicit loss or damping term, for example,

$$-kx - \eta \dot{x} = m\ddot{x}. \tag{7.11}$$

This equation states that the total force is the sum of the spring force, $-kx$, and the damping force, $-\eta \dot{x}$, where $\eta > 0$ such that drag works in the opposite direction of the velocity. The solution to this model is a damped oscillator rather than a perpetual oscillator (Benedek and Villars 2000b). Depending on the scale of damping, the solutions to this dynamical system can be underdamped (i.e., including diminishing oscillations) or overdamped (i.e., relaxing back to the rest point without oscillations). There is even a critical damping level that separates these two regimes (Figure 7.4 top). In this example, the particle begins to

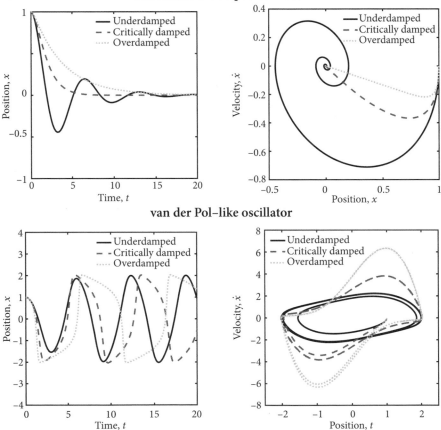

Figure 7.4: Dynamics in time and in the phase plane for a conventional damped oscillator (top) and a van der Pol–like oscillator (bottom). (Top) The conventional damped oscillatory dynamics are $-kx - \eta \dot{x} = m\ddot{x}$, with $k = 1$, $m = 1$ and $\eta = 0.5$, 2, and 4 for the underdamped, critically damped, and overdamped cases, respectively. The phase plane dynamics plot x versus $\dot{x}$. (Bottom) The van der Pol–like dynamics are $-kx - \eta(x^2 - 1)\dot{x} = m\ddot{x}$, with $k = 1$, $m = 1$, and $\eta = 0.5$, 2, and 4 for the underdamped, critically damped, and overdamped cases, respectively. The phase plane dynamics plot x versus $\dot{x}$.

accelerate given the initial extension, relaxing back toward its equilibrium at $x^* = 0$. However, because of the damping, the velocity decreases and the initial potential energy is not converted strictly into kinetic energy. Instead, the energy is dissipated and the system loses velocity such that subsequent oscillations are ever smaller. Past the point of critical damping, the system relaxes strictly back toward the rest state without oscillations. The relationship between position x and velocity $\dot{x}$ is apparent in the phase plane portrait of Figure 7.4 (top).

But what if an oscillatory system could sometimes have feedback that meant that the "drag" was either positive (inhibitory) or negative (excitatory). Although such feedback seems implausible in the spring or pendulum example, it is not infeasible for active, living systems. This is precisely the approach taken by Balthasar van der Pol, who proposed a variant of the following model (van der Pol 1920; Guckenheimer 1980):

$$-kx - \eta(x)\dot{x} = m\ddot{x} \tag{7.12}$$

where $\eta(x) = C(x^2 - 1)$. In this conceptualization of drag, there are regimes where drag is positive and other regimes where drag is negative. Hence, when there are large deviations such that $|x| > 1$, then the system has positive drag, which reduces the velocity; whereas when $|x| < 1$, then the system has negative drag and the system is excitatory. The result of the inclusion of state-dependent drag can be seen in Figure 7.4 (bottom). Insofar as the system starts with an initial condition that is sufficiently "stretched" (i.e., given $|x^2| > 1$), then the system behaves similarly to a conventionally damped oscillator, relaxing back toward equilibrium. However, once the system's position drops below 1, then the velocity increases, kicking the system away from the rest state and back out where it again can relax toward an equilibrium that it never reaches. The result is oscillation for the position, and more generally in the phase plane. Hence, this change in sign underlies a radical change in the outcome of the dynamics.

The actual van der Pol model is slightly different:

$$\ddot{x} + C(x^2 - 1)\dot{x} + x = 0 \tag{7.13}$$

where the system has been transformed to one with units such that the intrinsic frequency is 1. This nonlinear second-order differential equation can be written as a coupled system of nonlinear differential equations in terms of the position x and velocity $v = \dot{x}$:

$$\dot{x} = v \tag{7.14}$$

$$\dot{v} = -x - C\left(x^2 - 1\right) \tag{7.15}$$

However, instead of working with the position/velocity, consider the transformed variable $y = \dot{x}/C + x^3/3 - x$, which combines both position and velocity into an effective term (admittedly opaquely so). The technical appendix shows how, after some algebra, this can be written as

$$\dot{x} = C\left(y + x - x^3/3\right) \tag{7.16}$$

$$\dot{y} = -x/C \tag{7.17}$$

Either formulation is equivalent, yet this second way of presenting the coupling lies at the heart of the FitzHugh-Nagumo (FN) model, which itself is meant to be a model of the more complex HH dynamics of neuronal excitations. In essence, FN takes the standard oscillator augmented by a drag that could be inhibitory or excitatory and then, using the transformed variables of van der Pol (1920), augments the basic equations so that they can be written in a slightly different form:

$$\dot{x} = C\left(y + x - \frac{x^3}{3} + Z\right) \tag{7.18}$$

$$\dot{y} = -\frac{1}{c}(x - a + by) \tag{7.19}$$

where x is the voltage, y is the recovery or gating variable, and Z is the time-dependent perturbation, e.g., as due to an applied current. There are many variants of this model. The one used for the bulk of analysis in this chapter is

$$\dot{v} = v(a - v)(v - 1) - w + I_a \tag{7.20}$$

$$\dot{w} = bv - \gamma w \tag{7.21}$$

This is what we refer to as the Fitzhugh-Nagumo model, or the FN model. But, before analyzing the FN model, we need to take one more step—understanding a new class of dynamics in the van der Pol case, i.e., relaxation oscillations. The hint that such relaxation oscillations are apparent in Figure 7.4 (bottom) in the overdamped case. In that case, the system relaxes slowly like a conventional spring oscillator until reaching the critical point $x = \pm 1$, at which point the system rapidly moves to the other side, then slowly relaxes back, before rapidly moving again.

7.3 RELAXATION OSCILLATIONS—A GENERALIZED VIEW

Fast-slow dynamical systems denote dynamics in which a subset of the variables change far faster than another set. Such time scale differences can sometimes enable useful approximations to system state (e.g., what are termed *adiabatic approximations*). More relevant to the study of neuron and cardiac excitations, the separation of time scales can also drive the kind of excitability—including filtering and refractory periods observed in electrically excitable cells and tissue.

Generalized fast-slow dynamical systems can be written as

$$\epsilon \frac{dx}{dt} = f(x, y, \epsilon) \tag{7.22}$$

$$\frac{dy}{dt} = g(x, y, \epsilon) \tag{7.23}$$

where $f(\,)$ and $g(\,)$ are of order 1 and $0 < \epsilon \ll 1$ is the speed parameter. The fact that ϵ is very small implies that the rate of change of x is very large and therefore that x is the fast variable and y is the slow variable. For generalized models, x and y may denote *sets* of fast and slow variables, respectively. These equations are written in terms of the slow time scale, t, Hence, there is another way to write this same model, by rescaling time such that $\tau = t/\epsilon$, i.e., so that τ denotes the fast time scale:

$$\frac{dx}{d\tau} = f(x, y, \epsilon) \tag{7.24}$$

$$\frac{dy}{d\tau} = \epsilon g(x, y, \epsilon) \tag{7.25}$$

In this fast time scale, x changes on the order of 1, whereas y changes very slowly. These two ways of writing the same set of equations are the basis for understanding singular limits, in particular the limit where $\epsilon \to 0$. Analyzing the dynamics of fast-slow systems in these singular limits provides a practical way of stitching together expected dynamics in the case of finite ϵ.

7.3.1   Fast flows in the van der Pol oscillator

The van der Pol (VDP) oscillator is the basis for the FitzHugh-Nagumo model. It can be written as

$$\epsilon \dot{x} = f(x, y, \epsilon) = y - \frac{x^3}{3} + x \tag{7.26}$$

$$\dot{y} = g(x, y, \epsilon) = a - x \tag{7.27}$$

where $0 < \epsilon \ll 1$ is the speed parameter and a is the forcing rate. In the limit that $\epsilon = 0$, the singular limit of the VDP model is

$$\frac{dx}{d\tau} = y - \frac{x^3}{3} + x \tag{7.28}$$

$$\frac{dy}{d\tau} = 0 \tag{7.29}$$

which is known as the *fast subsystem*. Because $dy/dt = 0$ in this singular limit, then the value of the slow variable y acts as a parameter of the system, such that x should converge to an equilibrium point that satisfies

$$y - \frac{x^3}{3} + x = 0. \tag{7.30}$$

The set of points that satisfies this condition is called the *critical set*, i.e., $C_0 = \{(x, y) : f(x, y, 0) = 0\}$; and in this case, because the set is smooth, it is referred to as the *critical manifold*. The stability of any point in the critical manifold can be analyzed locally, i.e., by calculating the eigenvalues of the Jacobian evaluated at points along the critical manifold:

$$\mathbf{J} = \begin{bmatrix} \frac{\partial f}{\partial x} & \frac{\partial f}{\partial y} \\ 0 & 0 \end{bmatrix}_{C_0} = \begin{bmatrix} 1 - x^2 & 1 \\ 0 & 0 \end{bmatrix}. \tag{7.31}$$

The use of linear stability analysis was already introduced in previous chapters. Recall that the objective of a linear stability analysis is to identify whether or not small perturbations would diminish or increase, exponentially. Here, because y is a parameter in this singular limit, then the stability of the system must be evaluated at every point on the critical manifold. There are two eigenvalues, $\lambda_1 = 1 - x^2$ and $\lambda_2 = 0$. The zero eigenvalue denotes

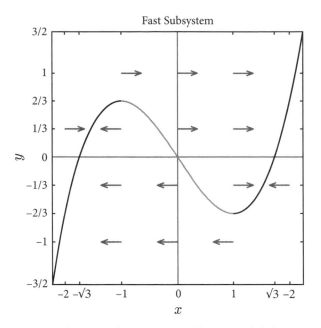

Fast Subsystem

Figure 7.5: Flows in the fast subsystem of the VDP model. The curve denotes the critical manifold, where black is attracting and gray is repelling. Arrows denote dynamics in x in the fast subsystem.

the fact that changes in the y direction are neither magnified nor diminished (at least not in the singular limit). In contrast, it is evident that points on the critical manifold where $|x| < 1$ have $\lambda_1 > 0$ and are unstable. In contrast, points on the critical manifold where $|x| > 1$ have $\lambda_1 < 0$ and are stable. The stability and instability of the critical set can be thought of in terms of three components: an attracting branch for $x < -1$, a repelling branch for $-1 < x < 1$, and another attracting branch for $x > 1$. Hence, when viewed in terms of the slow parameter y, the system exhibits two saddle-node bifurcations, at $y = -2/3$ and $y = 2/3$ (Figure 7.5). This means that for large values of y there is only a single, large positive x^* equilibrium, and for small values of y there is only a single, large negative x^* equilibrium. However, for intermediate values of y, the fast subsystem has three fixed points, two of which are stable and one of which is unstable (note the black versus gray shading to denote stability and instability, respectively). Yet the fast subsystem is only part of the story, because once the system relaxes to this critical manifold, then the slow variable can begin to change.

7.3.2   Slow flows in the van der Pol oscillator

The slow flows of the VDP model are described by

$$0 = f(x, y, \epsilon) = y - \frac{x^3}{3} + x \tag{7.32}$$

$$\frac{dy}{d\tau} = \epsilon g(x, y, \epsilon) = \epsilon (a - x) \tag{7.33}$$

This is a differential algebraic equation, in the sense that y is changing dynamically subject to the constraint that $y = \frac{x^3}{3} - x$. Hence, the dynamics are stuck to the critical manifold and, in the absence of forcing, $dy/d\tau = -\epsilon x$ such that y increases when $x < 0$ and y decreases when $x > 0$. These dynamics suggest there are four regimes. First, when the system is initiated for $x < -1$, then x and y increase toward $(-1, 2/3)$. In contrast, when the system is initiated for $-1 < x < 0$, then y increases and x decreases toward the same point $(-1, 2/3)$. Likewise, when the system is initiated for $0 < x < 1$, then y decreases and x increases toward the same point $(1, -2/3)$. Finally, when the system is initiated for $x > 1$, then x and y decrease toward $(1, -2/3)$. These dynamics are depicted in Figure 7.6. However, these dynamics raise an important question: what happens as the system approaches either $(-1, 2/3)$ or $(1, -2/3)$? Indeed, it seems something important might happen as dynamics converge to this point—yet the point is not an equilibrium of the system.

Indeed, in the slow flow, the constraint curve can be differentiated such that

$$\dot{y} = \frac{3x^2\dot{x}}{3} - \dot{x} \tag{7.34}$$

$$\dot{y} = \dot{x}\left(x^2 - 1\right) \tag{7.35}$$

$$-x = \dot{x}\left(x^2 - 1\right) \tag{7.36}$$

$$\dot{x} = \frac{x}{1 - x^2} \tag{7.37}$$

where the last two substitutions are enabled by recognizing that $\dot{y} = -x$. Hence, the dynamics along the constraint curve suggest the system will undergo a singularity at $x = \pm 1$. This singularity is in fact a "jump point" for relaxation oscillations. The jump point corresponds to a singular limit where the dynamics should rapidly move in the fast variable back to a stable branch of the manifold. In the oscillatory system, this corresponds to a rapid switch in dynamics. Further, to verify that this system of equations does in fact remain on the constraint curve if it begins there, note that

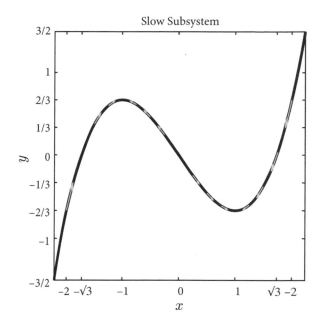

Figure 7.6: Flows in the slow subsystem of the VDP model. The curve denotes the critical manifold; the gray arrows inside the curve denote a change in y, wherein the overall dynamics are fixed along the constraint curve.

$$\frac{dx}{dy} = \frac{-1}{1 - x^2} \tag{7.38}$$

such that $dx\left(1 - x^2\right) = -dy$. Integrating both sides yields

$$x - \frac{x^3}{3} = -y + \text{const} \tag{7.39}$$

and for any value on the critical manifold C_0, this reduces to $y + x - \frac{x^3}{3} = 0$—precisely the constraint curve of the slow subsystem! This slow subsystem is only part of the story. The full dynamics require stitching both slow and fast dynamics together.

7.3.3 Stitching fast and slow flows together

The dynamics in the singular limits suggest that a van der Pol oscillator should exhibit a particular kind of relaxation oscillation. First, if the system is initialized far from the critical manifold, then it should move rapidly in x toward a stable branch (as shown in Figure 7.6). However, by approaching a stable branch, the fast dynamics become slow, because by definition, $\dot{x} = 0$ on the critical manifold. More details on how close the fast variable must be to the critical manifold have been examined in a general sense in the nonlinear dynamics literature (Arnold et al. 1995; Desroches and Jeffrey 2011) and as applied to biological systems (Cortez and Ellner 2010). Once the fast variable is within "ϵ" of the critical manifold, then the system changes by following the slow subsystem in both x and y, following the critical manifold until it reaches a singularity. This singularity is a jump point where the system

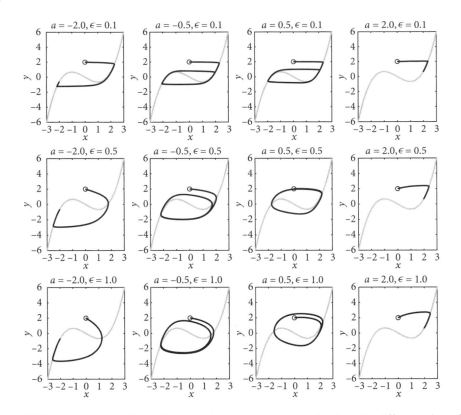

Figure 7.7: Flows in the van der Pol oscillator given variation in a and ϵ. Each column has different values of a, which determine the stability of dynamics, separating oscillatory dynamics (middle columns) from convergence to a fixed point (leftmost and rightmost). The fast-slow separation of scales is controlled by ϵ, where $\epsilon = 0.1$ (top row) $\epsilon = 0.5$, (middle row), and $\epsilon = 1$ (bottom row). Note that the full dynamics increasingly recapitulate a combination of singular limits as $\epsilon \to 0$.

departs from an unstable branch of the critical manifold and very rapidly moves in x to a stable branch. Then the cycle repeats, exhibiting rapid trajectories in the fast subsystem followed by slow trajectories in the slow subsystem (close to the critical manifold). Figure 7.7 shows three examples of these dynamics for the van der Pol oscillator, where the singular limit cases become an increasingly accurate representation of the full dynamics insofar as $\epsilon \to 0$. Critically, although relaxation oscillations are possible in this model, the long-term behavior depends on the value of a such that when $|a| < 1$ the system exhibits oscillatory dynamics and for $|a| > 1$ the system converges to a stable point on the critical manifold (see the technical appendix for more details).

7.4 PRINCIPLES OF EXCITABILITY: FROM CARDIAC CELLS TO TISSUE

The FitzHugh-Nagumo (FN) model is a two-dimensional "excitable" system meant to describe the coupled dynamics of cardiac cells. The FN model exhibits many of the features of the HH model. Yet, given that it is a lower-dimensional system, it enables interpretation in a way that provides insight into a broader class of models. The dynamics are of voltage v and a gating/recovery variable w:

$$\dot{v} = v(a-v)(v-1) - w + I_a \tag{7.40}$$

$$\dot{w} = bv - \gamma w. \tag{7.41}$$

This model is, in essence, a variation of the van der Pol oscillator with an applied current, and that analysis of it in the context of excitatory dynamics can borrow from the analysis of relaxation oscillations explored in depth in the prior section. As was noted in Chapter 2, a key step to analyzing coupled nonlinear dynamical systems is to identify the nullclines, find the fixed points, and then analyze the local stability of fixed points as a gateway to understanding the global behavior.

7.4.1 Synopsis of cardiac cell dynamics via the FN model

Recall that the FN model can be written as

$$\underbrace{\frac{dv}{dt}}_{\text{capacitive current}} = \overbrace{v(a-v)(v-1) - w}^{\text{current}} + \overbrace{I_a}^{\text{applied current}} \tag{7.42}$$

$$\underbrace{\frac{dw}{dt}}_{\text{channel dynamics}} = \overbrace{bv - \gamma w}^{\text{relaxation dynamics}}. \tag{7.43}$$

The nullclines of the FN system are

$$w = \overbrace{v(a-v)(v-1) + I_a}^{v \text{ nullcline}} \tag{7.44}$$

$$w = \overbrace{\frac{bv}{\gamma}}^{w \text{ nullcline}}. \tag{7.45}$$

Hence, there can be regimes where the cubic nullcline for $\dot{v} = 0$ intersects the linear nullcline for $\dot{w} = 0$ (Figure 7.8). In the regime that $0 < b, \gamma \ll 1$, the dynamics of the gating variable are much slower than those of the voltage. In that event, we have a "fast-slow" dynamical system, just as in the case of the VDP oscillator. If one takes a singular limit and assumes that w is slow, then the voltage v will change rapidly while w remains constant. This limit is appropriate so long as dynamics in the (v, w) plane remain away from the $\dot{v} = 0$ nullcline.

The dynamics unfold as follows. Given a sufficiently large excitation, the voltage will jump. The gating variable will increase, which can be thought of as a voltage-dependent switch in gating states. The slow increase in gating will, with time, lead to a switch in the dynamics of the voltage, slowly decreasing the voltage difference until the high-voltage state becomes unstable. Then the voltage rapidly declines to a low-voltage state, which enables the increase of the gating variable. These back-and-forth dynamics can lead to persistent "relaxation oscillations" in the FN model, just as an applied current can lead to beating in the HH model. Relaxation oscillations are periodic orbits characterized by fast excursions and slow relaxations in state space. The technical appendix provides additional detail on

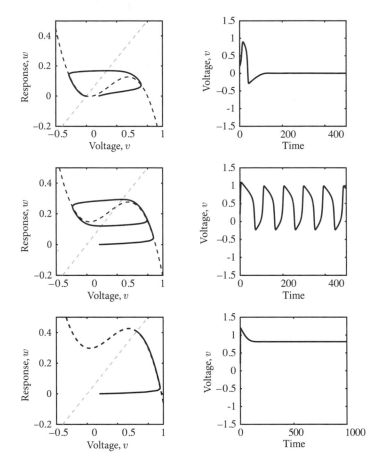

Figure 7.8: The FN model exhibits regular excitatory dynamics, including excursions between the critical manifold and dynamics that remain close to the critical manifold. Reading from top to bottom, parameters are $a = 0.1$, $b = 0.01$, and $\gamma = 0.02$ and the applied current $I_a = 0$, 0.15, and 0.3.

the analysis of the fast subsystem, in analogy to the analysis of the VDP oscillator. Hence, in this example, the FN model exhibits regular beating, just like the HH representation of a neuron given a constant applied force. The mechanisms include excitation, relaxation (along with filtering of "inputs"), and then a subsequent excitation. This behavior forms the basis for large-scale signal propagation. More detailed models of cardiac cells are reviewed elsewhere (Fenton and Karma 1998; Fenton et al. 2002).

7.4.2   Propagation of signals through tissue

Cardiac cells are excitable. But what about tissue? The dynamics of spatially extended excitable cells can be understood by coupling the reactions at one cell with others. In many systems, e.g., chemical reaction systems within cells or predator-prey dynamics at the population scale, all "species" can diffuse. That is not the case for nerve or cardiac cells. Instead, the gating proportions are local, implying that ionic concentrations will vary locally depending on the voltage-dependent change in gating. However, including an effective diffusion term in the voltage dynamics extends the core excitable model to a spatial version:

$$\frac{\partial v}{\partial t} = v(a-v)(v-1) - w + I + D\nabla^2 v \qquad (7.46)$$

$$\frac{\partial w}{\partial t} = bv - \gamma w \qquad (7.47)$$

where the Laplacian operator, ∇^2, is equivalent to $\frac{\partial^2}{\partial x^2}$ in one dimension, to $\frac{\partial^2}{\partial + x^2} + \frac{\partial^2}{\partial y^2}$ in two dimensions, and so forth. This equation is a reaction-diffusion model, insofar as each site has a local "reaction" (i.e., the changes in the spatially explicit $v(x,t)$ and $w(x,t)$) and "diffusion" (i.e., in this case, just the voltage).

The diffusion of voltage means that it is possible for an initial excitation to be transmitted spatially, akin to the movement of a signal through an axon, from the body of the nerve cell to others. Yet the challenge is that diffusion can actually work against the propagation of signals. Diffusion has the tendency to smooth out heterogeneities. Hence, it is worthwhile to consider two limits. First, in the limit that $D = 0$, then any excitation will be confined to a particular location and the signal will not propagate. However, in the other limit that $D \gg 0$, then any initial perturbation (e.g., as induced by an applied current) implies that the voltage will diffuse rapidly, to all cells. Because of the nonlinear nature of voltage-dependent feedback on gates, then such rapid diffusion can have the effect of eliminating the possibility of extended excitations. It is an intermediate regime where it is possible for an initial excitation to propagate.

Figure 7.9 shows how an initial excitation in a cell in a homogeneous cable can excite the entire cable. Initially, the diffusion begins to decrease the local voltage levels at the site of the excitation. However, the voltage-dependent gating response leads to a transient increase in the response variable, which in turn amplifies the voltage response in a nonlinear fashion leading to an amplification of the signal. Because of the no-flux boundary conditions, the excitatory dynamics complete a relaxation oscillation locally and then begin to excite new cells in the cable. A wave front develops and travels down the cable with both v and w increasing together, even though only v diffuses; it is the feedback between v and w in the FN model that amplifies local excitation, enabling the propagation of the wave (and the signal).

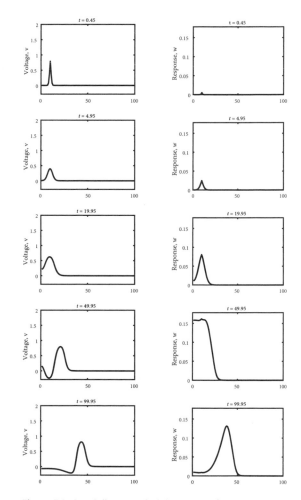

Figure 7.9: Spatially extended dynamics of the FN model in a 1D cable, given no-flux boundary conditions (i.e., $\nabla v = 0$ at the boundary of the domain). Complete details are available in the accompanying computational laboratories, including information on how to replicate these dynamics. Here $a = 0.1$, $b = 0.01$, $\gamma = 0.02$, $D = 0$, and there is no applied current. The reaction-diffusion model includes 100 cells, with normalized time and assuming $\delta t = 0.05$.

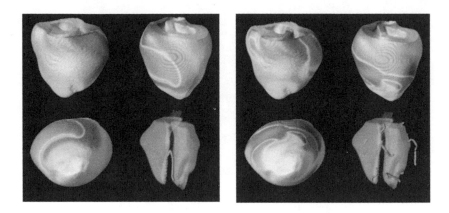

Figure 7.10: Spatially extended dynamics simulated in the rabbit ventricular geometry. (Left) A single spiral wave. (Right) Multiple spiral waves. Images reproduced from Figure 31 of Cherry and Fenton (2008).

7.4.3   From excitability to dysfunction

The excitability of cardiac tissue also implies that systems can initiate nontrivial, even unhealthy spatiotemporal dynamics. A key danger is that of spiral waves. A spiral wave impacts the heart organ's ability to pump blood. These waves can rotate and then break up, leading to fibrillation, which is disorganized and noncoordinated beating of cardiac cells in human tissue. The use of electric shocks to terminate arrhythmias can also be seen in this light: as an attempt to reset a spatially extended, excitable system to return to a different state in its bifurcation dynamics. The methods to develop whole-heart tissue simulations of cardiac dynamics are not altogether dissimilar from the simplified example in Figure 7.9. In essence, a spatially extended reaction-diffusion model can include realistic geometry, heterogeneities in diffusivity, and alternative reaction models, including early work (Noble 1962) to present-day extensions (Karma 1994; Fenton and Karma 1998; Fenton et al. 2002). These spatially extended systems reveal the potential for spiral waves and spiral wave breakup leading to arrhythmias (Figure 7.10 [Plate 6]). Intriguingly, viewing arrhythmias and atrial fibrillation through the lens of dynamical systems also raises new opportunities for therapeutic intervention: to apply lower-energy, albeit more nuanced, timing of electrical shocks to steer a dysregulated system back to a healthy state (Luther et al. 2011). With time, perhaps this integration of dynamical systems, computation, and experiment will become part of cardiac interventions.

7.5   TAKE-HOME MESSAGES

- Cardiac cells are excitable systems with similar excitability properties as nerve cells, including filtering, refractory periods, and the tendency to initiate large system excursions in response to sufficiently large perturbations.
- In aggregate, cardiac tissue has emergent properties, including the tendency to exhibit homeostatic patchiness where the system shifts between relatively constant rhythms that may then change over time in a nonstationary fashion.

- The dynamics of excitable systems can be understood as a type of relaxation oscillation.
- Classic models of relaxation oscillations extend conventional models of oscillatory dynamics with drag, and make the drag state dependent, including a potentially excitatory drag (rather than a strictly inhibitory feedback).
- Relaxation oscillations involve a slow and fast subsystem, with fast dynamics usually moving between pieces of critical manifolds and slow dynamics usually moving close to (or nearly on) critical manifolds.
- The FitzHugh-Nagumo model is a two-dimensional simplification of the Hodgkin-Huxley model—and itself is a variant of classic work by van der Pol on nonlinear oscillations.
- When coupled, excitable dynamics at one cell can propagate to nearby cells, forming tissue-level excitations.
- Although such dynamics can remain stable, they can also exhibit non-homeostatic spatiotemporal dynamics, including chaos and spiral waves that can underlie dysfunction like atrial fibrillation.
- The link between nonlinear dynamics and cardiac functioning remains a priority not just for understanding, but to develop system-informed approaches to drive a dysfunctional cardiac system back into a healthy dynamical regime.

7.6 HOMEWORK PROBLEMS

These problems leverage lessons learned in the accompanying computational laboratory guide, including a focus on the following practical toolkits:

- Analyzing the FitzHugh-Nagumo model
- Integrating nullcline and phase space dynamics
- Developing a core reaction-diffusion model in one dimension
- Computationally implementing boundary conditions in spatially extended dynamics
- Developing the start of a robust, spatially extended simulation
- Building 'movies' for spatiotemporal simulations

The computational lab provides an extended introduction to simulating the FN model and the spatiotemporal application.

PROBLEM 1. FitzHugh-Nagumo and Excitability

The FitzHugh-Nagumo model is

$$\dot{v} = v(a - v)(v - 1) - w + I$$

$$\dot{w} = bv - \gamma w$$

where v is the membrane voltage and w is the response. In this problem, use $a = 0.1$, $b = 0.01$, and $\gamma = 0.02$.

a. Determine a "bifurcation diagram" for the FN model as a function of variations in the driving current I from 0 to 25 in increments of 0.01.

b. Identify the three relevant regimes in your bifurcation diagram, explain their biological meaning, and provide a single example of qualitatively different dynamics in each regime.

c. Choosing $I = 0.125$, modify the response of the gating variables, increasing them by a factor of $\epsilon > 1$, i.e., $b(\epsilon) = 0.01\epsilon$ and $\gamma(\epsilon) = 0.02\epsilon$. In doing so, identify a critical speed beyond which the beating of cardiac cells disappears. How do you interpret this finding?

PROBLEM 2. Propagating Information

The computational laboratory extends the FitzHugh-Nagumo model to a 1D spatial system, i.e.,

$$\frac{\partial v}{\partial t} = v(a - v)(v - 1) - w + I + D\nabla^2 v \tag{7.48}$$

$$\frac{\partial w}{\partial t} = bv - \gamma w \tag{7.49}$$

where the Laplacian operator, ∇^2, is equivalent to $\frac{\partial^2}{\partial x^2}$ in one dimension. When discretized, the model is

$$\frac{v_i^{t+\Delta t} - v_i^t}{\Delta t} = v_i^t(a - v_i^t)(v_i^t - 1) - w_i^t + I + D\frac{v_{i-1} - 2v_i + v_{i+1}}{\Delta x^2}$$

$$\frac{w_i^{t+\Delta t} - w_i^t}{\Delta t} = bv_i^t - \gamma w_i^t$$

In this problem use $a = 0.1$, $b = 0.01$, and $\gamma = 0.02$, with $I = 0$ (no applied current other than the initial perturbation), $w_i(0) = 0$, a fixed time step of 0.025, and Neumann boundary conditions.

a. Using the 1D model, measure the traveling wave speed over a cable of length 100, given $D = 1$ and an initial perturbation, of 2.25 in the eleventh cell and $v_i = 0$ otherwise. Provide evidence for or against a claim that the speed remains uniform over the time measured.

b. Change the diffusion to be $D = 0.5$, 0.75, 1.25, and 1.5. Using all other default parameters, identify which cases enable a traveling wave and which do not.

c. Discuss why and how the diffusion rate can control whether or not a wave propagates. Did the results meet your prior expectation?

7.7 TECHNICAL APPENDIX

Derivatives of trigonemetric functions Consider a time-dependent function, $x(t)$, that satisfies $\ddot{x} = -ax$ where a is some constant and the double dot notation denotes a second derivative with respect to time. The solution to these systems is a combination of sine and cosine functions, precisely because the second derivative is proportional to the function. Recall that

$$\frac{d\sin t}{dt} = \cos t \tag{7.50}$$

and

$$\frac{d\cos t}{dt} = -\sin t. \tag{7.51}$$

The chain rule can be used to take derivatives when including an additional prefactor, e.g.,

$$\frac{d\sin(\omega t)}{dt} = \cos(\omega t)\frac{d(\omega t)}{dt} \tag{7.52}$$

$$\frac{d\sin(\omega t)}{dt} = \omega \cos(\omega t) \tag{7.53}$$

The same logic implies that

$$\frac{d\cos(\omega t)}{dt} = -\omega \sin(\omega t). \tag{7.54}$$

Hence, when faced with an equation like $\ddot{x} = -ax$ where a is some constant, the solution can be of the form $x(t) = sin(\omega t)$ such that

$$\dot{x} = \omega \cos(\omega t) \tag{7.55}$$

and

$$\ddot{x} = -\omega^2 \sin(\omega t). \tag{7.56}$$

To satisfy $\ddot{x} = -ax$ implies that

$$-\omega^2 \sin(\omega t) = -asin(\omega t) \tag{7.57}$$

such that $\omega = a^{1/2}$. Note also that sine and cosine functions can be used interchangeably given that $\cos(\theta + \pi/2) = \sin(\theta)$, they are equivalent, albeit phase-shifted functions.

Rescaling the van der Pol oscillator Consider an oscillatory system with state-dependent drag in terms of the position x and velocity $v = \dot{x}$:

$$\dot{x} = v \tag{7.58}$$

$$\dot{v} = -x - C\left(x^2 - 1\right)v \tag{7.59}$$

Instead of working in the $x - v$ space, consider a transformed variable,

$$y = \frac{\dot{x}}{C} + x^3/3 - x, \tag{7.60}$$

which combines both position and velocity. As such, then the derivative of y can be evaluated:

$$\dot{y} = \ddot{x}/C + \dot{x}x^2 - \dot{x} \tag{7.61}$$

$$= \ddot{x}/C + \dot{x}\left(x^2 - 1\right) \tag{7.62}$$

$$= \frac{-C(x^2 - 1)\dot{x} - x}{C} + \dot{x}\left(x^2 - 1\right) \tag{7.63}$$

$$= -(x^2 - 1)\dot{x} - \frac{x}{C} + \dot{x}\left(x^2 - 1\right) \tag{7.64}$$

$$= -\frac{x}{C} \tag{7.65}$$

As a result, the full transformed system can be written as

$$\dot{x} = C\left[y\frac{x^3}{3} + x\right] \tag{7.66}$$

$$\dot{y} = -\frac{x}{C}. \tag{7.67}$$

by using the definition of y from above.

Stability and bifurcations of the van der Pol oscillator Consider the van der Pol oscillator:

$$\dot{x} = f(x, y) = \left(y - \frac{x^3}{3} + x\right)/\epsilon \tag{7.68}$$

$$\dot{y} = g(x, y) = a - x \tag{7.69}$$

where a is a parameter and ϵ controls the relative speed of dynamics. The stability of a fixed point can be evaluated locally by linearizing the system near the fixed point:

$$x^* = a \tag{7.70}$$

$$y^* = \frac{a^3}{3} - a \tag{7.71}$$

The Jacobian for this system is

$$J = \begin{bmatrix} \frac{\partial f}{\partial x} & \frac{\partial f}{\partial y} \\ \frac{\partial g}{\partial x} & \frac{\partial g}{\partial y} \end{bmatrix}, \tag{7.72}$$

which becomes

$$J = \begin{bmatrix} \frac{1-x^2}{\epsilon} & \frac{1}{\epsilon} \\ -1 & 0 \end{bmatrix}, \tag{7.73}$$

which when evaluated at the fixed point becomes

$$J = \begin{bmatrix} \frac{1-a^2}{\epsilon} & \frac{1}{\epsilon} \\ -1 & 0 \end{bmatrix}. \tag{7.74}$$

The eigenvalues associated with this Jacobian can be solved in the standard way; they are solutions to $\mathrm{Det}(J - \lambda I) = 0$ such that

$$\lambda = \frac{\left(\frac{1-a^2}{\epsilon}\right) \pm \sqrt{\left(\frac{1-a^2}{\epsilon^2}\right) - \frac{4}{\epsilon^2}}}{2}. \qquad (7.75)$$

Note that for $|a| < 1$ both eigenvalues have a positive real part and an imaginary component and the fixed point is unstable (the full dynamics can exhibit relaxation oscillations, as explained in the chapter). In contrast, for $|a| > 1$ both eigenvalues are strictly negative and the fixed point is stable (the full dynamics converge to this fixed point on the critical manifold, as explained in the chapter).

Fast subsystem dynamics for the VDP oscillator In the fast subsystem, the van der Pol (VDP) oscillator reduces to

$$v' = v(a - v)(v - 1) - w + I_a \qquad (7.76)$$

where both w and I_a are parameters and $v' \equiv dv/d\tau$. This system is a cubic. Hence, there can be up to three equilibria, depending on the values of the gating variable, w. For example, examine the consequences of setting $w = 0.35$, 0.2, and 0.05. This is equivalent to asking: what happens to dynamics if we were to fix w at different values? In doing so, the answer may depend on the *initial* value of v. Figure 7.11 depicts the value of v' as a function of v, similar to ways of analyzing other 1D dynamical systems, like the logistic equation. Hence, the way to interpret these curves is that the fast dynamics of the voltage variable will converge to those stable values of v in this quasi-1D model. As is apparent, there is only one value of v^* such that $v'(v^*) = 0$ when $w = 0.05$ and when $w = 0.35$. In contrast, there are three fixed points of v^* when $w = 0.2$, but only two of them are stable.

The use of a fast-slow dimensional system provides a framework to analyze dynamics. Let $\mathcal{S}_0$ denote the critical set of the system, i.e., the set of values of (v, w) such that $v' = 0$. So long as the dynamics of the system are not close to this critical set, then the dynamics of v are much faster those that of w. Once the dynamics

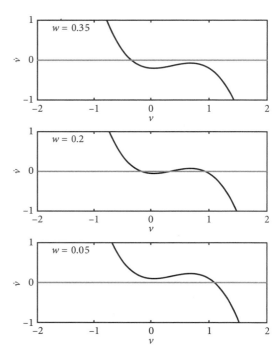

Figure 7.11: FN fast subsystem given parameter values $w = 0.05, 0.2$, and 0.35. Here parameters are $a = 0.1$, $b = 0.005, \gamma = 0.01$, and $I = 0.15$.

approach this critical set, they tend to move along it, i.e., by changing in both v (the fast variable) and w (the slow variable) until the stability of the critical set changes. Indeed, we can parameterize the critical set in terms of attracting sets and repelling sets. The term *attracting sets* denotes the fact that some portions of $\mathcal{S}_0$ are attracting fixed points in the 1D fast subsystem determined by the dynamics of v'; *repelling sets* denotes the fact that some portions of $\mathcal{S}_0$ are repelling fixed points in this subsystem.

The attracting and repelling fixed points also point to the existence of two bifurcations of the 1D fast subsystem given changes in w. Note that there are three solutions to the equilibrium condition

$$v(a - v)(v - 1) - w + I_a = 0. \tag{7.77}$$

These can be identified by finding intersections systematically as w increases. Figure 7.11 shows that the number of interesections changes from 1 to 3 back to 1, corresponding to saddle-node bifurcations near $w = 0.05$ and $w = 0.35$. These intersections suggest that the full dynamics would include rapid switching between high and low v, followed by slow relaxation along the critical manifold until the system reaches a jump point.

Cardiac rhythms—from healthy to diseased This chapter text includes a challenge to identify the healthy heart time series from four measurements of heart rate (beats per minute) over a 30-minute interval. The following image provides the time series, along with the correctly identified physiological state (reproduced from Goldberger et al. 2002). Critically, healthy heart rhythms exhibit homeostatic patchiness, which is not equivalent to a precisely fixed interbeat interval of 60 bpm. In the classroom, most students do not tend to identify the healthy heart time series (B) and often conflate it with (C), which is in fact an example of severe congestive heart failure. Notably, most students find the time series (A) and (D) to be suspect, even if they are not quite sure why.

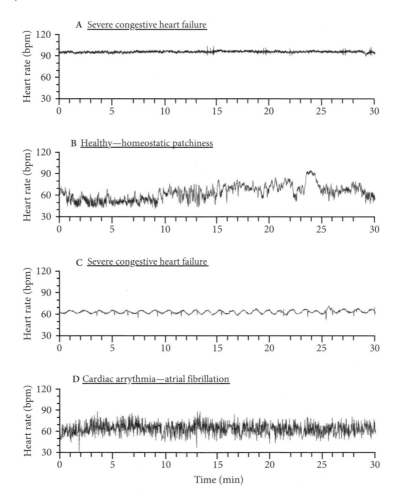

Chapter Eight

Organismal Locomotion through Water, Air, and Earth

8.1 MOVEMENT FROM WITHIN

The study of movement should be simple. At its core, understanding how organisms move would seem to be the logical extension of applying Newton's laws: (i) an object in motion will stay in motion, and one at rest will stay at rest; (ii) $F = ma$, that is to say that the change in momentum $\mathrm{d}(mv)/\mathrm{d}t$ is proportional to the applied force; (iii) for every action there is an equal and opposite reaction. In practice, the "object" in motion in Newton's laws is often idealized—a point mass that external forces act upon. Indeed, applying Newton's laws to the problem of the force acting upon celestial bodies—the Sun and Earth, for example— provides a principled route to recovering Kepler's laws of planetary motion and far more.

Yet the movement of living organisms is shaped by both external forces (e.g., gravity) and internally generated forces. Moreover, the idealization of forces acting upon a single, homogeneous object must also be adapted to the interactions between spatially extended components. Hence, even if we were to understand the parts list—the neurons that sense the environment, the electrophysiological transmission of signals from sensory neurons to the central nervous system and back again, the conversion of signals into firing of muscles, and the interaction of muscle contractions and extensions with tendons, joints, and limbs (the entire musculoskeletal system)—understanding the collective movement of a body would require that we grapple not only with Newton's laws but with the constraints imposed by the evolved structures and interactions with the external environment. Such a task is far from simple.

The task is far from simple precisely because, as Charles Darwin wrote at the end of his on the Origin of Species, "endless forms most beautiful and most wonderful have been, and are being, evolved" (Darwin 1859). These endless forms also bring with them seemingly endless movements. Studying endless movements most beautiful requires grappling with the flying motions of insects and birds, the ability of bipedal and multilegged animals to traverse heterogeneous and rough terrain, and the (sometimes) efficient swimming motions of cells and organisms across many scales of life, from *E. coli* to ciliates to electric eels to blue whales and beyond, including the movements of plant and fungi (Skotheim and Mahadevan 2005). These organisms span more than 7 orders of magnitude in length and more than 20 orders of magnitude in mass (Milo et al. 2010). Aristotle (ca. 350) first tried to conceive of how movement arose via the integration of parts in his *De Motu Animalium*, noting:

> The animal organism must be conceived after the similitude of a well-governed commonwealth. When order is once established in it there is no more need of a separate monarch to preside over each several task. The individuals each play their assigned part as it is ordered, and one thing follows another in its accustomed order.

This concession to organization has carried through to the present day. Lewis Thomas (1978) in his *The Lives of a Cell* views the dynamics of internal physiological regulation through a similar lens:

> Nothing would save me and my liver if I were in charge. For I am, to face the facts squarely, considerably less intelligent than my liver. I am, moreover, constitutionally unable to make hepatic decisions.

These quotes suggest that the answers to understanding organismal locomotion lie not from the imposition of external forces (or external will) but rather from within. In order to begin to make sense of animal movement requires switching perspectives. As Aristotle envisioned, movement arises not strictly from external forces (e.g., gravity) but from external and internal forces (e.g., those internally generated via force-generating mechanisms). Hence, the study of biomechanics requires a different starting point than Newton. Instead, we turn to an Italian physicist: Giovanni Borelli.

Borelli wrote what is now considered the first treatise on biomechanics—the two-volume *De Motu Animalium*—published posthumously in 1680. His work preceded that of Newton and followed upon that of Galileo—who discovered the concept of inertia (subsequently codified in Newton's first law). That is, an object at rest will tend to stay at rest and an object in motion will tend to stay in motion unless acted upon by another external force (e.g., gravity for a falling object or friction for an object sliding across a surface). Yet Borelli had a different starting point, in which internal forces rather than external forces mediated object movement. These internal forces are generated by contracting muscles. The consequence is that the center of mass of an organism is displaced as a result of the forces generated by contracting muscles and the interaction of the musculoskeletal system with the environment.

A few centuries later, in 1867, Mark Twain wrote what was to become the first of many celebrated short stories, essays, and novels—on the jumping frogs of Calaveras County. Of course, in that story, the celebrated jumping frog, Dan'l, didn't actually jump at the critical moment in the competition, to his owner's chagrin. Modern analysis of anuran jumping (i.e., that of frogs or toads) has documented a rapid takeoff with jump speeds on the order of $v = 2 - 4$ m/sec (note that Usain Bolt's top speed is ~ 12.2 m/sec, only a few times faster, despite the noted size advantage). Record anuran jump distances span 1–2 m (Astley et al. 2013). This distance is consistent with that expected from projectile motion (i.e., v^2/g), equivalent to ≈ 1.6 m for a launch speed of 4 m/sec, given a launch angle of 45 degrees and $g = 9.8$ m/sec^2 (see the technical appendix for a refresher on projectile motion dynamics).

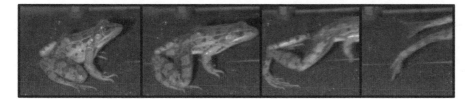

Figure 8.1: Catapult-like mechanisms enable rapid, long-distance jumps of frogs and toads (anurans). The initial jump is stimulated by the off-image presence of an individual who moves their fingers close to the posterior end of the organism. During the jump, the joint angle rapidly increases while the muscle contracts and the tendon stretches. The anurans are approximately 9 cm in length, with ∼ 0.1 sec intervals between images. Still images from the work of Astley and Roberts (2012).

These distances are remarkable in another sense: they seem to be farther than that theoretically expected from the power generated by skeletal muscles alone (Astley and Roberts 2012). A high-speed video capture augmented by X-ray motion analysis showed that jumping frogs contract muscle fibers even before their joints move. Then rapid movement of joints enabled by the recoil of tendons adds more power to what the skeletal muscles can generate on their own. Figure 8.1 provides a series of time captures of the jumping process separated by ∼ 0.1 sec, culminating in a record-like jump, unlike Dan'l in the Twain story.

But how and why did the anuran jump? Movement emerges from the interaction of musculoskeletal and nervous systems in complex environments and in response to external stimuli. Sensory neurons that detect change in the local environment send signals through the peripheral nervous system via afferent pathways toward the central nervous system, which then sends signals back outward to motor neurons via efferent pathways to trigger the release of stored energy and the integrated motion that spans molecules to tissue and limbs. The rapid release of stored energy in compressed muscles is accompanied by changes in limb position and a jump. Energy is stored within skeletal muscle fibers via the contraction of individual muscle cells made up of proteins organized into filaments, which can themselves contract and expand by sliding myosin and actin filaments along each other. The integrated system of movement spans scales from microns to centimeters in the case of jumping frogs and microns to meters for larger organisms. And yet, even if we were to identify and characterize the parts, our work would be far from complete.

Richard Feynman famously said "What I cannot create, I do not understand." Perhaps the most salient test for the state of biomechanics and neuromechanics lies in efforts to develop robots for traversal of heterogeneous terrains (Ijspeert et al. 2002; Kim et al. 2013; Aguilar et al. 2016; Yang et al. 2018). Despite advances, the results have been mixed—with lessons to be learned from both successes and failures. In the early 2010s, the US-based Defense Advanced Research Projects Agency (DARPA) aimed to stimulate the advancement of robotic movement through a series of DARPA Robotics Challenge events, including human-assisted navigation by bipedal robots of terrain. The event accelerated the use of high-performance video, microcontrollers, and computing to reshape an engineering-driven field—even as scientific principles provided routes to improve the fundamental design of robots (Scott 2004; Collins et al. 2005; Ting and McKay 2007). Yet the challenge also revealed how far the field has to go. The failures of certain bipedal robots have been set to music and are widely available online. At the time, what would have been an elementary task for a healthy human—walking straight, turning, stepping across a small barrier—proved challenging. But the robots are improving. Contrast the performance of DARPA Robotics Challenge entrants in 2015 (Krotkov et al. 2017) with bipedal and tetrapedal robots from Boston Dynamics and related humanoid and bioinspired robots just half a decade later (Nelson et al. 2019; Kim et al. 2021). Given the intent to establish principles underlying the interdisciplinary field of movement science, this chapter prioritizes a series of representative examples that link simple biophysical principles with organismal behavior. Reconciling form and function helps to explain both organismal evolution and successes, as well as gaps remaining in recapitulating (or even extending) such successes in physical models (e.g., in robots) (Aguilar et al. 2016). In doing so, the examples and dynamical principles discussed here focus on (i) movement of organisms off the ground, even if briefly, as a means to link internal forces with interactions with the environment; (ii) micron-sized traversals of *E. coli* and other microscopic swimmers that operate in the limit where inertia is irrelevant; (iii) extending principles at the microscopic scale to those governing movement of larger organisms over rough, granular terrain where dissipative forces still predominate. In some cases, these simple models are sufficient to engage directly with data, whereas in others, simple models provide a basis for deciding what constraints and principles must be considered when someone—perhaps you the reader—decides to create.

8.2   MOVEMENT WITH BRIEF MOMENTS IN AIR

The dynamics of aboveground motion involve movement at a ground-air interface. This interface is particularly important insofar as it has not always been self-evident that animal gaits ever lose contact with the ground. To resolve the dilemma, in the latter part of the nineteenth century, Leland Stanford funded a series of high-speed "chronophotographs" working with the photographer Eadweard Muybridge. Muybridge installed wire-tripped cameras along the Palo Alto track in the summer of 1878 to capture the movement of a racehorse, albeit in a sequence of still lifes. At the time, it was not yet understood whether or not all four legs of a horse lifted off the ground during a full gallop. Perhaps one of the legs was always in contact. The images put the matter to rest (Figure 8.2)—the gait of a horse includes a ground phase and an aerial phase in which all four legs are no longer in contact with the surface. This visual depiction also encapsulated the notion of a *gait*—a periodic repetition of body movements that leads to horizontal and/or vertical movement.

Strikingly, many features of gaits have turned out to be universal, including that of an alternation between an aerial phase and a ground phase. Precisely because of that alternation, Muybridge's images raise questions on how much energy is needed to enable an organism to self-propel. For example, where does the energy come from to generate the force necessary for the next aerial phase? It should be apparent from lived experience that organisms do not have the physical properties of a hard circular object bouncing off another hard surface. Contact with the ground leads to changes in foot positioning and leg extension, along with rapid changes in tendons, muscles, and joints. Yet the bounce back does not in fact require as much energy as an initial jump. Instead, the aggregate effect of these actions taken before, during, and after liftoff has the net result of leading to relatively efficient energy exchange between hops. This section aims to identify common

Figure 8.2: The Horse in Motion, Library of Congress Prints and Photographs Division (public domain). The images were taken at a distance of 27 inches and 0.04 seconds apart, with an exposure time of less than 0.002 seconds. The horizontal lines at the bottom of each image represent elevations of 4 inches; hence, at a gallop, all legs are elevated at least 1 foot above the ground, with an aerial distance of approximately 81 inches ($\approx$ 2 m).

principles of walking, running, and other gaits by addressing an even simpler version of the ground-air interface: hopping at zero horizontal speed. The analysis of even this apparently simple problem reveals principles that explain more complex gaits underlying animal locomotion.

8.2.1   Hopping mechanics

Hopping features two key elements: a ground phase in which the organism is in contact with a hard surface and an aerial phase. As such, analyzing hopping serves as a critical prelude to analyzing more complex gait dynamics, including walking and running, that have a horizontal component and either a full or partial aerial phase (for foundational work in the field, see Blickhan 1989; Farley et al. 1993; Blickhan and Full 1993). To begin, consider a simplified representation in which an organism of mass m interacts with its environment via an idealized spring of rest length L and spring constant k subject to gravity, g. The spring represents the integrated leg system. The position of the hopper is characterized only by its vertical position y. We use the convention that $y = 0$ denotes contact with the ground in which the spring is at rest (hence, one can think of 0 as the midpoint of an actual body). When the hopper hits the ground, the spring will compress, y will become negative, the compression will slow down the vertical velocity, and eventually the spring force will lead to a sufficient acceleration to alter the direction of the hopper's movement, subsequently leading to leg extension and then liftoff. The key difference between conventional spring-mass problems and that of a hopper is that once $y \geq 0$ the spring is no longer in contact with the ground and the only action of the system is due to gravity. The force is therefore piecewise continuous but not fully differentiable. As a result, the system dynamics can be written as

$$\dot{y} = v \tag{8.1}$$

$$m\dot{v} = -\overbrace{ky\Theta(-y)}^{\text{spring force}} - \overbrace{mg}^{\text{gravity}} \tag{8.2}$$

where Θ is the Heaviside function, such that the spring force is only relevant when $y \leq 0$. This can be transformed into a second-order differential equation, of the kind typically seen in classical mechanics contexts. For the ground phase the dynamics are

$$m\ddot{y} = -ky - mg, \tag{8.3}$$

whereas for the aerial phase the dynamics are

$$m\ddot{y} = -mg. \tag{8.4}$$

The hopping gait therefore consists of two portions—a ground oscillation involving the reaction of the spring with the surface and an aerial phase in which the hopper travels upward and downward in a standard projectile motion.

The ground dynamics can be solved by noting that the interactions during the ground phase should be oscillatory, albeit interrupted, beginning with ground contact (where $y(0) = 0$ and $\dot{y}(0) = -v_a$) and ending with liftoff (where $y(t_c) = 0$ and $\dot{y}(t_c) = v_a$). The period of time in contact with the ground is t_c—as yet to be determined. Hence, consider the solution ansatz

$$y(t) = a\sin\omega t + b\cos\omega t + c. \tag{8.5}$$

At $t = 0$, due to the contact $b = -c$ such that $y(0) = 0$. Further, because the hopper is moving downward at the initial contact, then $\dot{y}(0) = -a\omega = -v_a$, implying that $a = v_a/\omega$. The full equation must satisfy

$$m\left[-a\omega^2\sin\omega t + c\omega^2\cos\omega t\right] = -ka\sin\omega t + kc\cos\omega t - kc - mg, \tag{8.6}$$

which is only possible if $\omega^2 = k/m$ and $c = -mg/k$, implying that $b = mg/k = g/\omega^2$. Putting all these pieces together implies that the full ground solution is

$$y(t) = \frac{v_a}{\omega}\sin\omega t + \frac{g}{\omega^2}\cos\omega t - \frac{g}{\omega^2}. \tag{8.7}$$

This solution can also be used to calculate the duration of the ground contact phase, noting by symmetry that $\dot{y}(t_c/2) = 0$, such that

$$v_a\cos(\omega t_c/2) = \frac{g}{\omega}\sin(\omega t_c/2) \tag{8.8}$$

and

$$\frac{v_a\omega}{g} = \tan(\omega t_c/2). \tag{8.9}$$

This ground contact period can be solved numerically in terms of the driving parameters of the system, m, k, and v_a.

The aerial phase follows standard solutions to projectile motion (albeit only in the vertical direction; see the technical appendix for more details). That is, given liftoff at a velocity v_a, then the full solution for the velocity is linear in time:

$$v(t) = v_a - gt \qquad (8.10)$$

such that the aerial phase is of duration $t_a = 2v_a/g$. The position is quadratic in time:

$$y(t) = v_a t - \frac{1}{2}gt^2 \qquad (8.11)$$

leading to a hop height of $v_a^2/(2g)$. This hop height is equivalent to finding the height in which the kinetic energy at liftoff, $mv_a^2/2$, is equal to the potential energy at the top of the hop, mgh. As a result, we can combine the contact and aerial phases to estimate the gait duration as $T = t_a + t_c$, or

$$T = t_c + \frac{2v_a}{g} \qquad (8.12)$$

such that the hopping frequency, $f = 1/T$, and the fraction of time spent in contact with the surface is

$$p_c = \frac{t_c}{t_c + \frac{2v_a}{g}}. \qquad (8.13)$$

These two phases fully specify the system dynamics, in theory. They also provide some guidance as to the controls in the limit of small and large hops. Yet the question remains: to what extent can such insights provide information on hopping and running, in practice? It would seem that the many ways in which organisms interact with ground surfaces through posture changes, skeletomuscular changes, and complex organismal-surface interactions might limit the value of such simplified model insights. However, as it turns out, actual walking, running, and hopping are built on simple dynamical templates—including those akin to the interactions of a spring with a surface, as we explore next.

8.2.2 Hopping rhythms and the inverted pendulum

The dynamics of a 1D hopper can be explored numerically by converting Eqs. (8.1)–(8.2) into a computational framework, both to help visualize outcomes and to help explore the limits of the model's viability in light of physiological constraints. Following Blickhan (1989), we evaluate the model dynamics for humanlike hopping, assuming $m = 70$ kg, $L = 1$ m, $k = 10$ kN/m, and a vertical leap velocity of $v_a = 2.5$ m/sec. Figure 8.3 shows the alternation between aerial and ground phases along with revealing the gait's rhythm. Upon landing, the spring compresses, leading to an increasing force on the object that eventually reverses the direction of motion, initiating a new jump that culminates in the exit of the hopper from the ground phase back into the aerial phase with an exit velocity of the same magnitude at which it arrived. The model suggests that the compression of the spring given these parameters is on the order of 0.3 m—significantly further compression would not be physically possible. For example, the compression cannot exceed approximately one-half of the spring length. Hence, although the dynamics can be solved arbitrarily for choices of parameters, we must also take into account physiological constraints to consider plausible hopping dynamics.

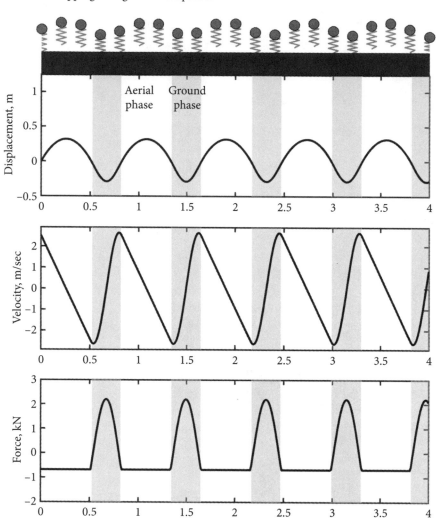

Hopping configuration snapshots

Figure 8.3: Hopping dynamics assuming a vertical leap velocity of $v_a = 2.5$ m/sec. Note that the panels depict the hopping configuration, displacement (m), velocity (m/sec), and force (kN). Here $m = 70$ kg, $g = 9.8$ m/sec^2, $k = 10$ kN/m, $L = 1$ m, and the dynamics are evaluated for a total of 5 second. The height of the jump is approximately 0.32 m (roughly 1 foot).

The driving parameters underlying one-dimensional hopping are the liftoff velocity v_a, the spring length L, the mass m, and the spring stiffness k. Note that if one uses a simplified assumption that lengths scale like $m^{1/3}$ (i.e., isometrically), then this provides at least some basis to explore the limits to hopping in humans and even across species. To begin, note that the period of time in the air for a hop with liftoff velocity v_a is $2v_a/g$, which for hops with liftoff speeds between 0.5 m/sec and 2.5 m/sec is approximately 0.1 to 0.5 seconds, with hop heights spanning 1.2 cm to 32 cm. These are close to physiological limits. However, the total hop time (and therefore the corresponding hopping frequency) also includes the

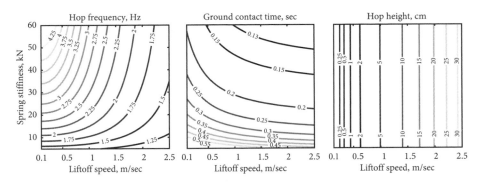

Figure 8.4: Hopping dynamics as a function of liftoff speed and spring stiffness. In all cases, $m = 70$ kg, $g = 9.8$ m/sec^2, and $L = 1$ m. Theory is calculated as described in the text. The panels depict contour lines for hop frequency (Hz), ground contact time (sec), and hop height (cm).

ground phase. Figure 8.4 uses the theoretical expectations described in the prior section to jointly visualize hopping features for combinations of liftoff velocity and spring stiffness. The three panels depict contour lines for hop frequency (Hz), ground contact time (sec), and hop height (cm)—while exploring a range of takeoff velocities from 0.1 to 2.5 m/sec and stiffness values from 5 kN/m to 60 kN/m. Notably, the inclusion of the ground phase means that there are joint changes in different features of hops. For example, increases in hop frequency come with a trade-off in terms of decreasing hop height as well as decreasing contact with the ground. These phase plots also point out fundamental constraints, e.g., too small a liftoff speed would be excluded given minimal physiological requirements for lift (try hopping with less than 1 cm of lift), and higher liftoff speeds would require deeper compression than possible (i.e., exceeding 1/3–1/2 of the spring length). For example, the compression length can be calculated as

$$y(t_c/2) = -\frac{|v_a|}{\omega} \sin(\omega t_c/2) + \frac{g}{\omega^2} \cos(\omega t_c/2) - \frac{g}{\omega^2} \qquad (8.14)$$

where t_c is the ground contact time and therefore $t_c/2$ corresponds to the time at peak compression. This particular set of phase plots includes ranges between approximately 3 and 30 cm of compression where the higher stiffness results in lower compression and vice versa.

Yet, notably, human hopping does have emergent regular patterns. Figure 8.5 shows patterns of hopping corresponding to individuals whose hopping dynamics were measured on force plates. The individuals matched their hopping frequency to a metronome over a range of approximately 2 to 5 Hz (try it for yourself). As is apparent, actual hopping involves modulation of ground surface dynamics such that the contact time remains nearly invariant; the mean is approximately 0.15 seconds, as shown in the dashed line in the top left panel of Figure 8.5. How then do humans modulate their interactions with the ground so as to maintain a fixed contact time while also modulating the frequency of jumps? The analysis of joint dynamics shown in the middle panel of Figure 8.4 provides a clue. If one were to follow an invariant contact time contour, then hop frequency could increase insofar as the spring stiffness increases and the liftoff speed decreases. This is precisely what human hoppers do. The bottom left panel of Figure 8.5 compares the measured spring stiffness against the

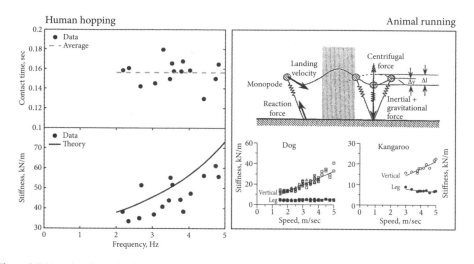

Figure 8.5: Hopping dynamics in humans and animals. (Left) Human hopping dynamics, with data digitized from Blickhan (1989). The theory is calculated using Eq. (8.9) and m = 70 kg, g = 9.8 m/sec^2, and L = 1 m. The top panel shows that contact time with the ground can remain invariant over a wide variation in hopping rates. This invariance is made possible by the change in leg stiffness over the same range of frequencies (bottom panel). (Right) Schematic of running like a monopode along with animal measurements, reprinted from Blickhan and Full (1993). The data shows changes in vertical stiffness increasing with increasing speed of horizontal motion in both dogs and kangaroos.

theory from this simplified spring-mass model. The solutions shown correspond to cases where the ground contact time is fixed at ≈ 0.15 seconds (the average of the measured hoppers) and the hopping frequency increases from 2 to 5 Hz. Combining the ground and aerial phases implies that the takeoff velocity must satisfy $v_a = \frac{g}{2}\left(f^{-1} - t_c\right)$ where $f = 1/T$ and t_c is the contact time. Then substituting in the experimentally measured value for $t_c = 0.15$ sec, the spring stiffness can be solved directly using Eq. (8.9). The agreement between the spring-mass model of human hopping and measurements is notable (see bottom left panel of Figure 8.5, albeit with some minor discrepancies). The theoretical curve has no additional fitting parameters. The significance of this relationship means that simple physical models can serve as dynamical templates for animal hopping—in fact for much more.

The 1D spring-mass hopping dynamic model forms the basis for understanding jumping and running—using the lens of "running springs" (sensu Farley et al. 1993). An illustration of the concept is shown in the right panel of Figure 8.5. As a monopode comes into contact with the ground, the interactions in the vertical direction can be approximated as those of a spring-mass system. The individual moves forward due to the horizontal component of the landing velocity while contracting the spring system, rotating forward, and experiencing an outward force on the body—similar to what one experiences on a merry-go-round, albeit for a much shorter ride. Finally, as the spring extends, the system moves back into the aerial phase governed largely by projectile dynamics (though of course organisms do change body configurations during flight as part of takeoff and landing dynamics). Notably, the effective vertical stiffness of an animal spring-mass system also increases with speed (Figure 8.5, right panel). In summary, building models of movement based on spring-mass interactions is a gateway to understanding natural hopping gaits and even running gaits for humans and a wide range of animals. These simplified models can be thought of as dynamical templates (Full and Koditschek 1999) that provide the basis for more complex, integrated control. They also operate in a particular regime—where inertia predominates. And yet for some organisms—small and large—that is not necessarily the case.

8.3 PRINCIPLES OF SLOW SWIMMING

Individual rod-like *E. coli* cells, each approximately 1–2 μm in length and less than 1 μm in diameter, move via a series of runs and tumbles. The runs represent near-constant velocity excursions of approximately 20–30 μm/sec enabled by the coordinated winding motion of multiple, surface-attached flagella. These self-assembled flagella, approximately 10 μm in length, are distributed uniformly on the cellular surface. When rotating in a counterclockwise motion, the flagella push fluid away from the cell body, propelling the cell forward. In turn, the drag leads to posterior positioning of flagella with respect to a cell in motion (Berg and Anderson 1973). These runs are then interrupted by "tumbles," short (< 1 sec) periods in which the flagella motors no longer turn counterclockwise, the flagella separate from the bundle, and the *E. coli* stops moving (Figure 8.6). This simple start-stop motion is capable of moving to and away from chemoattractants through a simple principle: a signal transduction pathway mediated by chemoreceptors at the cell surface extends the periods of runs when moving toward chemoattractants or away from chemorepellants, and decreases the periods of runs when moving away from chemoattractants or toward chemorepellants (see Chapter 5). This gradient sensing is an example of robust adaptation, as rapid changes in background concentrations of chemicals do not permanently shift the duration of runs and tumbles (at least within certain limits) (Barkai and Leibler 1997; Alon et al. 1999). Whereas Chapter 5 focused on the basis for signal transduction, this chapter focuses on the interaction of the moving bacteria with the environment. Specifically, one of the facts previously taken for granted in Chapter 5 is that when the flagella of *E.coli* stop turning in a counterclockwise fashion, then the organism stops moving—nearly instantly, on the order of a few microseconds. The near-instant cessation of movement seems counter to the first of both Galileo's and Newton's principles: that a body in motion tends to stay in motion. These principles come with a caveat—unless acted upon by another force. In this case, the life of the very small, the microbe-scale small, is different than our own, so much so that for an organism like *E. coli*, the only way to move is to swim, i.e., there is

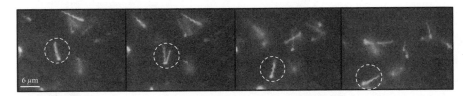

Figure 8.6: Still images during a few seconds of *E. coli* movement (Berg and Brown 1972; Turner et al. 2000; Berg 2006). A focal individual (circled) makes two tumbles separating out three runs. In the second panel, the flagellar bundle separates, corresponding to a brief cessation of movement. In the absence of a chemical gradient, each new run tends to be close to, but at a different angle than, the previous.

no gliding or coasting at these scales. The reasons, as will soon be explained, have to do with the relative importance of drag forces to inertial forces at micron scales—reasons that also explain principles of the motility "gait" of *E.coli*, cilia, and even far larger organisms when confronted with a medium that has many of these properties (e.g., lizards burrowing through sand).

8.3.1   Life at micron scales

The law $F = ma$ is introduced in the earliest of physics classes. Some readers may realize this formula recapitulates the principle of conservation of momentum in the absence of the application of an external force. Recall that momentum is the product of mass times velocity, $p = mv$, such that the derivative of momentum is $\dot{p} = m\dot{v} = ma$. In other words, the basic force principle can be written as $F = \dot{p}$. And, in most introductory physics classes, the trajectories of inanimate matter are often explored by replacing the force by that of gravity or a spring force and neglecting, at least at first, the concept of drag. Yet, at very small scales, drag cannot be ignored, not even to a first approximation. Consider the movement of a bacterial cell, like *E. coli* in water. Individual cells on the order of 1 μm in size are bombarded by collisions with water molecules, each about a quarter of a nanometer in size. Hence, if a bacterium were to stop rotating its flagella, then the interactions with these molecules would lead to a rapid decline in velocity—effectively instantaneously. This experience is wholly unlike what happens in "normal" swimming or movement where, even if a human swimmer (e.g., equipped with flippers) or a dolphin were to stop movement, the individual organism would still coast forward. They would coast precisely because inertial forces matter at such scales.

The extent to which drag and viscous forces dominate inertial forces can be understood by examining the governing equations of fluids—the Navier-Stokes equations (Tritton 1988). A full treatment lies far beyond the scope of this chapter. Nonetheless, certain limits are useful for understanding the movement of microscopic swimmers. Consider an incompressible fluid with density ρ and viscosity η with a velocity **u** (the bold denotes the fact that velocity is a vector field). The incompressibility condition that more fluid cannot be added into a particular volume implies

$$\overbrace{\nabla \cdot \mathbf{u}}^{\text{divergence of flow}} = 0. \tag{8.15}$$

This is just one part of the Navier-Stokes equations. The second equation must address Newton's second law, albeit for a three-dimensional fluid. The equivalent representation of Newton's second law for a fluid is

$$\overbrace{-\nabla p + \eta \nabla^2 \mathbf{u}}^{\text{pressure and viscous forces}} = \overbrace{\rho}^{\text{density}} \times \left(\frac{\partial}{\partial t} + \mathbf{u} \cdot \nabla \right) \mathbf{u}. \tag{8.16}$$

where over the parenthetical term: $\overbrace{}^{\text{acceleration}}$

The right-hand side of Eq. (8.16) denotes the mass density times acceleration, where the full time derivative in 1D can be written as $D/Dt = \frac{\partial}{\partial t} + \mathbf{u}\frac{\partial}{\partial x}$. The left-hand side of Eq. (8.16) represents the force acting on a particular parcel of fluid, including both pressure and viscous forces (and neglecting gravity). This system of equations can be nondimensionalized by rescaling in terms of a characteristic velocity v, length scale a, and time scale a/v (following the convention in Figure 8.7). Hence, the dimensionless variables can be written as $\tilde{\mathbf{u}} = u/v$, $\tilde{\mathbf{x}} = \mathbf{x}/a$, $\tau = t/(a/v)$, and $\tilde{p} = p/(v\eta/a)$. As shown in the technical appendix, rescaling the fluid equivalent of Newton's second law yields

$$\tilde{\nabla}\tilde{p} + \tilde{\nabla}^2 \tilde{u} = \text{Re}\left(\frac{\partial}{\partial \tau} + \tilde{\mathbf{u}} \cdot \tilde{\nabla}\right)\tilde{\mathbf{u}} \qquad (8.17)$$

where

$$\text{Re} = \frac{\rho v a}{\eta} \qquad (8.18)$$

is known as the Reynold's number. The Reynold's number is dimensionless (note that the fundamental units of η are $M \cdot L^{-1} \cdot T^{-1}$) and can also be written as $\text{Re} = \frac{va}{\nu}$ where $\nu = \eta/\rho$ is termed the *kinematic viscosity*. For water, $\nu \approx 10^{-2}$ cm^2/sec. Hence, the magnitude of the inertial terms is set by the length and velocity scales relative to that of the kinematic viscosity (Figure 8.7). Note that another interpretation of the Reynold's number is that it is the ratio of the inertial to viscous forces, i.e., $\rho \mathbf{u}\nabla\mathbf{u}$ divided by $\eta\nabla^2\mathbf{u}$. Assuming gradients operate over a characteristic distance a implies that inertial forces are of the magnitude $\rho v^2/a$ and viscous forces are of the magnitude $\eta u/a^2$. Taking the ratio of these forces yields $\frac{\rho av}{\eta}$, precisely the Reynold's number derived above.

Critically, the Reynold's number is dramatically different for small and large organisms. For very small organisms, the product of the length scale and the velocity is far less than 10^{-2} cm^2/sec. For example, consider a 2 μm *E. coli* (equivalent to 2×10^{-4} cm in length) swimming at 30 μm/sec (equivalent to 3×10^{-3} cm/sec). The Reynold's number for microbial organisms of this size and speed combination is

$$\text{Re}_{E.\ coli} \approx \frac{3 \times 10^{-3} \cdot 2 \times 10^{-4}}{10^{-2}} = 6 \times 10^{-5}. \qquad (8.19)$$

This value is so small in magnitude that it is possible to assume the Reynold's number is effectively 0. As a consequence, the swimming dynamics of *E. coli* and other microbes can be approximated as those in which inertial forces do not matter! In contrast, inertial forces certainly matter for swimming humans—relatively speaking, a speedboat compared to *E.coli*. A human swimmer moving at Michael Phelps-ian speeds of ≈ 10 km/hr (or ≈ 250 cm/s) given a length scale of 200 cm has an approximate Reynold's number of

$$\text{Re}_{Michael\ Phelps} \approx \frac{2.5 \times 10^2 \cdot 2 \times 10^2}{10^{-2}} = 5 \times 10^6. \qquad (8.20)$$

In this case, inertial forces are vastly greater than viscous forces—momentum matters. These are entirely different limits, representing more than 10 orders of magnitude difference in the relative importance of drag versus inertial forces (Lauga and Powers 2009).

To evaluate one particular consequence of these differences, consider how long *E. coli* would coast if it were to stop propelling itself forward via counterclockwise flagellar motion.

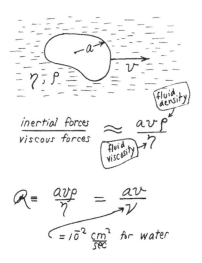

Figure 8.7: Life at a low Reynold's number by Edward Purcell (1977). Here a is the characteristic size of the organism, v is the velocity, ρ is the fluid density, ν is the kinematic viscosity, and $\eta = \nu/\rho$ is the dynamic viscosity.

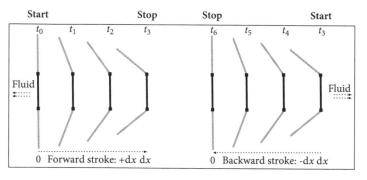

Figure 8.8: Symmetric paddling leads to no net displacement. In this example of a three-link swimmer, the symmetric gait includes a forward and backward stroke, denoted in the left and right frames, respectively. The forward stroke includes movement of two limbs backward, pushing fluid in the $-x$ direction and leading to forward propulsion in the $+x$ direction. The flow of time is marked by $t_0 < t_1 < t_2 \ldots < t_6$. The backward stroke is the reverse of this motion, pushing fluid in the $+x$ direction and leading to negative propulsion in the $-x$ direction. The net displacement is 0. In total, the forward stroke leads to a net displacement dx and the backward stroke leads to a net displacement $-dx$, i.e., there is no overall motion over the course of an entire gait.

Note that examining the Navier-Stokes equations reveals that there is a drag force proportional to velocity with a drag force constant on the order of $\frac{\eta}{a^2}$ per unit volume. This implies a drag force on an object of size a to be $k \sim \frac{\eta a^3}{a^2} = \eta a$. By analogy, if $F = -kv$, then scaling analysis implies that the characteristic coasting distance should be $l_c = vm/k$ (see technical appendix for derivation from first principles). Assuming that the density of *E. coli* is nearly that of water, then this can be rewritten as $l_c = \frac{\rho v a^3}{\eta a} = \frac{\rho v a^2}{\eta}$. Dividing both sides by the length scale, a, yields a coasting distance in terms of the relative size of the organism:

$$\frac{l_c}{a} = \frac{\rho v a}{\eta} = \mathrm{Re}. \tag{8.21}$$

Hence, the dimensionless coasting distance (i.e., coasting distance relative to the size of the organism) in a low Reynold's number regime is equivalent to that of the Reynold's number (Lauga and Powers 2009). For *E. coli* the implication is remarkable—if the bacterium were to stop moving, it would go only as far as approximately 6×10^{-5} its body length, and given that its body length is on the order of 2 μm, this implies that the coasting distance is less than a nanometer! Living at a low Reynold's number implies that to swim requires motion. Any cessation in motion would bring an individual bacterium to a nearly immediate halt, in which subsequent movement would be controlled strictly by the stochastic impacts of Brownian motion. Yet, beyond stopping on an angstrom (i.e., approximately the size of a DNA base), the fact that inertial forces can be neglected also implies that time (in the conventional sense) does not matter. As such, viewing the biomechanics of *E. coli* requires considering how microscopic organisms swim without coasting.

8.3.2   Purcell's three-link swimmer and moving at a low Reynold's number

Moving at a low Reynold's number requires navigating an environment where inertia is irrelevant. Given such limits, how should an organism like *E. coli* move? The answer lies in symmetry. Because $\mathrm{Re} \approx 0$, the constitutive equations corresponding to Newton's second law for fluids reduce in the limit to something akin to $F = 0$. That is, the net force on the organism at any given moment is zero. This is a remarkable statement. It implies that at a low Reynold's number organisms move in the absence of a force! This would seem to violate the very basis of Newton's laws.

In reality, in order for an organism to propel itself forward (or backward, or in any direction) in a low Reynold's number fluid, it must use internally generated forces to change its body configuration. Each change in body configuration leads to a corresponding change in the center of mass coordinates and orientation of the organism. The reason is that changes in body configuration come with an equal and opposite reaction from the environment. The accumulation of these infinitesimal changes in body configuration lead to an infinitesimal translation in the world frame—each small movement seems to come as if there were no inertia. Hence, if a microscopic organism had paddle-like appendages and used them to push fluid back (as in Figure 8.8, left), then the organism would move forward. Likewise, when the organism reversed the motion so as to return to its starting body configuration, then the organism would move backward (as in Figure 8.8, right). Over time, the organism would move back and forth, but in the low Reynold's number limit it would get nowhere at all. Given that the constitutive Navier-Stokes equations are time independent, then reversing a body configuration in the backward part of the gait simply undoes whatever gains were made in the forward motion. Moving forward requires that the organism deform its body over time without retracing its own "steps."

Figure 8.9 presents an illustrative example of a "model" organism—Edward Purcell's three-link swimmer (Purcell 1977). The swimmer has three links, representing a central body and two "limbs." These limbs can move independently, forming angles α_1 and α_2 with respect to the body. The organism is positioned at (x, y) and oriented at an angle θ relative to the world coordinates. The body coordinates (x_b, y_b) may also be used to denote the position of the body components relative to that of the world position (see associated computational guide for more details). This three-link swimmer provides a clarion example of the challenges of swimming in a low Reynold's number regime, as well as ways in which the addi-

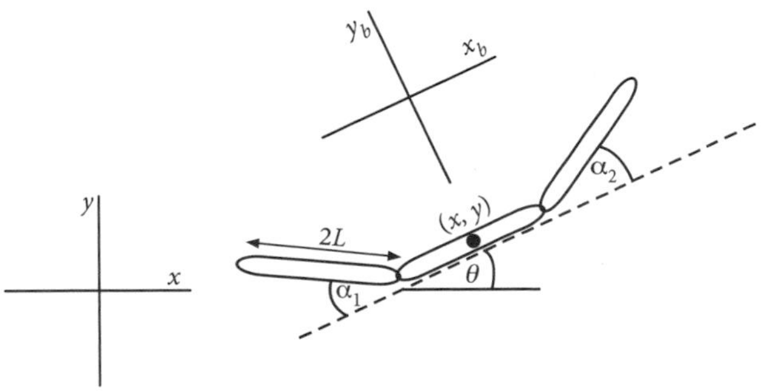

Figure 8.9: Edward Purcell's three-link swimmer (Purcell 1977). The swimmer has a body position (x, y), orientation θ, each of its three links (body and two limbs) have length $2L$, and the two limbs can freely swing about their point of attachment with angles α_1 and α_2, respectively.

tional degrees of freedom make swimming possible. Indeed, Figure 8.8 already revealed one way in which a three-link swimmer would not be able to successfully self-propel. If both limbs move in unison, one could envision this microscopic gait as akin to rowing, first backward, then forward, and again and again. Nonetheless, the symmetric swimmer does not advance, despite all the rowing. Alternatively, this is known as the scallop theorem (Purcell 1977), in that a scallop that opens and closes a single aperture so as to pull and push fluid away would not actually move ahead. Instead, movement at low Reynold's numbers requires artful use of asymmetries in traversing configuration space so that body deformations can repeat again and again, forming a gait while leading to a net translation in the intended direction.

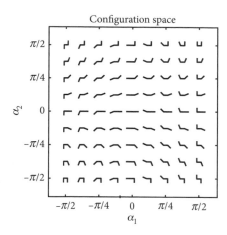

Configuration space

Figure 8.10: Configuration space for the three-link swimmer. The units of both axes are angles in the internal body space of the swimmer; each small image depicts the orientation of the three-link swimmer given the combination of α_1 and α_2.

What then are the possible mechanisms for a three-link swimmer to enable a net translation? In order to envision the possibilities, it is worthwhile to catalog the potential set of configurations. Figure 8.10 provides a map connecting the two degrees of freedom (α_1, α_2) to the resulting configurations. As is apparent, these shapes range from flat to U-like (upper right and lower left) to lounge chair–like configurations (upper left and lower right). In this space, a gait is defined as a closed loop, i.e., one that begins and ends at the same body configuration. A symmetric gait is one that reverses itself in the configuration space. For example, envision starting in the open position ⌣ and then progressively moving to ____ and then finally to ⌐⌐. This sequence moves from the upper right quadrant of the configuration space to the lower left quadrant and would push fluid downward and lead to the translation of the three-link swimmer upward. But, this is not yet a gait. A gait is defined as a closed trajectory in configuration space. Hence, when the three-link swimmer reverses this motion, moving from the lower left quadrant to the upper right quadrant, then fluid would be pushed upward and the swimmer would move downward, back to its starting position (just as shown in the schematic, albeit at a right angle, in Figure 8.8).

Forward motion is possible when a gait leverages a fundamental asymmetry. For the three-link swimmer, that asymmetry involves changes in configurations that form a closed loop without doubling back over the same configuration path. Figure 8.11 reveals one such trajectory in which the opening and closing of limbs occurs asynchronously and in staggered order, akin to a discretized version of a propagating wave down the length of the body. The result is that the swimmer is able to move forward more than it moves backward in each cycle, and then cycle after cycle allows a net translation in the forward direction. The computational guide associated with this chapter provides the basis for exploring configuration space and the link between gait and translation. Indeed, these principles have a direct bearing on the movement of *E. coli* and other microscopic swimmers. In the case of *E. coli*, the asymmetry is enabled through the handedness of the flagellar rotation. Rather than turning and reversing a flagellum in the counterclockwise and then clockwise direction, the flagellum only turns in the counterclockwise direction, leading to a net forward propulsion. (For more on fundamental principles underlying the swimming microorganisms in the limit of vanishing inertia, see Taylor 1951). Stopping involves unbundling the flagella, which leads to a near-instant halt to motion, as explained in the prior section. One can imagine other versions that leverage such principles—including macroscopic swimmers, albeit those that move in highly viscous fluids. A physical model of precisely this scenario is shown in Figure 8.12 (Plate 7), in which rubber band–propelled physical swimmers are immersed in a high-viscosity medium. Corn syrup has a viscosity a few thousand times higher than that of water, so using an object of approximately 2 centimeters implies that the Reynold's number is on the order of, if not less than, 1. These rubber band–powered swimmers are configured to move with a flapping machine such that a paddle at the end of the body moves back and forth, or via a corkscrew-like motion (left and right panels,

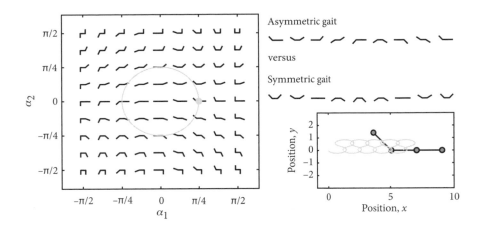

Asymmetric gait

versus

Symmetric gait

Figure 8.11: An asymmetric gait leads to forward propulsion of a 3-link swimmer. (Left) Asymmetric gait in configuration space; the sequence is recovered by following the gray trajectory counterclockwise. (Top right) Gait sequences. Forward translation (Bottom right) enabled by the asymmetric gait where the gray trajectory denotes the cumulative movement of the swimmer. The full simulation of the three-link swimmer is provided in the computational lab guide accompanying this chapter. In brief, this simulation uses a prescribed gait and couples changes in body configurations with drag forces proportional to the velocity of its limbs in order to calculate the acceleration on the swimmer. In this model, the asymmetric gait moves one limb at a time. The body configuration is updated based on changes in the velocity.

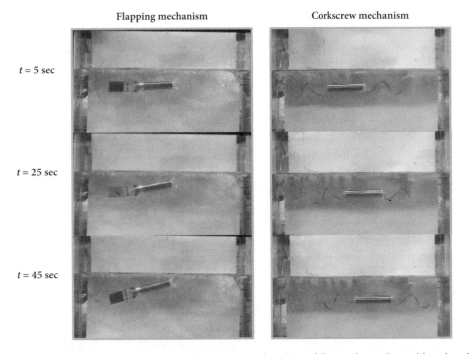

Figure 8.12: Examples of mechanical swimming at a low Reynold's number using rubber band–powered swimmers that include flapping (left) and corkscrew (right) mechanisms. The flapping swimmer does not move forward over time, whereas the corkscrew swimmer does. For videos see https://uwmadison.app.box.com/s/384lkhmyaj6jelnagl74vbbeusd83r7r. Videos and images courtesy of Michael Graham and Patrick Underhill, used with permission. Videos were taken by Patrick Underhill, Diego Saenz, and Mike Graham at the University of Wisconsin–Madison, with funding from the National Science Foundation. Swimmer designs were inspired by the work of G. I. Taylor (1951); additional context in Graham (2018).

respectively). In the case of the flapping mechanism, by flapping to the left and pushing fluid to the left, the organism would move to the right, and by flapping back to the right and pushing fluid to the right, the organism would to the left. But the flapping swimmer does not move forward, even if the same swimmer would in fact propagate forward in water while generating vortices behind its body. Instead, the corkscrew swimmer leverages the asymmetry of a chiral swimming motion to continuously drive its body forward, in much the same way that *E. coli* is able to propel forward. In corn syrup, we are able to get a glimpse of the environmental constraints driving the evolution of *E. coli*'s biomechanics.

8.4   TERRESTRIAL LOCOMOTION

Terrestrial locomotion involves more than short-term contact with a surface before liftoff. Hence, this section aims to unify the concept of dynamical templates with that of slow swimming by treating perhaps one of the most exciting frontiers in the study of animal locomotion: movement on and even through granular materials. The study of granular materials is fascinating its own right (Aranson and Tsimring 2006). Granular materials—whether sand, dirt, or immense piles of grain stored in silos—involve the interactions of large numbers of relatively small particles constrained to move due to hard body interactions with their neighbors. They can flow, like dry sand pouring out of a child's hand on the beach, or retain solid properties, like sand being formed into a castle wall with the addition of water from that same child's bucket. Unlike the dynamics of fluids, there is no equivalent set of constitutive equations, although phenomenological models are of significant use and applied relevance. Yet organisms and even robots must contend with such surfaces—they are the norm rather than the exception for moving across ground, debris, and complex terrestrial environments (Aguilar et al. 2016).

How then do organisms contend with moving on or through granular media? In brief, it appears that functional adaptations reflect particular properties of interactions with the environment. Even though there is not an equivalent of Navier-Stokes laws, the nature of hard body interactions with many small particles suggests that dynamics can be dominated by dissipative forces. As was explained, we can neglect inertia for *E. coli* swimming in water or for a flapping rubber band–powered swimmer moving in corn syrup. Likewise, an organism that tries to walk, burrow, or even swim through sand will not continue to move forward unless it continues to move its body. These body deformations lead to reaction forces from the deformable environment that enable a net translation. The accumulation of deformations represents a gait—one that can then lead to locomotion. Yet for granular media, the child's lesson that sand can be both a fluid and a solid is also a lesson underlying the evolutionary adaptation of organisms to movement through granular media. Indeed, the degree to which a fluid acts as a fluid or a solid depends on multiple factors—critically, both the packing fraction and the direction in which an object moves. Intuitively, the more highly packed a granular medium is, the more it resists motion through it. However, there is an asymmetry. Think how much easier it is to draw your hand through the top layer of sand than it is to try to compress that sand. In practice, movements through granular media that are perpendicular to the plane of motion are resisted more than those parallel to the plane of

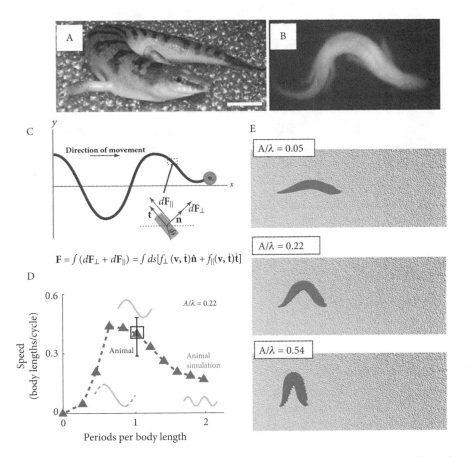

$$\mathbf{F} = \int (d\mathbf{F}_\perp + d\mathbf{F}_\parallel) = \int ds [f_\perp (\mathbf{v}, \hat{\mathbf{t}}) \hat{\mathbf{n}} + f_\parallel (\mathbf{v}, \hat{\mathbf{t}}) \hat{\mathbf{t}}]$$

Figure 8.13: Lessons from a sandfish lizard swimming through sand. The organism *S. sincus* is shown above the surface (A) and in the midst of subsurface locomotion (B) via X-ray tracking (reprinted from Maladen et al. 2011). (C) Schematic of resistive force theory; for origins, see Gray and Hancock (1955). (D) Comparison of propulsion speed in terms of body lengths per cycle versus the wave periods per body length (adapted from Maladen et al. 2011, with data from Maladen et al. 2009). (E) Discrete element method simulations of sandfish swimming with different aspect ratios for the amplitude of body oscillations relative to body length (courtesy of Daniel Goldman; images by Yang Ding).

motion (in practice, it turns out that the resistance force is twofold higher for perpendicular than for parallel motions (Zhang and Goldman 2014)).

With these preliminaries in mind, consider the motion of the sandfish lizard, *Scincus scincus*, as it moves underneath the sand's surface. This behavioral adaption has many reasons, not least of which is that it helps the organism and offspring avoid predators. Figure 8.13 (Plate 8) provides an essential compilation of key features of these dynamics. Notably, X-ray imaging and positionally associated markers have made it possible to track organismal behavior below the surface in fluidized beds in which the density of granular packing can be carefully controlled and reset (Maladen et al. 2009). In practice, the sandfish lizard self-propels by deforming its body via the propagation of a wave that traverses the body from head to tail with one oscillation per body length at any given time. Because dissipative forces dominate over inertial forces, a given infinitesimal deformation must lead to the paradoxical condition $F = 0$. This notion of self-propulsion at zero force is also central to the analysis of *E. coli* (a unification of these ideas can be understood in terms of the geometric phase; for more details, see Shapere and Wilczek 1987; Hatton and Choset 2013; Astley et al. 2020). In practice, this means that, depending on the body's configuration, the total force is the superposition of forces among body elements (while noting that there is an anisotropy between parallel and perpendicular directions; Figure 8.13C). Hence, if the organism were moving in a particular direction (forward, for example), then each body element would experience a net drag force proportional to the propulsion velocity—which itself depends on the wave speed.

The unique solution to this optimization problem can also be used to determine the optimal magnitude of wave amplitudes for self-propulsion. Figure 8.13E shows a comparison in a swimming simulation of a sandfish lizard using alternative gaits, in which A/λ denotes the ratio of the body wave magnitude to the length of the organism. As is apparent, there is a Goldilocks effect for efficient locomotion, such that too slender a wave ($A/\lambda = 0.05$) leads to inefficient propulsion as does too floppy a wave ($A/\lambda = 0.54$). The optimal ratio is $A/\lambda = 0.22$, close to the measured ratio of 0.3 for sandfish lizards—the solutions of which also recapitulate the observed propulsion speeds, as shown in Figure 8.13D. Notably, part of the propulsion of sandfish lizards through sand is realized by leveraging unusual features of granular media. Specifically, the noise of the sandfish lizard fluidizes the granular environment, unpacking it as it were, which reduces dissipative forces and allows the animal to move more efficiently through the environment.

In closing, the use of simple, physical models to explain evolutionary functional adaptations has a bearing on organisms ranging from *E. coli* (Purcell 1977) to animals, whether hopping or running (Farley et al. 1993), and even to robots (Aguilar et al. 2016). Indeed, more than three decades ago, Tad McGeer realized that the use of dynamical templates could provide an altogether different route to consider the development of robotic walkers (McGeer et al. 1990). Instead of overpowering each movement through burdensome feedback loops that connected sensors with central control, McGeer decided to let the natural dynamics of ground-air interfaces do much of the work. The result—a "passive dynamic walker"—was not altogether dissimilar to children's' toys that can be wound up and then walk on their own. The walker was approximately 1 m in height, had legs with mechanical knees and foot pads, and if given the right forward swing would manage to take multiple steps without failing or falling—in the absence of any cognitive control whatsoever: https://www.youtube.com/watch?v=WOPED7I5Lac. These ideas embody an attempt to understand the movement of organisms as passive dynamical systems, able to maintain robust motion in response to perturbations even without active, cognitive control. In essence, appropriate reductions of the complexity of extended systems can drive both biological understanding and perhaps even engineering of novel solutions to robophysical models (Collins et al. 2005). Beyond the development of ever more complex models, the use of the geometric phase to link internal changes in body configurations with net movement across a gait has the chance to unify the field and improve the engineered, built solutions that Feynman viewed as the ultimate hallmark of understanding.

8.5   TAKE-HOME MESSAGES

- Newton's laws are sufficient to describe component dynamics but do not—in and of themselves—provide an intuitive understanding of the movement of complex, extended organisms through the environment.

- Studying the movement dynamics of living systems requires linking changes in body configurations generated by internal forces to net changes in the position of an extended organism.

- Simplified dynamical templates, like a spring-mass system, can be used to explore movement at the air-land interface, including hopping and running.

- Changes in the effective stiffness of the body enable organisms to adjust their gaits even as they increase frequency or speed while keeping certain features constant (e.g., the contact time with the surface in the case of hopping).

- The dynamics of microscopic swimmers can be described via low Reynold's number flows in which dissipative forces dominate over inertial forces.

- When dissipative forces dominate over inertial forces, net translation requires leveraging asymmetries in body configuration changes, e.g., counterclockwise rotation of a flagellum or propagating waves down a body (as shown in the simplified three-link swimmer example).

- Swimming through granular media is analogous to swimming at a low Reynold's number for microscopic swimmers like *E. coli*.

- Because inertial forces are largely irrelevant, swimming through sand requires taking advantage of asymmetries so that gaits do not end up reversing back through the very path they carved.

- Studies of locomotion that leverage the power of geometric mechanics may also provide insights into the improved design and deployment of agile robots.

8.6   HOMEWORK PROBLEMS

This problem set builds upon the developing toolkit you have accumulated while leveraging principles of force-driven mechanics, following upon the templates in the computational laboratory guide associated with this chapter. The laboratory guide includes the following major concepts, while utilizing the three-link swimmer as the basis to explore the relationship between changes in body configuration and the dynamical concept of self-propulsion:

- Visualizing a configuration space
- Exploring the concept of a gait, i.e., a closed trajectory in configuration space
- Developing a dynamical system that incorporates drag-dominated force dynamics in an extended body
- Connecting transient motions to short-term dynamics
- Exploring the relationship between continuous dynamics and long-term propulsion

The overall objective of this problem set is to explore the principles by which gaits are connected to self-propulsion. The problem set also includes additional exploratory problems that help connect dynamical templates to understanding principles of simple locomotion (e.g., 1D hopping).

PROBLEM 1.   The Impact of Size on Propulsion

Using the self-propulsion code developed in the laboratory for the asymmetric gait, measure the self-propulsion speed. What is the speed in terms of body lengths advanced per cycle? Modulate the size of the object by factors of 0.5 and 2; how does the propulsion speed change with size?

PROBLEM 2.   The Impact of Speed on Gaits

Extend the self-propulsion code developed in the laboratory and evaluate the impact of slowing down or speeding up the default symmetric and asymmetric gaits by modulating the rate at which the organism goes through its gait cycle by factors of 0.5 and 2. In doing so, measure the self-propulsion speed and evaluate the extent to which changes in the speed of body deformations modulates propulsion. What is the theoretical expectation for dissipatively driven systems and what limits the applicability of the theory across scales?

PROBLEM 3.   The Greatest Gait of All

Explore and evaluate the relationship between the amplitude of asymmetric limb oscillations and the efficiency of self-propulsion. Are all oscillation amplitudes equally effective? Is there an optimal periodic gait? As a bonus, after you establish the relationship, evaluate whether or not you can improve upon the optimal periodic gait by modulating the limb motion to see if there are gait forms that outperform the optimal periodic oscillation.

Bonus Problem.   The Limits to Hopping

As a bonus, aim to recapitulate the ground-aerial dynamics of a one-dimensional hopper. In doing so, first recall that the force dynamics are continuous but not continuously differentiable (i.e., there is a shift at the point of contact). If your model is working, then you should be able to calculate a sequence of features across a ground-air cycle directly from your time series, including

- Aerial time of each hop
- Ground contact time
- Compression of the spring
- Maximum force

If your model is working, using the default parameters shown in Figure 8.3, modulate both the stiffness and the takeoff velocity to evaluate the maximum hop height—while keeping in mind the constraint that the spring should not compress more than one-half that of its length. What other constraints would limit the possible range of contact times, frequencies, and hop heights?

8.7 TECHNICAL APPENDIX

Reminders on simple matters of force As a reminder on the consequences of applying Newton's laws to macroscopic organisms, consider a canonical example related to projectile motion—calculating the horizontal flight for an object with mass m released at an absolute velocity v_0 given an initial inclination α in the absence of drag. Note that the dynamics of this system can be described in terms of trajectories in the x–y plane as follows:

$$\dot{x} = v_x \tag{8.22}$$

$$\dot{y} = v_y \tag{8.23}$$

$$\dot{v}_x = 0 \tag{8.24}$$

$$\dot{v}_y = -g \tag{8.25}$$

where $g \approx 9.8$ m/sec^2 is the force of gravity. Here the force acts strictly on the vertical direction, and given the absence of drag, the horizontal velocity $v_x = v_0 \cos \alpha$ remains constant throughout the entire flight. There are a few ways to get intuition on the horizontal distance. First, because the horizontal velocity does not change, the distance must be $d = v_0 t_f \cos \alpha$ where t_f is the time of flight (the product of velocity and time). Note that the only factors involved in setting the time of flight are the force of gravity g and the vertical velocity $v_0 \sin \alpha$. By dimensional analysis, we expect that the time for upward motion should be of the order $v_0 \sin \alpha/g$, which has units of time. This implies that the time of flight should be on the order of $2v_0 \sin \alpha/g$ and therefore that the total horizontal distance traveled should be of the order $2v_0^2 \cos \alpha \sin \alpha/g$, accounting for both upward and downward ascent periods. This scaling argument agrees in limiting cases, given that the horizontal distance will be 0 in the event of a strictly horizontal or vertical release, corresponding to $\alpha = 0$ and $\alpha = \pi/2$, respectively. The force diagram is shown below; in all cases, the force acts strictly in the vertical direction.

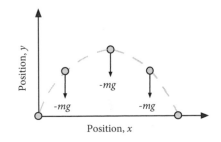

A complete solution can be obtained by explicitly solving the differential equations. First, note that $v_y(t) = v_y(0) - gt$ such that the object no longer has a vertical velocity when $t_{1/2} = v_y(0)/g$. And the object must have its maximal *negative* vertical velocity at $t_f = 2t_{1/2}$ when $v_y(t_f) = -v_y(0)$. Due to symmetry, the time at which the vertical velocity is 0 corresponds to the flight's apex. This time must also correspond to exactly half of the total time traveled, i.e., $t_{1/2} = t_f/2$. At that point, the object has traveled a horizontal distance $d_x = v_x(0) * t_{1/2}$. The total distance should be twice that, i.e., $2v_x(0)v_y(0)/g$, or precisely $2v_0^2 \cos\alpha \sin\alpha/g$, the value expected on dimensional analysis alone. Second, another feature of the trajectory—the maximal height—can be found by solving the differential equations or by leveraging principles of conservation of energy. The initial energy is $\frac{1}{2}mv_0^2$. At the point of maximal height, the vertical velocity is 0, but the total energy is a combination of kinetic and potential energies, i.e., $\frac{1}{2}mv_x(0)^2 + mgh$, where h is the maximal height. Note that conservation of energy, and recall that the initial kinetic energy, is $\frac{1}{2}mv_0^2$. At its highest point, the total energy must be equal to

$$\frac{1}{2}mv_0^2 = \frac{1}{2}mv_x(0)^2 + mgh \tag{8.26}$$

such that

$$h = \frac{\frac{1}{2}v_0^2 - \frac{1}{2}v_x(0)^2}{g} \tag{8.27}$$

$$h = \frac{\frac{1}{2}v_0^2}{g}\left(1 - (\cos\alpha)^2\right) \tag{8.28}$$

This value also has the correct limits; when $\alpha = 0$ then the maximal height is 0, and when $\alpha = \pi/2$ then all of the kinetic energy is translated into potential energy at the flight's apex, i.e., $h = \frac{\frac{1}{2}v_0^2}{g}$, although there is no horizontal displacement.

If one were to try to "throw" an object as small as *E. coli* at the speed at which *E. coli* itself moves, then it would not get very far. The expected maximal horizontal distance would be $2v_0^2 \cos\alpha \sin\alpha/g$, and assuming $\alpha = \pi/4$, then this is $v_0^2/(2g)$, which would be $\frac{30^2}{(9.8\times10^6)}\,\mu$m or $\approx 10^{-4}\mu$m. It may be tempting to try—but of course for objects of this size, neglecting drag is problematic. Indeed, the actual stopping distance is even smaller, as explained in the chapter.

Drag and terminal velocity The preceding discussion considered dynamics of projectiles subject to gravity but in the absence of other forces. In reality, drag is a factor even for objects moving in air. Drag (or what is commonly termed *wind resistance*) opposes motion. For an object in flight, it has the opposite sign of the current velocity, so that the vertical component of the drag force is downward in the upward component of the flight and upward in the downward component of the flight. Throughout the flight the drag force opposes the forward motion, as in the force diagram below.

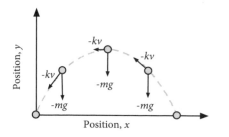

For projectile motion, the dynamics of this system can be described in terms of trajectories in the $x-y$ plane as follows:

$$\dot{x} = v_x \tag{8.29}$$

$$\dot{y} = v_y \tag{8.30}$$

$$m\dot{v}_x = -kv_x \tag{8.31}$$

$$m\dot{v}_y = -mg - kv_y \tag{8.32}$$

But, for the purposes of identifying how drag modifies dynamical outcomes, consider the contrasting dynamics of the vertical velocity for an object dropped from a sufficiently high level:

$$\dot{y} = v_y \tag{8.33}$$

$$\dot{v}_y = -g - kv_y/m \tag{8.34}$$

This dynamical system can be solved explicitly, noting that

$$\frac{dv_y}{g + kv_y/m} = -dt, \tag{8.35}$$

which after integrating by parts yields

$$\ln\left(g + kv_y/m\right) = -kt/m + \text{const} \tag{8.36}$$

and then

$$v_y = Ce^{-kt/m} - g/k \tag{8.37}$$

where the value C can be determined by noting that $v_y = 0$ when $t = 0$ such that the velocity follows

$$v_y = \frac{g}{k}\left[e^{-kt/m} - 1\right]. \tag{8.38}$$

This provides two insights. First, in the presence of drag, the velocity does not increase unabated but rather reaches a terminal velocity of $v_c = g/k$ (in the downward direction) over a characteristic time $t_c = m/k$. Hence, drag decreases the maximum velocity and also shortens the period over which the velocity increases. The generic relationship between scaled velocity and time is shown in Figure 8.14.

As an aside, in the case of an object falling due to gravity, the terminal velocity will be nearly reached as long as the object is dropped from a sufficiently high initial position y_0 that the time to reach the ground $t_f \gg t_c$. The final time corresponds to solving

$$\frac{dy}{dt} = v_c\left[e^{-t/t_c} - 1\right] \tag{8.39}$$

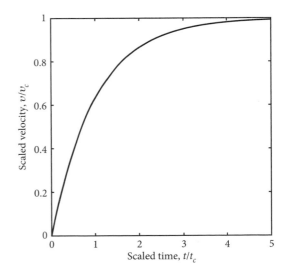

Figure 8.14: Scaled velocity as a function of scaled time for an object accelerating due to a constant force (e.g., gravity) and subject to a drag force proportional to velocity. In this example, the constant force is g (kg m/sec^2), the drag constant is k (kg/sec), and the mass is m (kg). The characteristic time is $t_c = m/k$ and the terminal velocity is $v_c = g/k$.

and finding the time t_f when $y(t_f) = 0$. One can already see that this distance should be greater than $v_c t_c$, or $y_0 > \frac{mg}{k^2}$ of $v_c = g/k$ (in the downward direction) over a characteristic time $t_c = m/k$. Integrating both sides yields

$$y(t_f) = \text{const} + v_c \int_0^{t_f} \left[e^{-t/t_c} - 1 \right] \tag{8.40}$$

$$y(t_f) = \text{const} - v_c t_c \left[e^{-t/t_c} \right]_0^{t_f} - v_c t_f \tag{8.41}$$

$$y(t_f) = \text{const} - v_c t_c \left[e^{-t_f/t_c} - 1 \right] - v_c t_f \tag{8.42}$$

and noting that $y(t = 0) = y_0$ implies that

$$y(t_f) = y_0 - v_c t_c \left[e^{-t_f/t_c} - 1 \right] - v_c t_f. \tag{8.43}$$

Now note that the final time must satisfy $y(t_f) = 0$:

$$y_0 = v_c t_c \left[e^{-t_f/t_c} - 1 \right] + v_c t_f. \tag{8.44}$$

Coasting distance Consider an object (or organism) moving at a speed v experiencing a drag force proportional to its speed such that the equations of motion are

$$\frac{dx}{dt} = v \tag{8.45}$$

$$m \frac{dv}{dt} = -kv \tag{8.46}$$

where k is the drag constant and m is the mass. How long will the organism coast?

To answer this question, first note that the velocity can be solved explicitly by integrating both sides of the force equation

$$\log v(t) = C - \frac{k}{m} t \tag{8.47}$$

where C is an integration constant. In the event that the organism has an initial velocity v_0, then

$$v(t) = v_0 e^{-t/\tau_c} \tag{8.48}$$

where $\tau_c = m/k$ is a characteristic coasting time. The position can also be solved, noting that

$$\frac{dx}{dt} = v_0 e^{-t/\tau_c}, \tag{8.49}$$

which can be integrated to yield

$$x(t) = v_0 \tau_c \left(1 - e^{-t/\tau_c} \right). \tag{8.50}$$

Hence, the total distance traveled is $v_0 m/k$, such that the coasting distance increases with initial velocity and mass and decreases with the drag force.

Rescaling Navier-Stokes equations The Navier-Stokes equations are

$$\overbrace{-\nabla p + \eta \nabla^2 \mathbf{u}}^{\text{pressure and viscous forces}} \quad = \quad \overbrace{\rho}^{\text{density}} \times \left(\frac{\partial}{\partial t} + \mathbf{u} \cdot \nabla \right) \mathbf{u} \qquad (8.51)$$

where p is pressure, $\mathbf{u}$ is velocity, ρ is density, v is kinematic viscosity, and $\eta = v/\rho$ is dynamic viscosity. This system of equations can be nondimensionalized by rescaling in terms of a characteristic velocity v, length scale a, and time scale a/v. The dimensionless variables are $\tilde{\mathbf{u}} = \mathbf{u}/v$, $\tilde{\mathbf{x}} = \mathbf{x}/a$, and $\tau = t/(a/v)$. Rescaling the Navier-Stokes equations yields

$$-\frac{1}{a\eta} \tilde{\nabla} p + \frac{v}{a^2} \tilde{\nabla}^2 \tilde{\mathbf{u}} = \frac{\rho v^2}{a\eta} \left(\frac{\partial}{\partial \tau} + \tilde{\mathbf{u}} \cdot \tilde{\nabla} \right) \tilde{\mathbf{u}}, \qquad (8.52)$$

which can be reduced to the following by multiplying all terms by a^2 and dividing them each by v:

$$-\frac{a}{v\eta} \tilde{\nabla} p + \tilde{\nabla}^2 \tilde{\mathbf{u}} = \frac{\rho v a}{\eta} \left(\frac{\partial}{\partial \tau} + \tilde{\mathbf{u}} \cdot \tilde{\nabla} \right) \tilde{\mathbf{u}}. \qquad (8.53)$$

Replacing $\tilde{p} = p/(v\eta/a)$ yields

$$\tilde{\nabla} \tilde{p} + \tilde{\nabla}^2 \tilde{u} = \mathrm{Re} \left(\frac{\partial}{\partial \tau} + \tilde{\mathbf{u}} \cdot \tilde{\nabla} \right) \tilde{u} \qquad (8.54)$$

where the dimensionless Reynold's number is

$$\mathrm{Re} = \frac{\rho v a}{\eta}. \qquad (8.55)$$

The interpretation of the Reynold's number is found in the chapter text.

Part III

Populations and Ecological Communities

Chapter Nine

Flocking and Collective Behavior: When Many Become One

9.1   LIFE IS WITH OTHER ORGANISMS

Collective behaviors are replete in natural systems, from starling flocks to bee swarms to ant trails. The term *collective* implies that the behavior is exhibited by more than one organism—indeed, usually many more than one. But the term *collective* requires more than simply the presence of many organisms doing something, somewhere. If it were not so, then any group of microbes, insects, birds, fish, or mammals (including humans for that matter) would inevitably exhibit collective behavior. To distinguish these ideas, it is worthwhile to consider two alternative concepts: (i) a fast herd of deer; (ii) a herd of fast deer. These descriptions may appear identical. Yet the subtlety in the placing of the adjective implies two alternative interpretations:

Fast herd of deer: The properties of the group are different than the individual properties.

Herd of fast deer: The properties of the group reflect the individual properties.

Given these definitions, the notion of "fast" modifies the herd, and one can see this notion conveyed in everyday parlance. Perhaps the most telling example is that of the wisdom of the crowd, i.e., the idea that groups can have more accurate decision-making abilities than do individuals. Yet collective behaviors need not always have a positive, normative benefit, e.g., the fallacy of the crowd is also possible. The ongoing spread of misinformation and conspiracy theories (in humans) is but one indication that crowds do not always make prudent choices (Kouzy et al. 2020). Irrespective of the normative value, the emergence of collective behaviors implies that no one individual need be particularly fast, wise, agile, prudent (or perhaps, slow, wise, or imprudent) for a group to have such properties. But how can new properties emerge among groups that are not necessarily contained in the individuals who comprise them?

The answer to this question takes some time to unravel in practice. As we will find, the transmission of local information between individuals can propagate to the group scale, enabling the emergence of novel patterns and dynamics. This concept is evident in the behavior of bird flocks, as well as in bees, fish, and humans. Figure 9.1 (Plate 9) presents a gallery of collective behavior spanning wingless locusts, army ants, golden rays, fish, starling

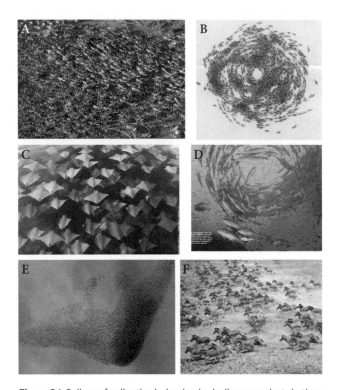

Figure 9.1: Gallery of collective behavior, including snapshots in time of (A) wingless locusts, (B) army ants, (C) golden rays, (D) fish, (E) starling flocks, and (F) zebra herds. Reproduced from Vicsek and Zafeiris (2012).

flocks, and zebra herds. These images are snapshots. Yet the snapshots have dynamic implications. Individual organisms moving in close proximity can exhibit collective motions, patterns, and order that seem to be not just a scaling up of the individual property but a new property altogether. Emergence is precisely that: the formation of collective behaviors that individuals do not possess on their own. This difference between individual and collective behavior is perhaps most evident in the case of bird flocks—with a salient example taken from flocks of starlings.

Flocks of the European starling are perhaps the "model" collective. Starlings can form coherent groups involving hundreds to thousands of organisms. Midair, the flocks seem to twist, wind, unravel, pull themselves into knots, and then untie themselves. Even after breaking apart, multiple flocks can fly close and then re-form a larger flock seamlessly. The flocks seem alive—in the sense that the flock, rather than the bird, appears to have its own dynamics. The static still image shown in Figure 9.2 (Plate 10) provides a sense of this coherence. In this example, a flock of over 1200 birds is moving nearly uniformly (in the frame). Yet the distances that span the flock, on the order of 36 m, exceed the potential range over which each individual bird is sensing, communicating, or responding directly to others. This long-range coherence implies an ability for this group to exhibit novel behavior not encoded in the local interaction rules or processes. But it is equally important to point out that the flock is not absolutely coherent (Figure 9.3A). Instead, when comparing an individual bird's velocity to its local neighbors, it is possible to identify domains of fluctuations. The coherent orientation domains in Figure 9.3B imply that fluctuations are correlated over distances significantly greater than the interaction range. Starlings typically interact directly with a handful of nearby birds (estimates are on the order of half a dozen birds). In contrast, the fluctuation domains are coherent over hundreds of birds. This gap and evidence of long-range order represents the key challenge: how do individual interactions lead to the emergence of new properties at the system scale?

The emergence of new properties at different scales not contained in the components is a hallmark of complex adaptive systems (Mézard et al. 1987; Parisi 1988; Gell-Mann 1994b) and was anticipated through the concept "more is different" (Anderson 1972). A complex adaptive system—sensu John Holland (1992), Murray Gell-Mann (1994a) and Simon Levin (2003)—is characterized by individual components and/or agents, each with localized interactions, in which an autonomous process selects a subset for replication. This definition of complex adaptive systems requires some notion of replication. However, the scale of replication is usually that of the individual agent and not the group itself—i.e., herds do not beget herds, but rather animals beget animals, which may have properties that tend to promote collective behavior at the group scale. It is this difference that underscores potential evolutionary misconceptions of a "fast herd"—of particular relevance to understanding multilevel selection (see Traulsen and Nowak (2006)).

This chapter aims to explore and explain how properties at the population scale emerge from local interactions between individuals. A central focus is on understanding the emergence, maintenance, and possible dissolution of dynamic flocks—like those of the European starling. But, to get there, this chapter first reviews mechanisms that can generate spatial patterning in ecological model systems—the rocky intertidal zone and semiarid grasslands. These two systems include a combination of gradients and feedback that determine which individuals predominate. Following these biological examples, the chapter introduces a more abstract model of spatial localization—the zero-range process model—that serves as a bridge to understanding the basis for

Figure 9.2: Flock of European starlings and emergent order. A snapshot of $N = 1246$ birds from a flocking event, which has a linear effective size of 36.5 m. Reproduced from Bialek et al. (2012).

dynamical order in moving flocks. Doing so also helps reveal what is different about having order in space, i.e., where organisms are, versus having order in dynamics, i.e., where organisms move. The combination of spatial and kinetic order has a wide range of applications, not only to flocks of starlings but to migratory herds, locusts, schools of fish, and even the swarming behavior of microbes.

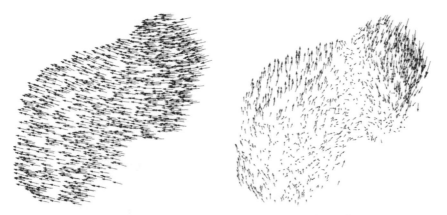

Figure 9.3: Velocity and velocity fluctuations in flocks of European starlings. The snapshot is taken from a flocking event with a linear effective size of 36.5 m. (Left) 2D projection of normalized velocities of $N = 1246$ birds, with arrow heads denoting direction of motion. (Right) 2D projection of the velocity fluctuations relative to the overall center of mass velocity of the flock. Reproduced from Cavagna et al. (2010).

9.2 ENDOGENOUS VERSUS EXOGENOUS DRIVERS OF SPATIAL ORDERING

9.2.1 The rocky intertidal: A motivating example

The start of this book introduced the canonical example of oscillations among fish populations in the Adriatic Sea. This observation, by Umberto d'Ancona, was the inspiration for Vito Volterra's proposal of feedback mechanisms that could lead to sustained oscillations in population abundances of both a predatory and a prey fish even without the inclusion of any exogenous variation. The study of predator-prey dynamics, as explored by both Volterra and Lotka and countless others since, has had an outsized influence in the study of biological dynamics generally. The idea is that oscillations of molecules, cell numbers, and populations can arise even in situations when the environment and associated environmental drivers are homogeneous and constant in time. The reason is that endogeneous, nonlinear feedback can lead to persistently oscillatory dynamics in populations. Here, the terms *endogenous* and *exogeneous* refer to whether the perturbation is internal to or external to the system, respectively. Given the balance between endogenous drivers and exogeneous drivers in shaping total population dynamics, it is worthwhile to ask a related question of spatially distributed populations: are exogeneous drivers required to impose spatial patterning on where organisms persist at any give moment in time? Or, instead, are there potential internal feedbacks between organisms that can lead to clustering or localization in certain parts of a habitat?

Environmental drivers can strongly shape where and when organisms can survive. These environmental drivers may be abiotic factors, e.g., temperature, light, or salinity. The relationship between abiotic drivers and species distributions is most apparent when crossing a gradient. For example, the density and composition of plants, shrubs, and trees changes markedly when traversing upward from base to summit. In the Appalachian Mountains of the eastern United States, there is a marked transition from hardwood-dominated forests (e.g., beech, oak, cherry) with bushy shrub undergrowth at lower elevations to spruce-fir dominated forests with moss dominated undergrowth at elevations exceeding 1000 m. In the western North American Rockies, the timberline (or tree line) demarcates the elevation above which trees cannot grow. Hence, if exogenous drivers were the only factor, then one could envision a process by which the environment "selects" those species whose habitat range coincides with the particular combination of abiotic factors present at a particular site.

However, what happens when the habitats of species are not mutually exclusive? Figure 9.4 illustrates this point by extending the ranges of each species so that there is a region of overlap. The gradient in this case is meant to be illustrative but raises important points. If many species can survive in one site, does that mean both will, or if only one is present, does that necessarily mean that no other species could have colonized and persisted at that site? If the outcomes depend on interactions, there must necessarily be a tension between exogenous and endogenous drivers. This tension underlies one of the most well-studied and influential of ecological examples of spatial patterning: that of the rocky intertidal zone.

Rocky intertidal zones are of interest for many reasons, not least of which is that there are multiple types of gradients, including wave energy, disturbances, and duration in which the surface sediment is exposed to air rather than submerged. Approaching the sea from land, one encounters a banded pattern with mussels, followed by starfish and then algae and other seaweeds (Figure 9.5 [Plate 11]). It could be that these banded patterns reflect exogenous factors, but the reality is more complicated. The exogenous explanation seems plausible, in that mussel attachment is facilitated by the presence of rock and not by sand alone. However, the banded nature of starfish is harder to explain.

In a foundational series of studies, Robert Paine set out to test whether exogenous or endogenous drivers were primarily responsible for the patterning of distinct species in a rocky intertidal ecosystem off the coast of Washington adjacent to Tatoosh Island (Paine 1974; Paine and Levin 1981). Paine did so by manipulating the system—literally yanking the starfish up and out from their established nooks at the water's edge and throwing them far out to sea, far enough that reestablishment was unlikely (Yong 2013). This species removal raised interesting questions: would the algae and mussels remain where they were or, instead, would, they take over habitats that had

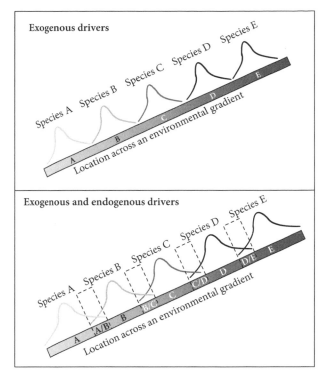

Figure 9.4: Zonation theory given environmental gradients, as inspired by the rocky intertidal. (Top) An example of gradients in environmental conditions such that distinct species only persist in specific zones in the environment. (Bottom) Alternatively, there may be overlap between species ranges given an environmental gradient such that feedback between organisms determines species ranges. In the case of the rocky intertidal, these interactions include feedback between mussels, starfish, and kelp (for additional context, see Paine 1974; Paine and Levin 1981).

Figure 9.5: Banded patterns of mussels (*Mytilus californianus*), starfish (*Pisaster ochraceus*), and seaweed/ kelp in the rocky intertidal. Photo by Dave Cowles, Goodman Creek, WA, July 2002, used with permission.

previously been occupied by starfish? As it turns out, something quite different happened. First, mussels did establish themselves in areas where the starfish had been. Yet, by reestablishing an expanded region of mussel beds, this reconfigured habitat was then colonized by a large diversity of other organisms. Hence, these newly available sites along the seashore were not fundamentally inhospitable to other species; rather, the combined effects of starfish predation and the absence of facilitation by mussel beds had made it impossible for other species to establish. As a result, it seemed that the presence or absence of mussels had a determining effect on the viability of other species. The story of how mussels were recognized as a keystone species is exceptional in its own right, and also transformed the way that ecologists view interactions in complex communities.

This early work helped establish the rocky intertidal as a model ecosystem for studying the combined effects of disturbance and of ecology, of keystone species and feedbacks, and of the emergence of patterns arising from interactions. Experimental work on algal zonation in the New England rocky intertidal, in particular by Jane Lubchenco (1980), generalized these findings. In essence, both abiotic and ecological feedbacks are essential to constraining species ranges at the broad scale, as well as spatial patterns at the small scale. Those areas most difficult for species survival may be set by abiotic factors. However, those areas ostensibly hospitable for the survival of multiple species may prove challenging due to competition, predation, parasitism, mutualism, and other ecological feedback. This general finding should also force us to confront a problem: when we see organism clusters in a particular location, is there something special about that location or, instead, is that "good" location a result of stochastically driven feedbacks that could have led to aggregation at another site? As we explore in the next example, the answer is often that internal feedbacks can give rise to what appears to be prioritized locations for organismal clustering.

9.2.2   Vegetation stripes: Making one's own luck

Spatial patterning and aggregation of organisms occur at many scales, from biofilms among microbes, to stalk formation within *Dictyostelium discoideum*, to labyrinthine patterns of mussels, to landscape scales in the case of vegetation stripes in semiarid habitats. Indeed, the emergence of regular vegetation stripes provides an ideal opportunity to illustrate how aggregation need not be indicative of an exogenous benefit associated with a particular location. Instead, vegetation stripes reflect feedback between plants and soil that enables vegetation to make its own luck.

The key observation from semiarid landscapes is that vegetation is not uniformly distributed (see Figure 9.6A for an example). Instead, given semisloped terrain and limited rainfall, the vegetation on a landscape can form regular patterns, including linear groves of acacia (in Australia), tiger bush (in Niger and Mauritius), and clusters of regularly spaced sage brush (in Utah) (Rietkerk et al. 2004). A characteristic in each is that these patterns emerge in semiarid or arid zones (e.g., typically no more than dozens of centimeters of rainfall per year). In addition, surface soil characteristics are often common; for example, compacted or sealed soil at the surface implies that deep penetration of moisture is unlikely, and instead, rainfall flows overland. This combination of factors also provides a clue for the emergence of inhomogeneity as vegetation and rainfall interact.

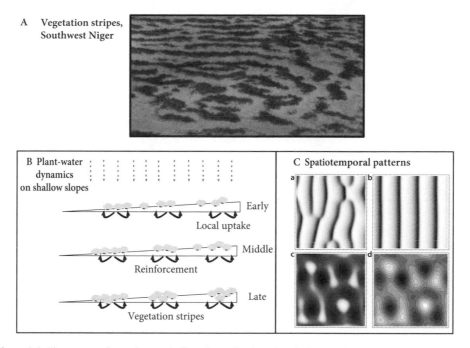

Figure 9.6: Plant-water dynamics on shallow slopes leads to banded vegetation patterns in simple, nonlinear PDE models with local facilitation. (A) Tiger bush vegetation stripes that can span many kilometers. (B) Schematic of how water-vegetation feedback leads to clustering. (C) Demonstration of the results of numerical simulation of the underlying PDE. Panels A and C reproduced from Klausmeier (1999).

Building upon these field-based observations, consider an environment with homogeneous rainfall on a sloped surface, involving a few simple rules (Figure 9.6B). First, vegetation takes up water, leading to less water and more vegetation. Next, vegetation "diffuses" locally as a consequence of growth and dispersal. Finally, water flows downhill. These rules form the basis for a partial differential equation (PDE) model of water-plant feedback on shallow landscape slopes (Klausmeier 1999):

$$\frac{\partial W(x,t)}{\partial t} = a - lW(x,t) - rW(x,t)N(x,t)^2 + v\frac{\partial W(x,t)}{\partial x} \quad (9.1)$$

$$\frac{\partial N(x,t)}{\partial t} = rjW(x,t)N(x,t)^2 - mN(x,t) + D\left(\frac{\partial^2}{\partial x^2} + \frac{\partial^2}{\partial y^2}\right)N(x,t) \quad (9.2)$$

where W denotes water concentration and N denotes vegetation. In this model, a denotes the uniform rainfall rate, l is the evaporation rate, v denotes the rate of downslope water movement, r is a coupling constant, j is an efficiency of water use, m is a basal plant loss term, and D is an effective plant diffusion term. Critically, the model assumes that there is no extrinsic variation in rainfall across the landscape and that all of the relevant kinetic rates are similarly homogeneous. Further, the x direction is oriented downhill, such that water flows downslope while vegetation diffuses locally via growth. Of note, the nonlinear feedback between water and plants reflects the facilitation of efficient water capture in semiarid environments by local root development that increases with plant biomass. This nonlinear feedback has dramatic consequences.

First, even in the absence of diffusion of vegetation matter and the overland flow of water, the model has the potential to exhibit alternative stable states. The technical appendix provides details on the distinct types of possible steady states—including a bare vegetation state ($W^* = a/l, N^* = 0$). Critically, the leading order of the growth term for vegetation is N^2, implying that small levels of vegetation do not increase in abundance. Instead, it takes more than $N_0 > \frac{m}{rja/l}$ for the density-dependent growth to exceed the density-independent mortality. Hence, given a sufficiently large perturbation, the system can move toward a vegetated state where both $W^* > 0$ and $N^* > 0$. In the absence of spatial coupling, each local site is distinct. Yet the underlying dynamical system suggests that alternative states are possible. Given sufficient density, the vegetation can grow, drawing down water levels. In a coupled system, those lower water levels, if flowing overland, may introduce a *shadow effect*—that is, local depletion of water by vegetation makes it harder for other vegetation to grow adjacent to an established patch. In contrast, when vegetation levels are low, then water accumulates closer to its baseline rate set by the exogenous rainfall, i.e., a/l, and may eventually provide enough water for the growth of new vegetation to take hold.

Indeed, local feedback between water and plants leads to marked divergences in a fully spatial model, even in an otherwise uniform landscape. To see how, consider a landscape with an initially homogeneous but random distribution of plants and water. Those regions where the local densities of plants are slightly elevated will increase their local water uptake compared to surrounding locations. As a consequence, less water will flow downhill (in the x direction of the model), leading to a shadow effect. Less water will further disfavor plant growth. The small in homogeneity in plant biomass will increase leading, to bare and occupied patches. Given the asymmetry imposed by the water flow, the coupled water-vegetation dynamics can generate banded stripes whose spacing reflects a combination of the kinetic constants (Figure 9.6C). These stripes are a steady state of the system—yet they also break its symmetry. If one were to run the same simulation again albeit with slightly different random configurations, then the local densities of plant biomass could lead to stripes in entirely different locations. Hence, the nonlinear feedback in the system leads to systemic differences in densities even if the expectation is that all sites are equivalent, given that each site receives the same light and water input on average. The robust feature of the vegetation stripes in semiarid landscapes is the characteristic distance between bands and not the location of the bands. This finding suggests that spatial localization can be driven by endogenous feedback even in the absence of exogenous differences.

9.2.3   Zero-range processes—a gateway to spontaneous symmetry breaking

As has been shown in this chapter, internal feedback can generate differences in the spatial distribution of populations, even in the absence of exogenous differences that might favor one location over another. Yet the prior examples include self-reinforcing mechanisms between a focal population and the environment. This raises questions on the extent to which microscopic details matter in driving the emergence of order. It seems useful to have an even simpler example of how strictly local interactions can lead to the emergence of coherence in the distribution of individuals in a population. The aim is to capture the essence of *spontaneous symmetry breaking* in a toy model of condensation, i.e., when many, if not nearly all, of the organisms in a population end up aggregated in a local site. As will be explained, which site becomes the gathering point is completely random given that all sites are in principle equally suitable for the population. This "broken symmetry" is also the hallmark of more realistic models of flocking that serve as the motivating series of examples in this chapter.

To begin, consider a population of size N where individuals can move between two connected sites. There are n_A individuals at site A and n_B individuals at site B. Because $n_A + n_B = N$, the system can be defined in terms of a single value: $n \equiv n_A$, the number of individuals at site A. The only rule is that the individual rate of movement between sites is a function of the local density: $u(n)$. If each individual's movement is independent of local context, then $u(n) = r$ and the long-term equilibrium must have the properties of a binomial distribution, which for large N leads to an ever-decreasing coefficient of variation and a nearly equivalent occupation at each site. But it is also possible that an individual's rate of movement decreases with density, such that $du(n)/dn < 0$. This decreasing movement rate could be a result of direct or indirect interactions between individuals, e.g., due to an increasing tendency to colocate next to other individuals, communication, or other forms of aggregation. Whatever the mechanism, the key feature is that the interactions are strictly local. The movement rate of individuals out of site A only depends on n_A, likewise the movement rate of individuals out of site B only depends on n_B. This class of models, known as *zero-range process models*, was originally introduced more than 50 years ago (Spitzer 1970) and is extensively used, sometimes implicitly, as a representative model of aggregation and condensation (Evans 2000; Großkinsky et al. 2003; Evans and Hanney 2005), including in biological transport models (Chou et al. 2011).

As an example of the potential dynamics, consider a generalized form $u(n) = rn^{1-\epsilon}$ where $0 \leq \epsilon$. Neglecting fluctuations, the rate equation for the population at site A should follow:

$$\dot{n}_1 = -rn_1^{1-\epsilon} + rn_2^{1-\epsilon}, \tag{9.3}$$

which can be simplified to

$$\dot{n} = -rn^{1-\epsilon} + r(N-n)^{1-\epsilon}, \tag{9.4}$$

recalling that $n_2 = N - n_1$ and $n \equiv n_1$. Given that the population size is fixed, it is possible to rewrite the dynamics in terms of the fraction of the population in site A, i.e., $x = n/N$, such that

$$\dot{x} = rN^{-\epsilon} \left[(1-x)^{1-\epsilon} - x^{1-\epsilon} \right]. \tag{9.5}$$

Irrespective of the value of ϵ, each of these rate equations is symmetric about $x = 1/2$ such that all have a fixed point at $x = 1/2$. But the stability of the fixed points depends on ϵ. To illustrate this point, consider three alternative exponents: $\epsilon = 0$, $\epsilon = 0.5$, and $\epsilon = 2$. As is apparent from the curves in Figure 9.7's top panels, the stability of the fixed point undergoes a change as a function of the inhibitory movement of exponent ϵ. For $\epsilon = 0$, each individual has the same rate of movement as any other, irrespective of group size. For $0 < \epsilon < 1$, the rate of a single individual's move is larger in large populations than in small populations. In contrast, for $\epsilon > 1$, individuals in smaller populations are more likely to move than individuals in larger populations. This effect intensifies with population size. As a result, once a population is large, individuals tend to stick there, a tendency that increases as more and

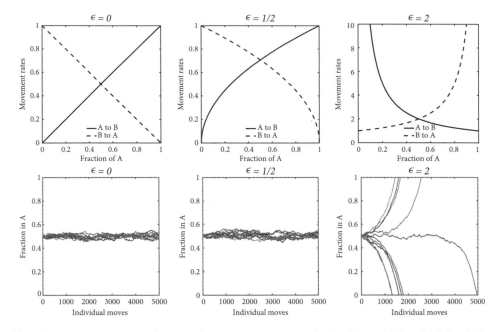

Figure 9.7: Zero-range process dynamics for a two-state metapopulation with $\epsilon = 0$, 1/2, and 2 (left, middle, and right columns, respectively). (Top row) Movement rates as a function of the fraction of the population x in site A. (Bottom row) Trajectories of a fully stochastic model. Each of the ten trajectories begins with $x = 0.5$ using a population of size $N = 1000$.

more individuals move there from the other site. This eventually leads to a critical outcome: the aggregation of (nearly) all individuals at a single site.

A fully stochastic implementation of this model is shown in the lower panels of Figure 9.7. As in the top panels, the two-state model is evaluated using different hopping rate exponents, ϵ. Insofar as $\epsilon < 1$, then an individual from a larger population is more likely to move out of its site than is an individual from a smaller population. In those cases, movement favors an equilibration of individuals that is equal between sites. But, when $\epsilon > 1$, then the dynamics include an initially stochastic period given that the populations are nearly equivalent. Once a sufficiently large fluctuation emerges, then individuals in the site with the larger population size have a decreasing movement rate, leading to increasing aggregation at that site and eventually a condensation such that all individuals are at a single site (see the technical appendix for the Master Equation of the system and additional evidence of aggregation in the stochastic model). This aggregation breaks the symmetry in the system, and as is apparent, the choice of which site "wins" is determined by chance.

The zero-range process model can be extended to multiple sites and dimensions. Here we consider a metapopulation framework, in which sites are connected along a one-dimensional lattice (or ring). Each site has n_i individuals such that $N = \sum_{i=1}^{L} n_i$. Alternative formulations of the zero-range process are possible; here we consider the movement of individuals in both directions so that when an individual leaves site i they have an equal chance of moving to site $i - 1$ or $i + 1$. We consider periodic boundary conditions such that site 1 is connected to site L. The snapshots of the zero-range process in the case of $N = 1000$

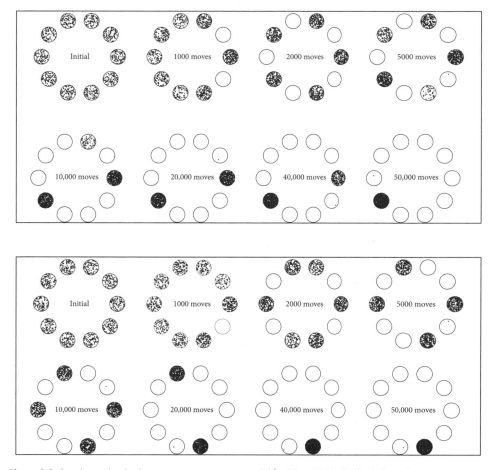

Figure 9.8: Condensation in the zero-range process model for $N = 1000$ individuals, hopping rate exponent $\epsilon = 2$, and $L = 10$ sites distributed on a ring with no preferred direction for movement. The snapshots are taken between 0 and 50,000 movements in the model, as noted in the center of each ring. The placement of individuals in a patch is randomized for visualization purposes; all individuals in the same patch have an equal effect exclusively on the local patch only (i.e., they have zero range). The top and bottom panels show two different instances of zero-range processes, each leading to the distinct emergence of an aggregated site.

and $L = 10$ are shown in Figure 9.8. In both instances shown in the top and bottom panels, a homogeneously distributed population begins to exhibit fluctuations that are then reinforced by the tendency for individuals in larger populations to remain fixed and individuals in smaller populations to move. Over time, a subset of sites becomes dominant, with individual sites becoming depleted until, eventually, a pair of sites competes, leaving only one aggregated site. The two instances also reveal the broken symmetry in the dynamics given that there is nothing special or unique about any of the L sites, yet ultimately one is selected and becomes the stable, dominant site. Although highly simplified, this zero-range process model provides insights into how local (indeed, zero-range) interactions can nonetheless lead to a long-range impact in the system, ultimately leading to population aggregation.

The next section addresses how to take this principle and extend it beyond static order to dynamic order as well.

9.3   VICSEK MODEL: UNITING STATIC AND DYNAMIC ORDER

9.3.1   Local rules and emergent outcomes

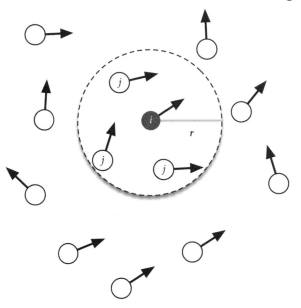

Figure 9.9: Schematic of reorienting particle motion in the self-driven particle model (Vicsek et al. 1995). The focal particle i is shown in black. Particles within a sensing radius r (dashed circle) are noted with the letter j.

In 1995, Tamás Vicsek and colleagues introduced a deceptively simple idea (Vicsek et al. 1995). The idea was that a minimal set of local interaction rules could give rise to a combination of spatial and temporal order. The prior sections of this chapter have already introduced examples of spatial aggregation and clustering. These examples have inspired a proliferation of generative mechanisms that could drive clustering. And even these examples are a small subset of the potential mechanisms that can lead to spatial order. However, understanding the emergence of joint clustering of localization and movement had defied simple, mechanistic explanations. The key challenge was to reconcile the local sensing and response behaviors of individual organisms with the manifestly long-range order inherent in the collective behavior of many organisms (refer to the gallery in Figure 9.1 for examples). Reconciling individual and collective behavior was also a key challenge of the then burgeoning field of complex adaptive systems.

Vicsek and colleagues proposed a model of "self-driven particles" to address the potential role of individual, local behavior in shaping large-scale collective motion. As presented, the model only has one rule (Vicsek et al. 1995):

> At each time step a given particle driven with a constant absolute velocity assumes the average direction of motion of the particles in its neighborhood of radius r with some random perturbation added.

Figure 9.9 provides a visual illustration of the basis for this update rule. In practice, there are certain subtleties involved in moving from that sentence to the generative mechanisms of self-driven particles, which is now known as the standard Vicsek model, or SVM for short.

The original SVM describes a system of N particles in a two-dimensional periodic domain of size $L \times L$. The location and velocity of each particle is denoted by $\vec{x}_i(t)$ and $\vec{v}_i(t)$, respectively. The magnitude of the velocity, v_0, remains fixed but the angle, θ_i, can change. The change is determined as follows. As a result of sensing the velocities of its neighbors, each particle reorients itself over a small time interval,

$$\vec{v}_i(t+1) = v_0 \frac{\langle \vec{v}_j(t) \rangle}{|\langle \vec{v}_j(t) \rangle|} + \text{perturbation},$$ (9.6)

and then advances in the new direction using a ballistic flight model:

$$\vec{x}_i(t+1) = \vec{x}_i(t) + \vec{v}_i(t+1).$$ (9.7)

It is possible to view these rules through the lens of updates to the current angle $0 \le \theta_i \le 2\pi$ in which a given particle i is moving, such that the reorientation rule can be written as

$$\theta_i(t+1) = \arctan\left[\frac{\langle v_{j,x}(t) \rangle}{\langle v_{j,y}(t) \rangle} \right] + \Delta_i(t)$$ (9.8)

where the x and y in the above equation denote the planar x and y components of the velocity vector. In either formulation, the average velocity direction of neighbors is taken over all particles j whose Euclidean distance to the focal particle i is less than or equal to r, i.e., $\|\vec{x}_i(t) - |\vec{x}_j(t)\|_2 \le r$ (Figure 9.9). The computational labguide shows how to resolve the ambiguity in the resulting angle based on the sign of the input velocity vectors. Critically, the set of particles that influences the focal particles always includes itself. Hence, the influence of the focal particle on itself is modulated by other particles in the focal neighborhood. In the SVM, individual particles track the aggregate direction of their neighbors the more crowded the local environment becomes. Finally, the noise term $\Delta_i(t) \sim U[-\eta\pi, \eta\pi]$ is drawn from a uniform distribution with strength controlled by $0 < \eta < 1$.

Hence, SVM does have one rule compressed into one sentence, even if takes a bit longer to explain and define (see Figure 9.9 for an illustration of the core concepts). The model is governed by a number of parameters, including the sensing radius r, the particle density ρ, the noise strength η, and the magnitude of velocity v_0. Surprisingly, this simple model yields a wide range of outcomes, spanning uncorrelated and seemingly random motion, formation of coherent flocks that move in different directions, and organization of large-scale patches. We explore these outcomes in the next section.

9.3.2   Self-propelled particles and the spontaneous emergence of order

The dynamics of the SVM can lead to the *spontaneous emergence of order* as well as to spontaneous symmetry breaking, as was already shown in the zero-range process model introduced in Section 9.2.3. These two terms are related.

First, consider a flock of many individuals at a high enough density that each individual tends to interact with a half-dozen individuals even as the flock contains hundreds. In that event, two individuals that are sufficiently close are likely to share many interacting partners. As such, their reorientation is likely to be in a more similar direction than when comparing the reorientation direction for individuals farther away in the flock. This local realignment reinforces itself. Hence, at the next time period, the mutual sharing of directions can lead to local patches of coherently moving individuals. Over time, the entire flock can be connected or, depending on the degree of noise, interaction rates, and density, the

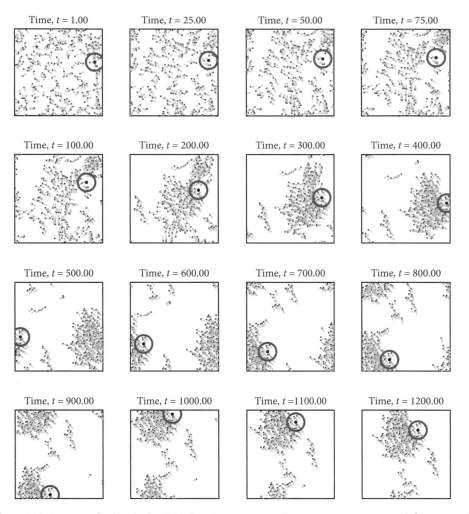

Figure 9.10: Emergent flocking in the SVM given $N = 100$, $\eta = 1$, $L = 5$, $v_0 = 0.03$, $r = 0.5$, and $dt = 0.2$, such that the density is $\rho = N/L^2$ or $\rho = 4$. The circle focuses on a particular individual for illustration of position in the flock—the individual is just as uninformed as any other individual with respect to a preferred direction. Note that these simulations use periodic boundary conditions (full details provided in the associated computational labguide).

flock can remain largely incoherent or characterized by distinct domains. Figure 9.10 illustrates the emergence of order from initially random orientations in the case of the SVM, using $N = 100$, $\eta = 1$, $L = 5$, $v_0 = 0.03$, $r = 0.5$, and $dt = 0.2$, such that the density is $\rho = N/L^2$ or $\rho = 4$. Using these parameters, then each individual should typically detect $\rho \pi r^2$ individuals if they are initially placed at random (a value close to 3).

Figure 9.10 reveals the rapid emergence of a preferred direction. This emergence implies two key facets of the SVM. First, despite each individual having no preference with respect to a direction of motion, the fact that each organism responds to the behavior of others reinforces local decisions, leading to collective behavior. Second, the direction itself is random. If one were to rerun the simulation (or observe a particular flock), the choice of direction at a given moment need not be somehow "better" or "optimal," but instead reflects the reinforcement of local decisions that break the symmetry of the system, i.e., the equivalency of all directions with respect to potential motion. Hence, although a direction is chosen, the hallmark of collective behaviors is not the particular direction, but the fact that there is a strong alignment of movement in a particular direction. This notion is an example of broken symmetry (Anderson 1972). Measuring that alignment requires a metric that is agnostic, i.e., rotationally invariant, with respect to a particular reference frame.

Quantifying the ordering requires an *order parameter*, i.e., a variable that is zero (in the infinite size limit) when there is random structure and then becomes nonzero when there is collective behavior. Hence, the order parameter is inherently a feature of the collective rather than the individual. The order parameter typically used in the SVM is the average velocity:

$$\bar{v} = \frac{1}{Nv_0} \left| \sum_{i}^{N} \vec{v}_i \right|.$$

The average velocity is equal to the sum of velocity vectors divided by the sum of the magnitude of each vector. An order parameter near 0 implies that the individuals in the flock are not correlated in their movement direction, and an order parameter near 1 means all the individuals are moving in nearly the same direction. Because of the dynamic feedback between current and future orientation, this order parameter, $\bar{v}$, is a dynamic feature of the flock. Initially, if the flock is initially randomly oriented, then $\bar{v}(t=0) \to 0$. Given suitable feedback, the order parameter can increase over time.

Figure 9.11 shows how the order parameter changes over time in two flocks that have the same number size and interaction radius but differ in one respect—the extent of noise. As is apparent, higher levels of noise in the update rule work against the emergence of local correlations that can then propagate over time to lead to patches and even coherence on the scale of the system. Instead, although local patches may form, sufficiently strong noise restricts the system to inhomogeneous motion. However, when noise is sufficiently low, spontaneous symmetry breaking occurs, and the flock moves largely (though not absolutely uniformly) in the same direction.

9.3.3 Informed leaders and collective behaviors in groups

Although many flocks do not have a preferred sense of direction, there are many examples in which informed individuals can shape the direction of animal groups.

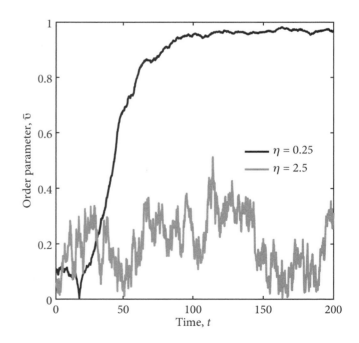

Figure 9.11: Order parameter in a flock with low noise, $\eta = 0.25$, and high noise, $\eta = 2.5$. In both cases, the flock contains $N = 250$ in a domain of size $L = 5$ giving a sensing radius of $r = 0.3$, implying a local sensing density of ≈ 3 individuals if initially distributed at random.

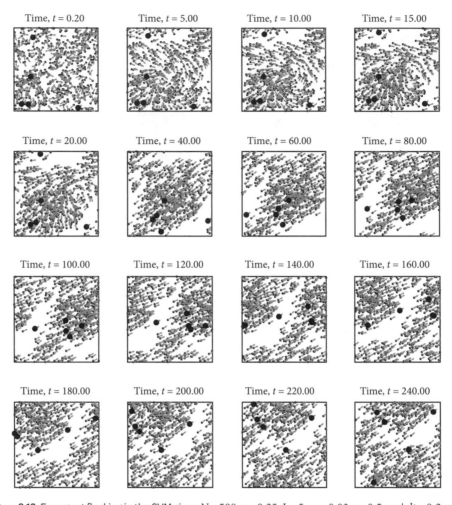

Figure 9.12: Emergent flocking in the SVM given $N = 500$, $\eta = 0.25$, $L = 5$, $v_0 = 0.03$, $r = 0.5$, and $dt = 0.2$, such that the density is $\rho = N/L^2$ or $\rho = 20$. The circles focus on a limited set of five informed leaders; their preferential direction of motion ends up determining the dynamics of the entire group. Full details provided in the associated computational lab.

Well-known examples include foraging individuals in social insect communities, including bees and ants. In these cases, informed individuals share chemical, visual, and/or tactile signals within the group that influence the choices of other members of their colony. Eventually, the reinforcement of informed individual preferences can lead to collective decisions, e.g., prioritizing trails to acquire food or identifying hive/nest locations. These same principles can be used to extend the SVM to include the impacts of informed subsets of individuals within a larger group.

Figure 9.12 depicts the results of a standard Vicsek collective flocking model given a higher-density flock than we have shown so far. There are 500 individuals in the flock, given $\rho = 20$, such that a typical individual interacts with $\pi r^2 \rho \approx 16$ individuals if distributed at random. In this case, the flock differs in one key respect from the SVM simulations.

Here 1 in 100 individuals is informed with respect to a particular preferred direction. These informed individuals can be identified by the larger black circles (in all of these panels, the position of the black circles with respect to the grey line denotes the direction of movement of informed individuals). Despite the fact that a very small number of individuals share the same preferred direction, these individuals are able to influence the entire group to go in a single direction, even though each interacts with a relatively small subset of individuals. The propagation of information is made possible by the influence of uninformed individuals via their neighbors. This remarkable level of influence by a small subset of individuals reinforces a point made at the beginning of this chapter: the emergence of collective behavior is not meant to have a normative value—it can lead to positive or negative impacts on a population. However, the evolution of such behavior in complex adaptive systems suggests that the net benefit outweighs the potential negative costs over evolutionary time periods.

9.4   COLLECTIVE DECISION MAKING AT THE FLOCK SCALE

9.4.1   Lessons from locusts

The emergence of order in animal groups presents certain challenges for the experimentalist. Yet, in recent years, studies of collective organismal behavior—including those of locust swarms—have been brought from the field to the laboratory. The movement of locusts is of particular concern from an environmental and human health perspective. Locust swarms can overwhelm efforts to prevent loss of crops and vegetation, with swarms exceeding tens if not hundreds of square kilometers km^2. Before going airborne, wingless juvenile locusts can also march, on land, over many kilometers—and early detection of the transition from individual to collective behavior is considered critical to intervention. The density of marching locusts is on the order of 50/m^2, suggesting that even relatively small densities (given their size) can lead to aggregate behavior.

In experimental work, marching locusts were challenged with a quasi-1D habitat—a circular terrain with a central excluded region. Hence, locusts could march at random, but to the extent that they formed a swarm, the quantification of the swarm would be in either a clockwise or counter-clockwise direction. Extending the principles from the SVM to 1D, the team used an alignment-based order parameter (Buhl et al. 2006):

$$\Phi^t = \frac{1}{m\pi/2} \sum_{i=1}^{m} \chi_i^t \tag{9.9}$$

where $\chi = \arcsin(\sin\theta - \alpha)$, α is the direction of motion in radians, and θ is the direction of the locust with the respect to the center of the arena, again in radians. Hence, when a locust is moving in a perfectly circular motion, the difference between its angle and that of the vector from the center of the area to it is either $\pi/2$ or $-\pi/2$. The use of the sin and arcsin functions then transforms this back into the domain $[-\pi/2, \pi/2]$, and therefore the normalization constraints of the order parameter are between -1 and 1. As before, an order parameter value of 0 denotes the absence of collective behavior.

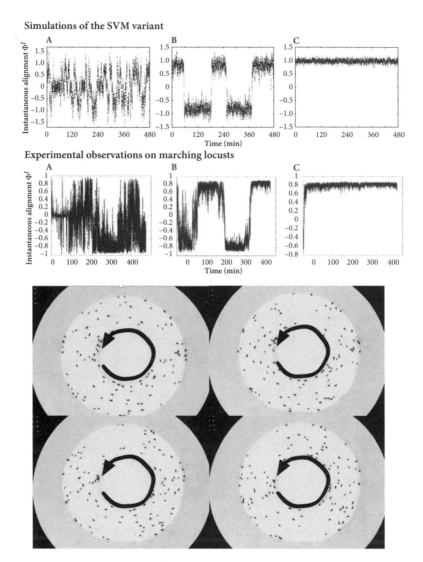

Figure 9.13: Simulation and experiments of marching locust alignment in a circular arena. (Top) Numerical simulation of a modified SVM; note that the instantaneous alignment is measured across three scenarios for 3, 11, and 47 locusts (A, B, and C, respectively). (Middle) Experimental observations given an average 3.5, 12, and 47 locusts in A, B, and C, respectively. (Bottom) Sequence of instantaneous snapshots of the experimental arena where the direction of motion (counter-clockwise) is noted in the middle of the arena—the excluded region. Images reproduced from Buhl et al. (2006).

The result of this experiment can be seen in Figure 9.13. Given expectations from field observations, typical marching locust densities of $50/m^2$ are equivalent to only ≈ 20 locusts in the ring-shaped arena used for the experiments. As long as there were 10 or more locusts in the arena, the locusts spontaneously began to march in the same direction. The outcomes are markedly different as a function of the density of locusts. The contrasting outcomes in the model (top panel) are for 3, 11, and 47 locusts, whereas the experiments (middle panel) show results for 7, 20, and 60 locusts, of which 3.5, 12, and 47 were moving on average. The salient point is that, although low-density systems can sometimes align, the overall movement remains largely disordered. In contrast, as densities increase, the system can alternate between ordered states, even as it spends the bulk of its time in a particular direction. Finally, at sufficiently high densities, the system moves to a near-uniform alignment that can last for many hours (note the time scale in hundreds of minutes on the axes).

This type of density-dependent transition is a hallmark of a phase transition, between disorder and order. Indeed, the SVM is expected to exhibit a transition between disorder and order in the limit of an infinitely large system, such that the average velocity $\bar{v}$ in the original model or the alignment ($|\Phi|$) is expected to grow like

$$\bar{v} \sim (\rho - \rho_c)^{\alpha}, \tag{9.10}$$

where ρ_c denotes a critical point in density demarcating the boundary between disorder and order; this holds for values of ρ close to and above ρ_c. The finite size of flocks will inevitably introduce corrections to this scaling. The critical value ρ_c is not universal, but the scaling exponent α is expected to hold across a wide range of problems. Indeed, this kind of phase transition is a hallmark of scale-free behavior, one that has direct relevance to asking questions about the correlation length in velocity fluctuations in large flocks of European starlings.

9.4.2 Lessons from starlings

It is now time to reinterpret the domains of coherent velocity fluctuations shown in Figure 9.3 To do so, we assess the relationship of fluctuations via a *spatial correlation function*, which typically measures the extent to which a property of a system is retained at different distances away from a randomly chosen center. Here the velocity correlation function is defined as the inner product of velocity fluctuations of birds given that they are separated by a distance r:

$$C(r) = \frac{1}{c_0} \frac{\Sigma_{ij} \vec{u}_i \cdot \vec{u}_j \delta(r - r_{ij})}{\Sigma_{ij} \delta(r - r_{ij})} \tag{9.11}$$

where δ is the Dirac function that is 1 when its argument is 0 and 0 otherwise (Cavagna et al. 2010). In this way, the correlation function only weights those pairs of birds that are separated by a distance r, and the constant c_0 is a normalization factor to ensure that $C(r) = 1$ (given the contributions from self-correlations). Critically, here $\vec{u}_j$ denotes the fluctuation of the individual starling velocity with respect to the flock average, i.e.,

$$\vec{u}_j = \vec{v}_j - \bar{v} \tag{9.12}$$

where $\bar{v}$ is precisely the form used above, i.e.,

$$\bar{v} = \sum_{k=1}^{N} \vec{v}_k. \tag{9.13}$$

Hence, this approach to analyzing fluctuations about the mean represents yet another step toward probing the extent of emergent order in the system (Bialek et al. 2012).

Like the use of an order parameter in the SVM, the spatial correlation function provides information on the collective behavior of the system. C is expected to be zero when velocities are uncorrelated, because each of the dot products, $\vec{u}_i \cdot \vec{u}_j$, ends up contributing in distinct directions and canceling out (similar to the principle of the order parameter in the SVM). In contrast, when a flock is nearly aligned the value approaches 1, and when the flock is antialigned the value can become negative. In practice, we expect that the correlation function should decay with distance and approach 0 (so that the relative fluctuations of two very distant birds will be uncorrelated). Perhaps such a decay might have a characteristic length, e.g., $C(r) \sim e^{-r/d}$, such that there is a typical distance over which birds are correlated irrespective of the size of the flock. Strikingly, this is not the case. Correlation lengths in flocks of different sizes can be measured via a motion tracking system. The system uses stereometric techniques to measure the 3D positions and velocities of birds in 24 distinct flocks at the Palazzo Massimo, Museo Nazionale Romano, in the city center of Rome—an inspired work that is part of a larger project led by Andrea Cavagna and Irene Giardina along with an international team of colleagues (Cavagna et al. 2010). One might expect that the velocity fluctuations would have a fixed domain size for both small and large flocks. Instead, the correlation length grew proportionately with the size of the system (24 observations in total, correlation coefficient of $r = 0.98$). For example, flocks of size 20 m, 40 m, and 60 m in extent had correlation lengths of approximately 7 m, 14 m, and 21 m, respectively. This growth of the correlation length with system size is a hallmark of a scale-free system, in which the apparent "scale" reflects artificial constraints because of the size of the flock rather than an inherent length scale. As such, the correlation length can be written using a conventional ansatz to describe the expected correlation, $C(r; L)$, that is conditional upon the size of the system:

$$C(r; L) = r^{-\gamma} f\left(\frac{r}{L}\right) \tag{9.14}$$

where γ is a scaling exponent and f is a scaling function. The scaling function should have certain properties; e.g., in the limit of an infinitely large system, $L \to \infty$, then one expects the size of the system not to matter, such that $C(r; \infty) = r^{-\gamma}$, implying that $f(0) \sim 1$. Although identifying the exact scaling function represents an empirical question, the finding that correlations scale with system size and not with an intrinsic characteristic length opens up new sets of questions that remain at the center of the field. Unlike in critical phase transitions where the appearance of long-range order is most relevant at the transition point, here the emergence of scale-free behavior reflects the spontaneous emergence of a new property, characteristic of flocks—the breaking of rotational symmetry. In the zero-range process model, symmetry breaking meants that one site is selected out of many equivalent sites. Here broken rotational symmetry means that one direction to move in is selected out of many equivalent directions.

There is still much to resolve in linking individual dynamics with those of the collective. For example, although our discussion uses a metric distance to calculate interactions, field observations suggest that interactions between starlings are limited not by distance but to a greater extent by information capacity. That is, rather than metric distance, it may be topological distance (i.e., a group of the 6–8 closest birds) that informs local updating and enables flock cohesion (Ballerini et al. 2008). Moreover, quantitative features of realized orientation waves in starling flocks suggest that extensions to the SVM are needed that include a notion of decision-making inertia (Cavagna et al. 2018). Nonetheless, precisely because of their simplicity, these findings (including simple models of flocking) reinforce the central point of complex adaptive systems: more is different.

9.5   TAKE-HOME MESSAGES

- Local interactions among a small number of agents can lead to coherence, structure, and order, in space and in orientation.
- The emergence of structure can include static features like the colocation of organisms, as well as static and dynamic coherence, e.g., the colocation of organisms that move in the same or similar directions.
- Collective groups on the move can perform without a leader or with very few "informed" individuals.
- A small number of informed individuals can have an outsized influence on the behavior of an entire group.
- A simple model of group formation—the standard Vicsek model—is ideal for exploring how a minimal set of local rules generates a large space of possible outcomes, including flocking.
- Local interactions drive broken symmetry and enable long-range order.
- Repeated local interactions can generate coherent and potentially durable spatial groups.

9.6   HOMEWORK PROBLEMS

This problem set builds upon the computational laboratory, which focuses almost entirely on the Vicsek model of self-propelled motion. Keep in mind that individual stochastic runs can differ in output and that robust answers sometimes depend on running replicates. The computational laboratory covers a number of topics, including

- Individual-based stochastic modeling
- Rule-based updating
- Periodic boundary-base computation
- Visualization methods to depict positions and velocities
- Order parameter calculations
- Distinguishing between leaders and the crowd in social groups

These computational methods, including solutions, provide the critical basis for the following problems.

PROBLEM 1. Critical Noise

Building on the self-propelled particle model or SVM in the laboratory, measure the order parameter $\bar{v}$ as a function of noise using a flock of size $N = 300$, $r = 0.3$, $v_0 = 0.03$, and $dt = 0.2$. Modulate η from 0.25 to 2.5 and try to identify a transition point between order and disorder. If possible, evaluate whether you have enough evidence to suggest a critical transition between order and disorder.

PROBLEM 2. Critical Density

Building on the self-propelled particle model or SVM in the laboratory, measure the order parameter $\bar{v}$ as a function of noise using a fixed noise level of $\eta = 1.5$, $r = 0.3$, $v_0 = 0.03$, and $dt = 0.2$. Modulate N from 10 to 200 and try to identify a transition point between order and disorder. If possible, evaluate whether you have enough evidence to suggest a critical transition between order and disorder.

PROBLEM 3. Leadership

Modify the simulation code developed in the computational laboratory to have a subset of 10 informed individuals that are moving in orthogonal directions (e.g., $\pi/2$ vs. 0 in radians). Using the same conditions as in the computational laboratory, evaluate whether or not one "wins" or if both groups "win" (i.e., with distinct patches of individuals moving in the direction of the two different groups).

PROBLEM 4. The Scale of Information Exchange

Modify the SVM to evaluate how patterns change with scale using a flock of size $N = 200$ and parameters $L = 2$, $r = 0.3$, $v_0 = 0.03$, $dt = 0.2$, and $\eta = 1.5$. Increase the size of the system from $L = 2$ to $L = 10$ while retaining the same density, i.e., increasing $N = 5000$. If possible, identify a characteristic length scale over which orientations are aligned and compare this to the size of the system. Is there a characteristic scale? Compare and contrast your findings with those of Cavagna et al. (2010).

PROBLEM 5. Aggregation in Zero-Range Process Models

Develop a simulation in a ring lattice of size L with population n_i such that $\sum n_{i=1}^{L} = N$ is fixed. Model the movement of individuals to adjacent sites such that per capita hopping rate is proportional to $n_i^{1-\epsilon}$. Exploring ranges of values close to $\epsilon_c = 1$, evaluate the number of moves it takes such that the system aggregates to a single site. How does the time change as a function of ϵ and N? Focus on relatively small systems for the purposes of the homework, e.g., N in the range of 100–1000.

- Develop your own approach to calculating the pair correlation function, e.g., by comparing all pairwise distances and then binning them into categories before averaging their orientation.
- Using your pair correlation function, estimate the decay of correlations in the SVM in the high and low noise conditions described in the chapter.
- Replicate the quasi-1D simulation of marching locusts using a banded periodic boundary condition simulation with hard boundaries on one set of edges. Using densities similar to those in the marching locust example, see if you can replicate the regimes of disorder, alternating directions, and coherent marching in a single direction as a function of increasing locust density.

9.7 TECHNICAL APPENDIX

Vegetation patterns—steady states without diffusion The coupled dynamics of water $W(x, t)$ and vegetation $N(x, t)$ are written in the chapter as

$$\frac{\partial W}{\partial t} = a - lW - rWN^2 + v\frac{\partial W}{\partial x} \tag{9.15}$$

$$\frac{\partial N}{\partial t} = rjWN^2 - mN + D\left(\frac{\partial^2}{\partial x^2} + \frac{\partial^2}{\partial y^2}\right)N \tag{9.16}$$

where the explicit location and time are suppressed for convenience. In this model, a denotes the uniform rainfall rate, l is the evaporation rate, v denotes the rate of downslope water movement, r is a coupling constant, j is an efficiency of water use, m is a basal plant loss term, and D is an effective plant diffusion term. In the limit of $D = 0$ and $v = 0$ (assuming no overland flow), the model becomes

$$\frac{dW}{dt} = a - lW - rWN^2 \tag{9.17}$$

$$\frac{dN}{dt} = rjWN^2 - mN \tag{9.18}$$

or

$$\frac{dW}{dt} = a - lW - rWN^2 \tag{9.19}$$

$$\frac{dN}{dt} = N(rjWN - m) \tag{9.20}$$

This system has multiple steady states in the form of (W^*, N^*), including a bare vegetation state $(\frac{a}{l}, 0)$ and a vegetated state $(W_c, \frac{m}{rjW_c})$ where the coexistence state includes both a high vegetation and low vegetation term. Note that when $W^* = a/l$ invasion by N requires that $N > m/rjW^*$ or $N > \frac{lm}{rja}$. Increasing vegetation decreases the water levels.

The solution to the coexistence state is

$$W_c = \frac{\frac{a}{l} \pm \sqrt{\left(\frac{a}{l}\right)^2 - \frac{4m^2}{j^2 rl}}}{2}.$$ (9.21)

Note that this equation is only valid insofar as

$$a^2 > \frac{4m^2 l}{j^2 r}.$$ (9.22)

Hence, the steady state analysis suggests the potential for a high and low state in water levels, which are then inversely correlated to vegetation levels because of the $1/W_c$ dependence for N^* in the vegetated regime. A high vegetation state draws down water, whereas a low vegetation state is associated with higher levels of water.

Master equation for the two-population, zero-range process Consider a metapopulation of two sites with n individuals on site A and $N - a$ individuals on site B. The per capita movement rate is proportional to $n^{-\epsilon}$ such that the aggregate movement rate out of site A is $n^{1-\epsilon}$ and the aggregate movement rate out of site B is $(N-n)^{1-\epsilon}$ where the rate constant prefactor is set to 1 without loss of generality. Define $p(n,t)$ as the probability that there are n individuals in site A at time t. Given the transition rates, the master equation can be written as follows for $0 < n < N$:

$$\frac{\partial p(n,t)}{\partial t} = -n^{1-\epsilon} p(n,t) + (n+1)^{1-\epsilon} p(n+1,t)$$
$$+ (N - (n-1))^{1-\epsilon} p(n-1,t) - (N-n)^{1-\epsilon} p(n,t)$$ (9.23)

where $p(-1,t) = 0 = p(L+1,t)$ can be used to enforce the restriction that $0 \le n \le N$. At the equilibrium, $\partial p(n,t)/\partial t = 0$, which means there are only two rates that balance for $n = N$:

$$N^{1-\epsilon} p^*(N) = (N - (N-1))^{1-\epsilon} p^*(N-1),$$ (9.24)

leading to

$$p^*(N-1) = N^{1-\epsilon} p^*(N).$$ (9.25)

This rate equivalence means that detailed balance holds. Detailed balance denotes the fact that $p(n)W_{mn} = p(m)W_{nm}$, where $p(n)$ and $p(m)$ are probabilities of being in states n and m, respectively, and W_{mn} and W_{nm} are the transition rates from n to m and from m to n, respectively. Via detailed balance, there is a recursion relationship for all the probabilities, e.g.,

$$p^*(N-2) = \frac{N^{1-\epsilon}(N-1)^{1-\epsilon}}{2^{1-\epsilon}} p^*(N).$$ (9.26)

and more generally to the midpoint (assuming N is even), then

$$p^*(N/2) = \frac{N^{1-\epsilon}(N-1)^{1-\epsilon}\dots(N/2+1)^{1-\epsilon}}{(N/2)^{1-\epsilon}(N/2-1)^{1-\epsilon}\dots 2^{1-\epsilon}} p^*(N).$$ (9.27)

Insofar as $\epsilon > 1$, then the values drop off as a near-geometric series (i.e., exponentially fast for large N). In contrast, when $\epsilon < 1$, the values increase as a near-geometric series toward a maximum at $p^*(N/2)$. Moreover, due to the symmetry in the problem, $p(0) = p(N)$. Overall, the equilibrium distribution has two peaks for $\epsilon > 1$ and a single peak in the middle for $\epsilon < 1$. In practice, this means that the two-site zero-range process leads to condensation at either site A ($n = N$) or site B ($n = 0$) insofar as the per capita rate of movement decreases faster than the local abundance.

Chapter Ten

Conflict and Cooperation Among Individuals and Populations

10.1   GAMES, RELATIVELY SPEAKING

For over a decade, I have started my game theory lectures in both undergraduate and graduate level classes by auctioning off a $5 bill. The rules are simple. Anyone in the class can bid and the highest bidder pays their bid and receives the $5 bill. So, if you were there and bid the highest amount, let's say $2, then you'd have made a $3 profit. Not bad. The only other rule is that the second-highest bidder also has to pay but does not get the $5 bill. Bids begin at $0.50 and can go up in $0.50 increments. That's it.

The auction rules seem straightforward enough. So what do you think happens? And perhaps more importantly, what would you do if you were sitting in the third row of class on a crisp Thursday morning? Would you bid, and if so, how much? And if others outbid you, what would you do then? For those who have seen or participated in live auctions, whether of refurbished cars or Old Masters paintings, or who just like to watch reruns of *Storage Wars*, you are familiar with the addictive nature of getting involved in the bidding excitement. Yet here the stakes are lower. And surely these highly trained students in highly competitive undergraduate and graduate programs will remain calm in the midst of a low-stakes auction. Perhaps. Instead, despite significant variation in class size, disciplinary training, and background, the same remarkable thing has happened in this auction every year. I will not give away what that remarkable thing is because it is worth trying it for yourself, ideally in a group with 10 or more, to set the stage for how multiplayer games can lead to unexpected outcomes.

I have recently added another game—the ultimatum game—to the start of class. For this game, I break the class into as many groups of two as possible and offer one of the two individuals in each group a virtual amount of cash, usually $2 to start. Let's call this individual Fortuna because she has the initial good fortune to receive the money in the first place. Fortuna then has a choice: she can offer whatever fraction of the $2 to the opponent, at which point the opponent has a decision to make. Let's call this opponent Solomon, because he has a challenging decision to make. If Solomon accepts the offer, then he receives what Fortuna offered and Fortuna receives what is left. However, if Solomon rejects the offer, then neither he nor Fortuna receive anything. For example, if Fortuna offers $1 it is likely that

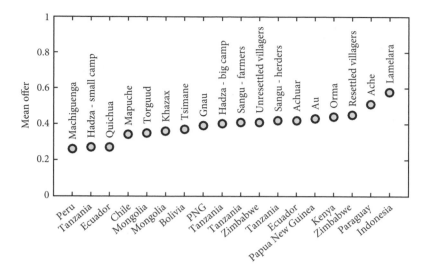

Figure 10.1: The ultimatum game across cultures—evidence for (relatively) fair offers (adapted from Henrich et al. 2001).

Solomon will say yes, and both will receive an equal share of $1. But what if Fortuna offers just $0.50, in which case Solomon receives only one-third of what Fortuna gets ($1.50), or even worse, a single cent, in which case Fortuna holds on to $1.99 of the original pot?

All of these scenarios unfold in a classroom setting. What is remarkable is how much variation there is among the decisions that the Fortunas and the Solomons make. Whenever Fortuna makes a "fair" offer of $1 or nearly so, then the Solomons inevitably say yes and accept. It seems fair. Later, when asked, the Fortunas reason that they wanted to make sure they received the $1 and so offering an equal share certainly guarantees it. Others reason that they simply perceived that offer as fair in the first place. Yet there are some Fortunas, year after year, who offer a single penny. In many cases, those Fortunas have taken a game theory class, usually in the economics department. We will put aside what that says about such classes (and economics) for a different day. What is even more remarkable is the fact that although many (if not most) Solomons say no, there are some Solomons who accept the offer! They will take a penny while the Fortunas hold on to $1.99. Those Solomons are often veterans of economics classes and when asked they have a perfectly reasonably explanation for their decision. If they accept the offer, then they will have $0.01 more. If they decline the offer, they will have nothing. And they reason being a penny richer is better. Perhaps—but does having one penny more really make one better off if others receive nearly 200 times as much? Notably, when such games are played in many different small-scale societies, the actual offers are usually far more substantial and far more equitable (Figure 10.1).

There is an old adage, penny-wise and pound-foolish. This adage applies here. To understand why, imagine that this good fortune is no longer $2 but instead represents access to a watering hole for parched wildebeests, the remains of a half-eaten gazelle discovered by two vultures, a food item collected by a hunter-gatherer, or an enzyme that can transform an inaccessible, scarce nutrient into a bioavailable form. In that case, what are the consequences for agreeing to a severely "unfair" offer? If you are Solomon, what are your odds of surviving and reproducing if you take 1 part in 200 of the gazelle, 1 part in 200 of the water, or 1 part in 200 of the hunted animal or farmed vegetable, while Fortuna takes nearly all? Not good. Although the economists might say that an increase, however small, is always better than staying where you are, evolution tells a different story. Fitness is relative. What good is being 1 unit ahead in an absolute sense when you have fallen 199 behind in relative terms? Instead, when strategic decisions involve increases and decreases in reproductive success, then there is the real possibility that enforcing fairness makes more sense. This is particularly true given access to limited resources.

The concept of relative fitness is one that we have already discussed and so, before replaying what is widely known as the ultimatum game, I revisit this concept of fitness. And then we play again. This time the virtual stakes are higher, $20. (For any administrator reading, keep in mind that no legal tender is used to dish out in class.) And this time, although some Fortunas offer 1 cent, I have never seen a Solomon accept. This is not to say that such things cannot happen, but the point of playing this game is to begin to build intuition on the link between strategic choices, games the organisms play across all scales of life, and ultimately the feedback between games and evolution.

How should organisms interact with each other given multiple "behaviors"? The answer to this question depends on context, in two senses. The first sense is literal. Organisms display a myriad number of different kinds of behaviors, so interactions necessarily stem from the particular kind of behaviors available. Meerkats may defend or retreat from territory; organisms of all sizes—from sea lions to large ungulates to birds to reptiles—engage in complex mating displays that can often end in fights when displays alone do not resolve the potential conflict. Predators must choose between hunting and waiting, just as prey must decide between continuing to search for food or hiding from potential predators (for a classic example of such refuge dilemmas in marine bristle worms, see Dill and Fraser 1997). In the microbial world, such dilemmas abound. When essential nutrients are scarce, some microbes release extracellular enzymes into the environment that can turn inaccessible nutrients (like iron) into accessible forms (Wilhelm and Trick 1994). Such public goods may end up benefiting those who do not contribute enzymes in the first place. And, as we explore later in this chapter, even viruses may exhibit socially influenced behaviors when coinfecting cells (Díaz-Muñoz et al. 2017).

Organisms interact over territory, reproductive opportunities, food, and more. They interact via production or exploitation of public goods, the use of chemical cues, decisions to look for resources or hide, flowering (or not), and a multitude of other mechanisms. Given that the payoff to the individual organism depends not only on its behavior but the behavior of another individual (often of a different species), we refer to these kinds of interactions as games, and begin to apply concepts from game theory to help resolve, interpret, and predict what happens when organisms interact. However, this chapter is not focused on classic game theory, in a narrow sense. Central to thinking about biology through the lens of games is the recognition that the fitness of one individual may be related to the payoff, but it depends on its relative value to others. The relative nature of fitness and the feedback between payoffs, fitness, and populations underscore the central aim of this chapter—to understand evolutionary games.

Evolutionary game theory explores the feedback between payoffs, relative fitness, and populations. A bridge between classic game theory and evolutionary game theory is helped

Hawks and Doves: Competing for nesting sites

Viral Games Inside Microbial Cells

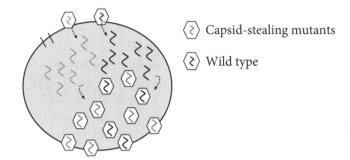

〈⋛〉 Capsid-stealing mutants

〈⋛〉 Wild type

Microbial Cheaters & Cooperators: Public Good Dilemmas

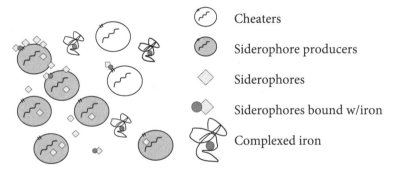

Cheaters

Siderophore producers

Siderophores

Siderophores bound w/iron

Complexed iron

Figure 10.2: Evolutionary game theory examples. Despite the differences in context, each represents examples in which individuals that belong to different types interact and, as a consequence, their reproductive success depends on the result of those interactions. In each of the cases, there is also the possibility of hostile action taking and/or cheating, i.e., not participating in some costly task. (Top) Hawks and doves competing for nesting sites. (Middle) Viral games inside microbial cells. (Bottom) Microbial cheating and cooperation arising from public good dilemmas.

by example—here the hawk-dove game. After introducing this game and the associated payoff concepts, the ideas are put into practice by examining archetypal examples: animal fights that don't end in harm, intracellular competition among viruses, and microbial interactions (Figure 10.2). In the classic example of animal fights, only one can win, but the conflict between hawk-like and dove-like strategies can be applied to a broad range of problems in which payoffs underlie interactions and coexistence of distinct strategies (Hofbauer and Sigmund 1998; Nowak 2006). Likewise, the use of intracellular resources by coinfecting viruses raises questions of whether viruses should invest in directing cellular machinery to produce structural components, or replicate and benefit from the products of other coinfecting viruses (Turner and Chao 1999). Finally, certain limited nutrients (like iron) may only be available for uptake given the release of extracellular enzymes (siderophores); however, cheating strains that don't produce such costly excreted enzymes can also benefit from the increased bioavailability of iron. Formative treatments of this and other excretion dilemmas in a sociomicrobiology context can be found in Xavier and Foster (2007); West et al. (2007); Nadell et al. (2008). To begin to reason through each of these examples and to place them in a comparative context requires a common conceptual framework. Hence, the next section starts with payoff matrices—and concepts from conventional game theory—as a means to address larger problems of evolutionary games.

10.2   PAYOFFS: A CLASSIC APPROACH

10.2.1   An illustrative example: The hawk-dove game

The hawk-dove game can be analogized to the interactions for nesting sites on a cliff face (Figure 10.2, top). These nesting sites are a critical resource for the reproduction and survival of offspring. Yet such sites are limited. In the hawk-dove game, there are more birds than sites, so each site may be contested by different combinations. Two hawks might meet, two doves, and even one hawk and one dove. The payoffs to each organism (formally, each player) in this interaction can be written as follows:

$$
\begin{array}{cc}
 & \text{Player 2} \\
 & \begin{array}{cc} \text{Hawk} & \text{Dove} \end{array} \\
\text{Player 1} \begin{array}{c} \text{Hawk} \\ \text{Dove} \end{array} &
\begin{array}{|c|c|}
\hline
\left(\frac{G-C}{2}, \frac{G-C}{2}\right) & (G,0) \\
\hline
(0,G) & \left(\frac{G}{2}, \frac{G}{2}\right) \\
\hline
\end{array}
\end{array}
$$

The format of this payoff matrix requires some explanation. The rows denote the actions of player 1. It is often helpful to think about the outcomes by putting yourself in this role, i.e., that of the focal player. The columns denote the actions of player 2. The elements of the matrix correspond to the payoffs to player 1 and player 2 before and after the comma, respectively. For example, the bottom left element, $(0, G)$, denotes the fact that if player 1 is a dove and player 2 is a hawk, then the payoff to the dove is 0 and the payoff to the hawk is G. This also explains why the payoffs are listed as $(G, 0)$ in the upper right element. Explaining the factors of 1/2 requires a bit more explanation. In this game, doves do not fight for a nesting site. Hence, in one interpretation the doves share the resource, each receiving one-half of the benefit, i.e., $G/2$. In an alternative interpretation, when two doves meet, one (at random) remains and the other flies away, hence, on average each player receives one-half of the benefit. When two hawks meet, the hawks fight and one loses, incurring a cost C, while the other wins, receiving a benefit G. Because either could be the winner (or loser), the average net payoff is $(G-C)/2$. Given the symmetric nature of the game, it is possible to rewrite the payoff matrix while forgoing the second value as follows:

		Player 2	
		Hawk	Dove
Player 1	Hawk	$(G-C)/2$	G
	Dove	0	$G/2$

Given these payoffs, what is the best thing to do? In other words, if you had to choose, would you be a hawk or a dove? There are certainly advantages to both. For example, hawks always outcompete doves, seemingly favoring the choice of hawks. However, hawks fight. That fights are costly, $C > 0$, means that being a hawk in a hawk world is a dangerous proposition. It is worthwhile to consider two limits: one in which fights are very costly, i.e., $C > G$, and the other in which fights come with little to no cost, i.e., $C \to 0$. In these two limits, the payoffs look like this:

Fights of limited cost, $C \to 0$

		Player 2	
		Hawk	Dove
Player 1	Hawk	$\approx G/2$	G
	Dove	0	$G/2$

Costly fights, $C > G$

		Player 2	
		Hawk	Dove
Player 1	Hawk	$-$	G
	Dove	0	$G/2$

In the first example, hawks receive approximately half of the gain ($G/2$), whereas in the second example the $-$ sign denotes that the average payoff to a hawk is negative and is worse than not having fought at all.

These examples can be analyzed in turn. First, when fights come with minimal costs, then it is possible to reason strategically regarding which animal type will be better off. Think of yourself as player 1, the focal player, and ask what you would do if the opponent, player 2, was a hawk. In that scenario, a dove gets a payoff of 0 while a hawk gets a payoff of $G/2$. Hence, it is better to be a hawk. Similarly, if the opponent was a dove, then it is still better to be a hawk because $G > G/2$. Hence, in the event that fights are not costly, then it is always better to be a hawk than a dove. And so it would seem that hawks should dominate the cliff face. Moreover, if these animal types were strategies that animals could adopt, i.e., a hawk-like or dove-like strategy, then for any set of costs $0 < C < G$, the same conclusion applies: it is better to be aggressive, irrespective of the choice of the opponent. Paradoxically, if all adopt a hawk-like strategy, then all will receive less payoff than had all adopted a dove-like strategy. Yet, without some kind of coordination, cooperation, or other dynamics, this would seem to be the rational choice (an issue that we examine in greater detail in the prisoner's dilemma framework in Section 10.3.2).

Next, when fights come with significant costs, i.e., $C > G$, then the benefits of the respective strategies change. If the opponent is a dove, then it is better to be a hawk (because

$x = 0$ $0 < x < 1$ $x = 1$

Figure 10.3: Illustration of the range of social contexts from all doves, $x = 0$, all hawks, $x = 1$, and a mixture of hawks and doves, $0 < x < 1$.

$G > G/2$). However, if the opponent is a hawk, then it is better to be a dove. The payoff is zero, but zero is better than negative; i.e., losing a nest site is bad, but not as bad as losing a wing or death, which could be the consequence of a costly fight. Hence, it would appear that if you know your opponent is a hawk, then it would be better to adopt a dove-like strategy, yet if you know your opponent is a dove, then it would be better to adopt a hawk-like strategy. But which one is your opponent going to be? There are many ways to embed uncertainty into game-theoretic frameworks, yet instead, we take an intentionally evolutionary perspective and directly connect payoffs to the social context, i.e., the fraction x of players that are hawks and the fraction $1 - x$ of players that are doves (Figure 10.3).

10.2.2 The social context of payoffs

The payoff to individual players depends on the social context, i.e., the fraction of individuals who are hawks versus those who are doves. To build up this idea more formally requires that we specify precisely how individuals interact. At first, we make the simplest assumption, namely, that the interactions are random. Hence, if x is the fraction of hawks, then an individual will have an x probability of encountering a hawk at a nesting site and a $1 - x$ probability of encountering a dove. The resulting payoff depends on what type of strategy the focal player utilizes. For example, doves receive a payoff of

$$P_{\text{dove}} = \overbrace{x \times 0}^{\text{dove–hawk}} + \overbrace{(1 - x) \times G/2}^{\text{dove–dove}}, \tag{10.1}$$

which is equivalent to

$$P_{\text{dove}} = \frac{G}{2} - \frac{G}{2}x. \tag{10.2}$$

In essence, the payoff to doves is maximal when there are no hawks and declines to 0 in the limit that the environment is (nearly) filled with hawks. In contrast, hawks receive a payoff of

$$P_{\text{hawk}} = \overbrace{x \times (G - C)/2}^{\text{hawk–hawk}} + \overbrace{(1 - x) \times G}^{\text{hawk–dove}}, \tag{10.3}$$

which is equivalent to

$$P_{\text{hawk}} = G - \left(\frac{G + C}{2}\right)x. \tag{10.4}$$

Again, the hawk payoff is at a maximum in a dove world and declines linearly with hawk frequency. The question underlying what is optimal depends on the comparison of these two frequency-dependent payoffs.

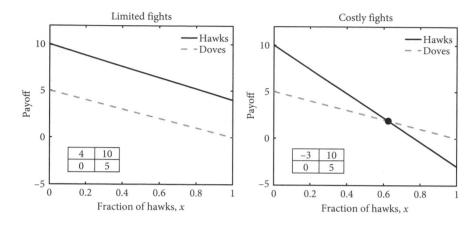

Figure 10.4: Payoff matrix in the hawk-dove game given fights with limited and significant costs (left and right, respectively). In these games, $G = 10$; $C = 2$ (left) and $C = 16$ (right).

To add greater specificity, consider two cases corresponding to the limited and costly fight examples. In both cases, $G = 10$; however, in the limited cost example $C = 2$, such that the payoff matrix is as follows:

		Player 2	
		Hawk	Dove
Player 1	Hawk	4	10
	Dove	0	5

In the costly fight example $C = 16$, such that the payoff matrix looks like this:

		Player 2	
		Hawk	Dove
Player 1	Hawk	−3	10
	Dove	0	5

The frequency-dependent payoff curves for these two examples are shown in Figure 10.4. Note that when fights are limited, i.e., $C < G$, then the hawk fitness always exceeds the dove fitness. We can prove this by noting that $P_{hawk}(x = 0) > P_{dove}(x = 0)$ and $P_{hawk}(x = 1) > P_{dove}(x = 1)$ when $C < G$, and recognizing that both x-dependent fitness values depend linearly on x. Hence, the two fitness values must be lines, where the endpoints of one exceed the endpoints of the other—and therefore this must hold for all intermediate values of x. This is not the case when $C > G$. Instead, when $C > G$, then the lines intersect when $P_{hawk} = P_{dove}$, which occurs when $x = G/C$. In the specific example, this corresponds to $x^* = 5/8$, which is denoted by the black circle in the right panel of Figure 10.4. Hence, when fights are costly, then being a hawk is advantageous when doves are prevalent and being a dove is advantageous when hawks are prevalent. From an evolutionary perspective, payoffs determine the demographic (or net growth) rates of individuals, which suggests that an evolutionary model might lead to a novel outcome: a mixed population of hawks and doves.

10.3 FROM PAYOFFS TO POPULATIONS

10.3.1 Replicator dynamics: Hawks and doves

This section introduces evolutionary game theory in the context of the hawk-dove game as a means to motivate a broader class of problems and challenges. Consider a two-player game involving two strategies, H and D, denoting hawks and doves (or aggressive and cooperative behaviors), respectively. The payoff matrix for the case $G = 10$ and $C = 16$ is

$$\mathbf{A} = \begin{bmatrix} -3 & 10 \\ 0 & 5 \end{bmatrix}. \tag{10.5}$$

In evolutionary game theory, such payoffs can be coupled to the changes in population or strategy frequencies, x_1 and x_2, e.g., where x_1 and x_2 denote the frequency of hawks and doves such that $x_1 + x_2 = 1$. The coupling is expressed via replicator dynamics. The standard replicator dynamics for two-player games can be written as

$$\dot{x}_1 = \left[r_1(\mathbf{x}, A) - \langle r \rangle (\mathbf{x}, A) \right] x_1 \tag{10.6}$$

$$\dot{x}_2 = \left[r_2(\mathbf{x}, A) - \langle r \rangle (\mathbf{x}, A) \right] x_2 \tag{10.7}$$

where r_1, r_2, and $\langle r \rangle$ denote the fitness of player 1, the fitness of player 2, and the average fitness, respectively, all of which depend on the frequency of players and the game-theoretic payoffs. This formula can be written generally as

$$\dot{x} = x(1-x)\left(r_1(x, A) - r_2(x, A) \right) \tag{10.8}$$

such that the dynamics of hawks obey a logistic equation whose growth rate (in parentheses) is positive when r_1 exceeds r_2 and negative otherwise. This use of replicator dynamics was already introduced in the context of mutation-selection balance in Chapter 4. Here the payoffs in Eq. (10.1) and Eq. (10.3) denote contributions to an effective per capita growth rate, weighted by the relative proportion of interactions. In this convention, the fitness values for the HD game are

$$r_1 = \frac{(G-C)}{2} x_1 + G x_2 \tag{10.9}$$

$$r_2 = 0 \times x_1 + \frac{G}{2} x_2 = \frac{G}{2} x_2 \tag{10.10}$$

and the average fitness is

$$\langle r \rangle = r_1 x_1 + r_2 x_2 \tag{10.11}$$

$$\langle r \rangle = \frac{(G-C)}{2} x_1^2 + G x_1 x_2 + \frac{G}{2} x_2^2. \tag{10.12}$$

Note that this system remains constrained by $x_1 + x_2 = 1$ (that is, $\dot{x}_1 + \dot{x}_2 = 0$). Hence, we can focus on a single variable and rewrite the dynamics of $x \equiv x_1$ and $1 - x \equiv x_2$, after some algebra (see technical appendix), as

$$\dot{x} = x(1-x)\left[\left(\frac{G-C}{2} \right) x + \frac{G}{2} (1-x) \right]. \tag{10.13}$$

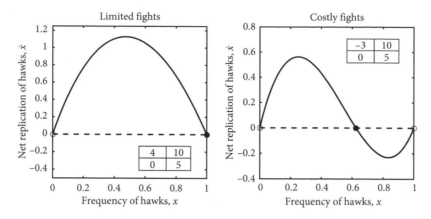

Figure 10.5: Hawk dynamics as a function of the social context, x, for limited (left) and costly (right) fights. In these games, $G = 10$; $C = 2$ (left) and $C = 16$ (right), using Eq. (10.14).

This can be seen explicitly by rewriting the dynamics as

$$\dot{x} = \frac{G}{2}x(1-x)\left(1 - \frac{C}{G}x\right). \tag{10.14}$$

The replicator dynamics for the hawk-dove game in Eq. (10.14) has up to three equilibria, including $x^* = 0$ and $x^* = 1$ as well as one more in the domain $[0, 1]$, namely, $x^* = G/C$ (when $C > G$). The stability can be identified by the sign of the cubic form on the right-hand side of Eq. (10.14), illustrated in Figure 10.5. When $C > G$, the leading term is positive, the extremal equilibria are unstable, and the internal mixed equilibrium is stable. Therefore, any initial condition not on a fixed point will converge to $x^* = G/C$. This mixed equilibrium denotes the coexistence of hawks and doves, with an increasing fraction of doves as costs increase. In contrast, when fights are limited and $C < G$, then $x^* = 0$ is unstable and $x^* = 1$ point is stable. Therefore, the system will converge to $x^* = 1$, corresponding to a hawks-only system. In summary, the more costly the fight is, the more likely it is that doves and dove-like strategies will predominate and that the community will include a mix of hawk-like and dove-like strategies.

10.3.2 Replicator dynamics: Prisoner's dilemma

The hawk-dove game is one of a family of two-player symmetric games (Hofbauer and Sigmund 1998; Nowak 2006) of relevance to understanding living systems. We can leverage the replicator dynamics to consider dynamics given another game with the following payoff matrix:

$$\mathbf{A} = \begin{bmatrix} 3 & 0 \\ 5 & 1 \end{bmatrix}. \tag{10.15}$$

This game is known as the prisoner's dilemma (PD), where the first strategy denotes cooperators and the second strategy denotes defectors. The dilemma is meant to represent the benefits to two individuals who have, in fact, committed a crime, but have been separated and are interrogated separately. Their options are to cooperate (with each other) by remaining silent or to confess and blame their coconspirator (thereby defecting from their

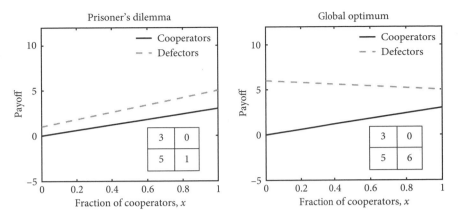

Figure 10.6: Payoff matrix in the prisoner's dilemma game (left) versus payoff matrix in a two-player game where "defection" leads to a global optimum. (Left) $R = 3, S = 0, T = 5, P = 1$. (Right) $R = 3, S = 0, T = 5, P = 6$.

partnership). The payoffs are such that, if both cooperate, they get the largest benefit. However, as is apparent from the analysis of the payoff matrix, the best that each can do is to defect while the other cooperates. Perversely, defection is the preferred strategy irrespective of what the the opponent does. Hence, both individuals reason this same way and choose to defect, receiving a worse payoff of 1 than had they decided to cooperate and receive a payoff of 3. This form has the same structure as the limited HD game analyzed in the prior section.

More generally, the payoff matrix for the PD game can be written as

$$\mathbf{A} = \begin{bmatrix} R & S \\ T & P \end{bmatrix} \tag{10.16}$$

where R denotes the reward for cooperation, T denotes the temptation to cheat, S denotes the sucker's payoff, and P denotes punishment. The key conditions for the PD game is that $T > R > P > S$. These conditions ensure that defection is favored for players irrespective of the opponent's strategy. The other key condition is that $R > P$. This ensures that when players choose to defect they are, in fact, worse off than had they chosen to cooperate. This reasoning implies that individuals acting in their self-interest are not necessarily better off. The fallacy that individuals acting in their own self-interest will actually make decisions that collectively benefit each of them is just that: a fallacy.

Figure 10.6 illustrates this point. The two panels include contrasting payoff matrices:

$$\mathbf{A}_{PD} = \begin{bmatrix} 3 & 0 \\ 5 & 1 \end{bmatrix} \qquad \mathbf{A}_{optimal} = \begin{bmatrix} 3 & 0 \\ 5 & 6 \end{bmatrix} \tag{10.17}$$

The $\mathbf{A}_{PD}$ payoff matrix is the prisoner's dilemma case. In it, defection is favored at all frequencies of cooperators. This implies that defection should increase relative to cooperation. Yet the tragedy of the PD game is that as $x \to 0$ the system is dominated by defectors, even though the payoff (equal to 1) is less than if all had cooperated (equal to 3). This gap can be seen directly in the figure by contrasting the position of the dashed line at $x = 0$ with value 0, to the solid line at $x = 1$ with value 3. The issue of a tragedy—and the tragedy of the commons—will be revisited at the end of this chapter. In contrast, the $\mathbf{A}_{optimal}$ matrix also has defectors doing better than cooperators at all frequencies; however, the payoffs to defectors increase with decreasing x, suggesting that the outcome here is a global optimum. There are a limited number of possible combinations of the entries of a 2×2 payoff matrix, and those correspond to a small number of different generic dynamics and of different relationships between local and global maxima (Hofbauer and Sigmund 1998).

As before, we can move from games to evolutionary dynamics by recognizing that the social context–dependent fitness is $r_1 = 3x$ and $r_2 = 1 - 4x$ such that

$$\dot{x} = -x(1-x)(1+x). \qquad (10.18)$$

Again, the stability can be identified from the sign of the cubic, i.e., $x^* = 0$ is stable and $x^* = 1$ is unstable. The payoffs have changed such that defection becomes the new equilibrium, and a population with a majority of C players will, over time, change to one with a majority of D players and, eventually, the elimination of C players altogether. Defection is universally favored. And the dilemma ends in tragedy. The broader themes connecting evolutionary dynamics and games require discussion not only of payoffs but also of the stability of action profiles with respect to perturbations—the topic of the next section.

10.3.3   Evolutionary Dynamics and Stable Strategies

There are a few remaining concepts relevant to understanding evolutionary game dynamics. For those familiar with conventional game theory, the central concept that is meant, putatively, to explain game dynamics is that of the Nash equilibrium. The Nash equilibrium concept pertains to action profiles; e.g., in a two-player game it refers to the strategy of both players, and in a multiplayer game it refers to the strategy of $n > 2$ players (Nash 1950). Formally, a Nash equilibrium is an action profile $a^* \in \mathbf{A}$ such that $\forall i \in \mathcal{N}$ where $\mathcal{N}$ are the player identities; then

$$u_i(a_i^*, a_{-i}^*) \geq u_i(a_i, a_{-i}^*). \qquad (10.19)$$

The notation a^* denotes an action profile—a vector—whereas a_i is a scalar. As such the notation a_{-i}^* denotes the set of actions taken at the equilibrium strategy profile by players *other* than i. Here u_i denotes the payoff associated with individual i given the action profiles. Hence, $u_i(a_i^*, a_{-i}^*)$ denotes the payoff to individual i insofar as both the focal individual and all other individuals use a Nash equilibrium strategy. In contrast, $u_i(a_i, a_{-i}^*)$ denotes the payoff to individual i insofar as that player deviates from the Nash equilibrium and when other players continue to play the Nash equilibrium strategy. In words, this means that the payoff that any player has when playing the Nash equilibrium cannot be unilaterally improved by changing their strategy while the other players continue to play the Nash equilibrium. Put another way, when all players play the Nash equilibrium strategy profile, then any player who changes their strategy will be no better off (i.e., will have the same payoff or worse). There can be more than one Nash equilibrium, so that if multiple players coordinate, then it would be possible to change many strategies at once and reach a better outcome. The PD game illustrates this point. The action profile (D, D) is the Nash equilibrium. Both players receive a payoff of 1. However, if either changes their strategy, then the profile would become (C, D) or (D, C). For the player that changes from $D \to C$, then their payoff goes down from 1 to 0. It would take both players to jointly change their behavior to C in order to improve their payoff. Despite its ubiquity in studies of game theory, and its critiques, it is not the only relevant concept for applications to living systems.

Consider the following three payoff matrices, where the convention remains that the values denote payoffs to the focal player (in rows) when playing against the opponent (in columns):

$$\text{Case 1}: \begin{bmatrix} 3 & 1 \\ 7 & 5 \end{bmatrix} \quad \text{Case 2}: \begin{bmatrix} 2 & 5 \\ 1 & 7 \end{bmatrix} \quad \text{Case 3}: \begin{bmatrix} 1 & 8 \\ 3 & 4 \end{bmatrix}$$

These three cases will help illustrate a number of concepts, starting with one from standard game theory and then moving into the realm of evolutionary game theory:

Invader strategy: A strategy that beats all others when another strategy is resident.

Unbeatable strategy: A strategy that cannot be beaten when it is dominant.

Coexistence: The dynamical regime where strategies can mutually invade one another.

Evolutionarily stable strategy (ESS): A strategy that is both an invader strategy and an unbeatable strategy.

In our three cases, case 1 is an example in which the second strategy is both an invader and an unbeatable strategy, and therefore it is an ESS (and also a Nash equilibrium). This means that the replicator dynamics for the payoff matrix in case 1 should end with the second player dominating. Case 2 is an example in which both strategies are unbeatable. Hence, neither player can be invaded if they are the resident. In this example, both pure strategies are Nash equilibria, such that we expect that the final outcomes depend on initial conditions. This phenomenon leads to alternative stable states. Finally, case 3 is an example in which both strategies can invade the other; neither is unbeatable. This case is equivalent to the HD game in which we expect the system to converge to an intermediate, mixed equilibrium. The reader may want to explore whether this mixed equilibrium is also Nash.

10.4 GAMES THAT REAL ORGANISMS PLAY

10.4.1 Animal fights

The interaction between animals does not invariably lead to costly conflicts. Animal behavior is replete with examples by which individuals fight over territory, resources, food, and access to reproductive opportunities (e.g., mates) in such a way that neither individual is physically harmed—or at least not critically so (Smith 1973). In fact, organisms have evolved multiple levels of sensing and behavior that are linked to assessment of the likely outcomes of such conflicts and that may provide a benefit to both. A classic example is that of red deer stag interactions when competing for access to female deer.

Figure 10.7 illustrates two of the key behaviors that can precede a potentially costly stag fight: roaring and parallel walking. These challenge type interactions occur at a distance, e.g., the challenger may roar hundreds of meters away from the target stag who is defending

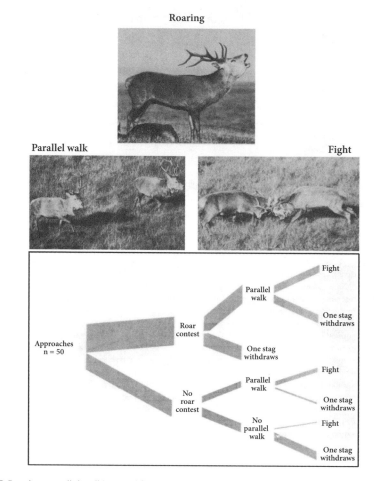

Roaring

Parallel walk

Fight

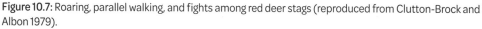

Figure 10.7: Roaring, parallel walking, and fights among red deer stags (reproduced from Clutton-Brock and Albon 1979).

access to females. The roaring between individuals often ends with the withdrawal of the challenger. However, further escalation is possible—including in the absence of the primary roaring round. Following roaring, about 50% of the contests lead to the withdrawal of the challenger. Parallel walking involves both defender and challenger, and can also precede a further escalation between individuals. The escalation proceeds when one stag (usually the challenger) lowers its head and a fight ensues. In the vast majority of cases, fights are averted.

This example suggests that, in the event of potential costly fights, other less costly displays might "advertise" the likely outcome of the fight, providing a signal to both individuals. For example, if the extent of roaring is an honest indicator of the health of an individual in the fight, then both animals can assess their likely outcome and proceed or withdraw accordingly. Such prefighting behaviors also suggest opportunities for cheating or "dishonest" displays—given the potential asymmetry in the interpretative value of behavior displays that precede fighting. A key take-away from this behavioral study is the extent to which individuals have tiered approaches to interact with potential competitors, assess their viability of success, and often avert a direct encounter. If the rule from the hawk-dove game is any guide, then sometimes the best outcome is not to fight at all.

10.4.2   Prisoner's dilemma in an RNA virus

Viruses are organisms too and worth a long read in their own right, not only because of SARS-CoV-2 but because of their global diversity and impact (Zimmer 2012) in shaping dynamics of cells, populations, and ecosystems (Suttle 2005; Weitz 2015; Breitbart et al. 2018). As organisms, viruses also have "behaviors," albeit the bulk of their behavior and potential relevance to game-theoretic questions emerges inside cells. Inside a cell, viral genomes can redirect host machinery to replicate viral RNA/DNA, synthesize structural components of the virus capsid, and then guide self-assembly of mature virions to be released back into the environment. Yet, when more than one virus infects the same cell at (more or less) the same time, then a new challenge is relevant, given that the outcome of the infection depends on what both viruses do.

In a now classic work, Paul Turner and Lin Chao investigated the dynamics of viruses competing for intracellular resources inside bacterial cells (Turner and Chao 1999). They did so by considering the dynamics of the RNA bacteriophage $\phi6$ and mutants. Phage $\phi6$ have a double-stranded RNA genome of ≈ 13.5 kb (comprising of three segments), and their genome is surrounded by a lipid membrane, which itself is surrounded by a protein capsid. Bacteriophage $\phi6$ can infect *Pseudomonas aeruginosa*, inject their genetic material inside bacteria, and then redirect cell machinery to produce more viruses, leading to the release of mature virions and death of the cell. This exploitation of cells by viruses has been well understood for more than half a century (see Chapter 1 on Luria and Delbrück's foundational work on the nature of mutation and selection). Yet, in certain conditions, for example, when virus abundances are relatively high, then multiple viruses—either of the same or different strain—infect the same cell. When this happens, then the viruses can be viewed as players in a game, with the cell setting the context and the payoffs determined by the intracellular strategies of the viral types.

Inside cells, viruses redirect cell machinery to produce capsids and genetic material—both of which are required to generate an infectious virion. Yet, when multiple viruses infect a cell, the production of a common pool of proteins could set the stage for defection. In an example of genomic downstreaming, some viruses have lost their ability to produce certain proteins (or have diminished expression of these proteins), while exploiting the ability of other viruses that can produce them (Huang 1973). These "defective interfering particles" (or DIPs) can flourish, as long as they are not alone. Such cheating can be partial. Viruses may produce fewer essential proteins, but nonetheless insert their genetic material into capsids whose components are produced from proteins expressed by other coinfecting viruses. This notion of stealing resources of other organisms can be a form of parasitism or predation, and is known as kleptoparasitism or kleptopredation, respectively. Examples of the former include cuckoo bees, which lay their eggs in the nests of other bees, with similar examples in the case of cuckoo birds (Brockmann and Barnard 1979). Yet, in an evolutionary sense, there is a risk in such kleptoparasitic strategies. Without the producer strains, the cheaters will fail. This is the tension underlying Turner and Chao's study: do they see examples of such game play, and if so, what kind of games do viruses play?

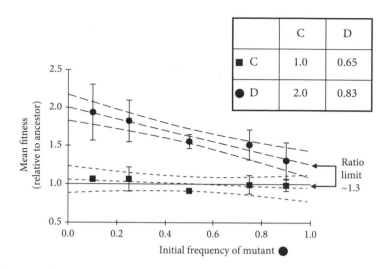

Figure 10.8: Interpreting the experiments of viral competition in a game-theoretic framework, adapted from Turner and Chao (1999). The estimated game-theoretic payoffs are shown in the matrix, where $\phi6$ is the cooperator and ϕH2 is the defector.

To examine this, the ancestral $\phi6$ phage and a mutant clone ϕH2 that was suspected of having a "cheater" phenotype were grown at high densities. A point mutation on $\phi6$ enabled a marked clone to grow on *Pseudomonas pseudocaligenes*. These phage cultures were then mixed together at different densities, and the fitness of viruses was measured in terms of the relative change in phage density ratio, i.e., $W = R_1/R_0$, where R_1 and R_0 were the ratios of the wild type:mutant phage densities in the collected lysate (at time point 1) versus those collected initially (at time point 0). Critically, all experiments were conducted at high multiplicity of infection such that most infections occurred with more than one phage. As shown in the top panel of Figure 10.8, the evolved phage, ϕH2, had a higher replication rate (proxy for fitness) relative to the ancestral type, $\phi6$, particularly when it was rare (i.e., low initial frequency in competition). The key point to recognize is that the x axis denotes the initial frequency in competition of the mutant. Hence, the fitness of the mutant is greater in an environment dominated by the wild type $\phi6$ rather than in that dominated by the mutant ϕH2. Taking the limit of the lines yields the values in the payoff matrix (Figure 10.8) of a reward of 1, a temptation to cheat of 2, and the ratio of the sucker's payoff to punishment of approximately $0.83/0.65 \approx 1.3$.

Additional experiments showed that ϕH2 replicated more slowly when alone than did $\phi6$, suggesting that the viruses ϕH2 and $\phi6$ exhibit a prisoner's dilemma that would favor the evolution of collectively "less fit" types. To prove that this game is in fact a prisoner's dilemma requires a different experiment. One needs to establish that wild-type (WT) viruses that coinfect single cells do in fact replicate faster than mutant viruses that coinfect single cells. To do so, Turner and Chao mixed WT viruses at high multiplicity of infection (MOI) with cells and separately mixed mutant viruses at high MOI with cells. They found that the defector phage replicated more slowly than did the cooperating phage by a factor of 0.83, thereby demonstrating that when phage evolved to defect, all were worse off.

This rationale seems counterintuitive. Why is that populations can evolve to be worse off? Why do individuals behave in a way that leads to a tragedy of the commons? The answers to both questions lie, at least in part, in the way that social context can provide opportunities for behavior (or strategies) that may be in the self-interest of each player (or population) but not necessarily in the interest of the community as a whole.

10.4.3 Microbial games

Oil and water don't mix. This lesson can be illustrated in simple experiments, amenable to elementary school science fairs. Yet the principle of the separation of immiscible liquids is also relevant to microbial populations. What happens when two distinct microbial populations are mixed together in the event that they differ in some key strategy to interacting with their neighbors? To explore games at the microbial scale requires that payoffs vary as a function of interactions. Here we turn to the type VI secretion system, or T6SS, as an archetype of how distinct outcomes in local interactions can give rise to emergent properties at the scale of populations.

The T6SS is a bacterial defense mechanism via which individual bacteria release toxic proteins into the cytoplasm of adjacent cells. The delivery mechanism is analogous to a nanomachine, insofar as a transmembrane syringe can transport individual proteins through the surface of adjacent cells and then release these proteins into the periplasm (Basler and Mekalanos 2012). Interestingly, the T6SS system is evolutionarily related to the contractile tail structure of bacteriophage. In the case of interactions with conspecifics, the release of these proteins—although costly—does not come with a toxic effect. Bacteria that utilize T6SS mechanisms as a form of competitive inhibition also produce antitoxins that inhibit negative effects to the producing bacteria themselves. In the case of *Vibrio cholerae* species, many strains possess distinct T6SS systems such that strains can be mutual killers of one another. This mutual killing can be thought of as a two-player game, in which all players pay a cost C for maintaining and utilizing the T6SS system but have a gain G via the elimination of nearby cells of different strain types.

The spatiotemporal consequences of T6SS have been explored in the interactions between variants of *V. cholerae* that function as mutual killers (McNally et al. 2017). Figure 10.9 (Plate 12) shows the dynamics of a red and blue labeled clone in the event of interactions between T6SS-deficient mutants (top) and WT strains with a functional T6SS system (bottom). The red and blue colors are distinct strains of *V. cholerae* (strains C6706 and 692-79, respectively). The mutants have a specific deletion that renders their T6SS system nonfunctional. In the T6SS-deficient case, the initially random mixture of clones remains spatially homogeneous without any apparent spatial pattern. In contrast, in the WT case, mutual killing by strains that have functional T6SS systems leads to the emergence of patches from an otherwise homogeneous starting condition. This emergence of patchiness was replicated in the original work through a series of mathematical models, including an individual-based model, a partial differential equation model, and variants that account for killing and for public good extraction. All of these models yield the same kind of coarsening behavior in which homogeneous distributions start to form patches and those patches grow larger over time—formally coarsening in the sense that the patches grow

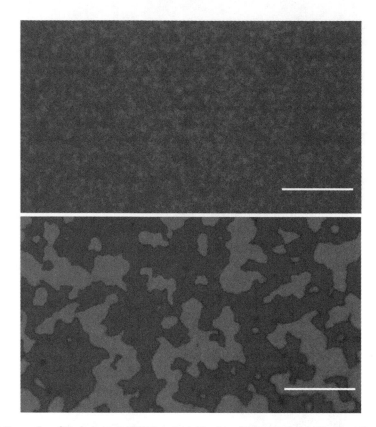

Figure 10.9: Dynamics of *V. cholerae* in T6SS⁻ mutants (top) and in interactions with a functional T6SS system (bottom). In the top panel cells cannot kill each other locally, while in the bottom panel cells can kill each other locally via the T6SS mechanism. The phase separation in the T6SS case in the bottom panel is a hallmark of coarsening and a direct result of the coordination nature of this bacterial game. Scale bars denote 100 μm. Reproduced from McNally et al. (2017).

like $t^{1/2}$, as expected in phase separation via the Cahn-Hilliard equations (Allen and Cahn 1972; Novick-Cohen and Segel 1984).

Here we take a slightly different approach to understanding the phenomenon of how mutual killing could lead to the emergence of patchiness and perhaps even encourage cooperation. Using the framework of this chapter, consider the strategy-dependent payoff of two *Vibrio* strains that compete in a two-dimensional spatial habitat, e.g., like an agar plate.

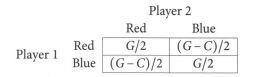

		Player 2	
		Red	Blue
Player 1	Red	$G/2$	$(G-C)/2$
	Blue	$(G-C)/2$	$G/2$

G denotes the gain associated with reproducing and C denotes the cost of the type VI killing machinery by a noncompatible type. Type VI secretion systems lead to the potential death of the target cell. Here the secretion is assumed to be costly to the opponent, i.e., $C > G$.

The expected dynamics of the red type with frequency x can be written in terms of the replicator dynamics for two-player games:

$$\dot{x} = r_1(\mathbf{x}, A)x - \langle r \rangle (\mathbf{x}, A)x \tag{10.20}$$

where $1 - x$ is the frequency of the blue type and the fitness values are

$$r_1(x) = Gx/2 + (1-x)(G-C)/2 \tag{10.21}$$

$$r_2(x) = (G-C)x/2 + (1-x)G/2 \tag{10.22}$$

$$\langle r \rangle = r_1 x + r_2 (1-x) \tag{10.23}$$

After some algebra (see technical appendix), the dynamics of the red type can be re-written as

$$\dot{x} = x(1-x)(r_1(x) - r_2(x)) \tag{10.24}$$

or, after some reduction,

$$\dot{x} = Cx(1-x)\left(x - \frac{1}{2}\right). \tag{10.25}$$

This cubic equation has three roots, at 0, 1/2, and 1. For any initial conditions where $x < 1/2$, the system will be driven into an "all-blue" state, $x^* = 0$, whereas for any initial conditions where $x > 1/2$, the system will be driven into an "all-red" state, $x^* = 1$, as in Figure 10.10. These equations describe the dynamics of frequencies in well-mixed populations, where the probability of meeting an individual of the other type is the same for every cell. However, the study of the T6SS system requires spatial organization, in which individuals interact with their neighbors, whether of the same or different strain type.

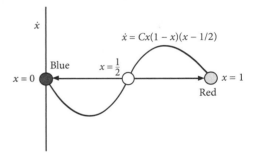

Figure 10.10: Sketch of the frequency-dependent growth rate, $\dot{x}$, of individuals playing strategy "red" given frequency x according to the replicator dynamics for T6SS-like games in Eq. (10.25). The fixed points occur for $\dot{x} = 0$, in this example when $x = 0$ (all blue), $x = 1/2$ (mixed), and $x = 1$ (all red).

Moving from two-player games to spatial extensions requires separating the game-play step from the replication step. In evolutionary game theory, payoffs accrue to replicators, and the replication of individuals in distinct populations changes the social context for subsequent interactions. By analogy, the procedure for considering a spatial game dynamic should follow along similar lines, while taking into account the localized nature of interactions. First, a focal player and a randomly selected neighbor are paired to interact, and the focal player accrues a payoff. This payoff is then directly tied to the reproduction or death of the focal player. The computational lab guide associated with this chapter provides both a pseudocode description of this procedure and a fully stochastic implementation. In brief, positive payoffs correspond to a per capita rate of reproduction, whereas negative payoffs correspond to a per capita rate of death. If the individual reproduces, then it takes over the location of a randomly selected neighbor. If the individual dies, then its location is taken over by a randomly selected neighbor (see Lin and Weitz 2019 for a similar implementation). In effect, this amounts to a coordination game. Red-red interactions lead to net positive payoffs, reinforcing the growth of patches. In contrast, red-blue interactions, which take place on interfaces, are

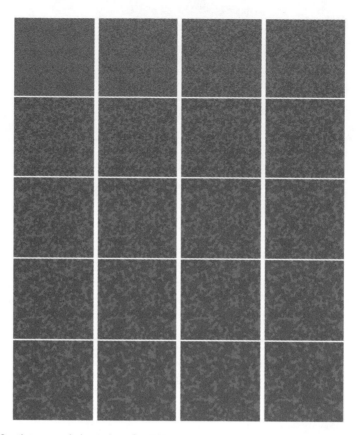

Figure 10.11: Spatiotemporal dynamics of a T6SS game in space, from random initial conditions to labyrinthine patterns. This game uses $G = 10$ and $C = 18$. The times shown here are after 0.2 time units (the equivalent of 20 iterations, each with a dt = 0.01). The time progresses across each row from left to right and then down from top to bottom. More details on how to develop a stochastic simulation of T6SS-like games are described in the computational laboratory associated with this chapter.

negative, leading to increased death and the expansion of patches. Figure 10.11 (Plate 13) shows a series of snapshots of a coordination game in space, beginning with homogeneously distributed red and blue types and leading to a gradual coarsening of the labyrinthine patterns. *Coarsening* is defined as the slow increase in the characteristic correlation length in the system, i.e., the length over which individuals in the population are likely to be of the same type. Critically, the coarsening increases with $t^{1/2}$, suggesting that the phenomenon of coarsening is a particularity not of mutual killing but of the larger class of coordination games (Cahn and Hilliard 1958; Cahn 1961). There are many ways for conspecifics to do better among conspecifics, and such types of asymmetries can, for better or worse, lead to the emergence of patchiness when individual games have population-level consequences.

In summary, qualitatively different games correspond to qualitatively different well-mixed dynamics. These well-mixed dynamics provide the basis for understanding "local" behavior of a spatially extended system. But, in spatially extended systems, there is feedback between the game and the composition of the local neighborhood. Thus, different games also give rise to different spatial patterns, including the growth of locally homogeneous patches in an otherwise heterogeneous population. Paradoxically, the adversarial nature of interactions between types leads to (potentially) cooperative local regimes that expand over time (McNally et al. 2017). This feedback between replication and local structure can lead to a wide variety of spatial patterns, which themselves may shape the evolution of new strategies (Nowak and May 1992; Hauert and Doebeli 2004; Doebeli and Hauert 2005; Roca et al. 2009; Helbing et al. 2010).

10.5   FEEDBACK BETWEEN STRATEGIES AND THE ENVIRONMENT

The replicator dynamics formulation of games can lead to a variety of outcomes, including single player dominance (as in the PD game), coexistence (as in the hawk-dove game), and alternative stable states (as in coordination games—the reason why we all agree to drive on the same side of the road, but the side can differ by countries). However, certain outcomes are not accessible. A two-player replicator dynamics system cannot oscillate given that the dynamics can be represented as a 1D system in terms of $\dot{x}$, whose nonlinearity depends on the payoffs and mapping from payoffs to fitness. The absence of oscillations is generic, so long as the payoff matrix and mapping of payoffs to fitness are fixed—because deterministic dynamics can only move in one direction at any given point in the domain $0 \leq x \leq 1$. But there is a problem with such an approach. Game outcomes depend on environmental context. Hence, replicator dynamics neglect the possibility that repeated actions eventually have an impact on the environment and, in turn, the associated payoffs.

This book has its origins in a class developed at Georgia Tech, in a state that has had a recurring set of water crises. Lake Lanier, the primary reservoir for the greater Atlanta metropolitan area, has experienced significant variation in water levels. The baseline surface of the lake tends to be at approximately 1072 feet above sea level in replete times, and drops below 1066 feet are cause for significant concern. In the summer of 2007, the state experienced an intense drought such that by November 2007 the water levels reached below 1058 feet. Severe water restrictions were imposed, and it took nearly two years before water levels again reached normal. Eventually, water restrictions were relaxed (e.g., in July 2009 the director of the Georgia Environmental Protection Division remarked that "our water supplies are flush"), and yet by the summer of 2012 the state was in another crisis, with the levels dropping below 1065 feet, again rebounding, and then experiencing another drought and levels of 1062 feet in 2017. This cycle of environmental scarcity leading to improved protections, followed by a restoration of a commons, relaxation of restrictions, and then another crisis recurred multiple times in approximately a decade in the state. Such coupling is hardly unique.

Consider the spread of an infectious disease in which a vaccine provides strong protection against infection and transmission. Yet perhaps an individual has some doubts regarding the safety of the vaccine. That person reason: if everyone has a vaccine, then why should I take it? That kind of question has a perverse soundness. In subsequent chapters, this book takes on the challenge of how herd immunity can be reached even if only a large portion of the population receives a vaccine. But, if everyone thinks that this level is reached and more people decide not to take a vaccine, then the risk can go up as a direct

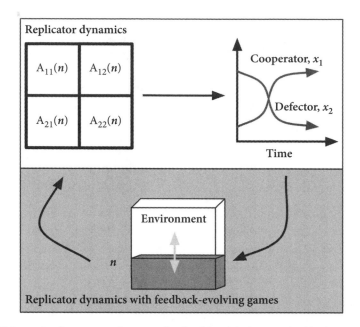

Figure 10.12: Schematic of a game-environment feedback model, characterized by frequency of cooperators, $x_1 \equiv x$, and defectors, $x_2 \equiv 1 - x$, as well as the environmental state n. The environmental state determines the payoff matrix (upper left), which influences the replicator dynamics (upper right). However, changes in the frequency of cooperators and defectors leads to a reshaping of the environmental state, which in turn changes the payoff matrix. This autonomous feedback system can give rise to game-environment dynamics different than classic replicator dynamics given a fixed payoff matrix. Reproduced from Weitz et al. (2016).

result of noncompliance, leading to greater incentives to comply. Yet, when compliance is high and the risk is low, the situation is reversed (Bauch and Earn 2004). The same concepts underlie the production of public goods and microbes. When public goods are replete, then individual microbes gain little benefit from enzyme production; yet if all microbes behave this way, then eventually the public supply can diminish, leading to an incentive for production, increasing the public supply, and thus restarting the cycle. This kind of coupling between game and environment warrants an extension of conventional evolutionary game theory.

In an effort to link game and environment, consider a modified version of replicator dynamics in which

$$\epsilon \dot{x} = x(1-x)\left[r_1(x, \mathbf{A}(n)) - r_2(x, \mathbf{A}(n))\right]$$
$$\dot{n} = n(1-n)f(x)$$

$$(10.26)$$

where x denotes the frequency of cooperators and n denotes the state of the environment. As before, the frequency of cooperators increases or decreases based on their fitness relative to defectors. Unlike conventional game theory models, the payoff matrix $\mathbf{A}(n)$ is *environment dependent* such that strategy and environmental dynamics are coupled (Figure 10.12) (Weitz et al. 2016). Here $f(x)$ denotes how players reshape the environment, and the term $n(1-n)$ in Eq. (10.26) restricts the environment between 0 (a depleted state) and 1 (a replete state). The players improve (degrade) the environmental state n when $f(x)$ is greater (less) than 0. Finally, ϵ denotes the relative speed of player and environment dynamics. When $\epsilon \ll 1$, the environment changes much more slowly than the frequency of players, and when $\epsilon \gg 1$, the environment changes much faster than the frequency of players.

It is worthwhile to explore this general model class via a particular choice of environment-dependent payoffs:

$$\mathbf{A}(n) = (1-n) \begin{bmatrix} 5 & 1 \\ 3 & 0 \end{bmatrix} + n \begin{bmatrix} 3 & 0 \\ 5 & 1 \end{bmatrix}. \tag{10.27}$$

This payoff matrix represents a linear combination of two scenarios, corresponding to $n = 0$ and $n = 1$. When the environment is depleted and $n = 0$, then the payoffs favor cooperation. Instead, when the environment is replete and $n = 1$, then the payoff matrix is equivalent to a PD game where defection is favored. As a result, incentives and behaviors change as the environment changes. Changes in the frequency of strategies in turn change the environmental state. Feedback dynamics are represented as follows:

$$f(x) = \theta x - (1-x) \tag{10.28}$$

such that cooperators enhance the environment and defectors degrade the environment given $\theta > 0$, the ratio of the enhancement rates to degradation rates of cooperators and defectors, respectively. Combining the elements together, this model of replicator dynamics with environmental feedback can be written as

$$\begin{aligned} \epsilon \dot{x} &= x(1-x)(1+x)(1-2n) \\ \dot{n} &= n(1-n)\left[-1 + (1+\theta)x\right] \end{aligned} \tag{10.29}$$

given the prior specification of $\mathbf{A}(n)$ (see the technical appendix for derivation).

The feedback between game and environment implies that there are now five fixed points where $\dot{x} = 0 = \dot{n}$, four of which correspond to "corners" in the plane. The four corners are (i) $(x^* = 0, n^* = 0)$—defectors in a degraded environment; (ii) $(x^* = 0, n^* = 1)$—defectors in a replete environment; (iii) $(x^* = 1, n^* = 0)$—cooperators in a degraded environment; (iv) $(x^* = 1, n^* = 1)$—cooperators in a replete environment. The last fixed point is $\left(x^* = \frac{1}{1+\theta}, n^* = \frac{1}{2}\right)$, corresponding to a mixed population of cooperators and defectors in an environment halfway between depleted and replete. These five fixed points guide the next question: what new dynamics might unfold in this two-dimensional system? Critically, each of the corner fixed points is unstable. For example, a degraded environment dominated by cooperators will experience an increase in the environmental state—driving the system away from the fixed point. Similar logic applies to each of the corner points. In contrast, a linear stability analysis of the interior fixed point yields two eigenvalues with zero real components and paired imaginary components (see the technical appendix for details). This finding implies that perturbations are neutrally stable—in fact, the model has a conserved quantity associated with a continuum of orbits in the plane (Figure 10.13). These orbits correspond to cycles in both the frequency of strategies and environmental state that differ only in their initial conditions.

The cyclical dynamics can be understood intuitively. First, consider a degraded environment dominated by defectors—but not entirely so. In this case, the payoff matrix given

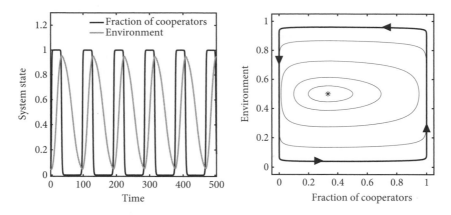

Figure 10.13: Cyclical dynamics of a game-environment feedback model in time (left) and in the phase plane (right). In this example, $\mathbf{A}(n)$ linearly interpolates between incentives for cooperation and defection at $n = 0$ and $n = 1$, respectively, as described in the text. Additional parameters are $\epsilon = 0.1$ and $\theta = 2$. The distinct curves in the phase plane correspond to different initial conditions—each is a closed orbit with counterclockwise dynamics. The asterisk denotes the neutrally stable fixed point ($x^* = 1/3$, $n^* = 1/2$). Reproduced from Weitz et al. (2016).

$n \rightarrow 0$ leads to the invasion of cooperators (recall when the environment is degraded, the payoff matrix corresponds to one where cooperation is incentivized and perhaps may even be required for survival). The invasion of cooperators is rapid and the dominance of cooperators enhances the environment. However, as the environmental state increases (e.g., $n \rightarrow 1$), so too do the incentives for defection (recall when the environment is replete, the payoff matrix corresponds to a PD game). The invasion of defectors is rapid and the dominance of defectors subsequently degrades the environment, leading back to the starting point of the cycle—which then repeats and repeats. These principles apply near the corners of the plane but also in the interior given alternative choices of initial conditions. Hence, inclusion of game-environment feedback leads to a new phenomenon: an oscillatory tragedy of the commons.

This example highlights the consequences of a joint change in strategies and environmental state. In the long term, this coupling of game and action can lead to highly unexpected outcomes. The particular example here is one in which the outcomes are oscillations between good environments marked by defection and bad environments marked by cooperation. As a final observation, the particular choice of symmetric payoff values given $n = 0$ and $n = 1$ is not generic—in fact, the quantitative values matter such that changing the game structure in the extremal environmental conditions leads to even richer dynamics, including cases where tragedies of the commons can be avoided altogether. As a result, this feedback can lead to intermediate outcomes: heteroclinic cycles in which the system hops between extremal states (Weitz et al. 2016), and true limit cycles once density-dependent (rather than frequency-dependent) environmental feedback is considered (Tilman et al. 2020). One of the major lessons from these simple models is that, when outcomes impact the environmental state, tragedies of the commons are not inevitable. Yet they can happen. In theory, averting a tragedy is possible when a group of individuals are willing to cooperate even when the environment is bad and the system is dominated by defection (Weitz et al. 2016).

Stepping back: The start of this chapter illustrated some of the ways in which these lessons may apply to humans and our use or misuse of the commons. As Figure 10.1 illustrates, individuals may be inclined to cooperate at levels far exceeding those expected through an examination of purely selfish incentives. Moreover, cooperation can be enhanced by incentives that restructure what happens over the long term. In that sense, being strategic does not always mean having an adversarial interaction with other players—since long-term viability depends on avoiding bad outcomes at least as critically as on making the most of good opportunities. Avoiding bad outcomes often requires cooperation. This means that it is not just idealism but also suitable incentives and context that can help redirect a system (or a commons) back in the right direction.

10.6   TAKE-HOME MESSAGES

- Individual organisms can display distinct strategies that impact their reproduction, survival, and fitness.
- From a game-theoretic perspective, the fitness of an organism depends on both what it does and what others do.
- The logic of animal thinking can be approached via the framework of evolutionary game theory.
- In evolutionary game theory, context-dependent strategies that are better than the average can invade, leading to mechanisms for domination and coexistence.
- The canonical example of the hawk-dove game is one in which costly fights lead to a coexistence of hawks and doves.
- The canonical example of a prisoner's dilemma game reveals how acting in one's self-interest need not lead to the best outcome for all (and in fact can lead to a worse outcome than had individuals not acted selfishly).
- Concepts of game theory can be seen in animal conflicts, virus-virus interactions, and microbial interactions.
- New extensions of game theory to spatially extended interactions and to game-environment feedback provide new opportunities to connect payoffs with the consequences of action taking for the system as a whole—including the environmental commons.

10.7   HOMEWORK PROBLEMS

The computational laboratory guide includes a range of material spanning the dynamical simulation of replicator dynamics and the stochastic and spatial implementation of games. As before, these exercises and toolsets are critical to addressing the homework problems. Topics covered include

- Replicator dynamic simulations for two-player games
- Stochastic, individual-based simulations for two-player games
- Birth-death simulations
- Development and implementation of pseudocode
- Spatial simulation, including periodic boundary conditions and neighbor selection

The code set for the stochastic spatial model is presented in its entirety in the solution set to the laboratory guide and is particularly relevant to addressing the problems here.

PROBLEM 1. Simulating Evolutionary Games

Consider the following sets of payoff matrices introduced in the chapter:

$$\text{Case 1}:\begin{bmatrix} 3 & 1 \\ 7 & 5 \end{bmatrix} \quad \text{Case 2}:\begin{bmatrix} 2 & 5 \\ 1 & 7 \end{bmatrix} \quad \text{Case 3}:\begin{bmatrix} 1 & 8 \\ 3 & 4 \end{bmatrix}$$

For each, develop a replicator model version of the game and evaluate the outcomes. Classify whether there is a single winning strategy, a mixed strategy, or alternative stable states. Compare and contrast your findings with those expected via a stability analysis of the 1D replicator dynamics.

PROBLEM 2. Stochastic Games

Using the payoff matrix introduced in Problem 1:

$$\text{Case 3}:\begin{bmatrix} 1 & 8 \\ 3 & 4 \end{bmatrix}$$

develop a stochastic model of replicator dynamics in a population of size 200. In doing so, consider starting conditions where the initial fraction of strategy 1 varies from 10% to 90%. How does the outcome of dynamics differ or agree with predictions from Problem 1? If it agrees, take an extreme case and initialize the population such that either 1% or 99% of the individuals are of strategy 1. Does the system inevitably follow the expected replicator dynamics? Why or why not?

PROBLEM 3. Spatial Games

Using the code base developed in the laboratory, evaluate the time scale of coarsening. Beginning with a homogeneously distributed environment of size 100×100, assess the extent to which patches grow in size and how that growth scales with the number of iterations. Develop your own approach to assessing the patch size, e.g., by leveraging two-dimensional spatial correlation functions or coming up with your own technique. (Advanced: Repeat this problem in an asymmetric game such that the gain for one type is 20 whereas for the other type it remains 16.)

PROBLEM 4. Game-Environment Feedback

Consider the coupled game-environment model presented in this chapter:

$$\epsilon \dot{x} = x(1-x)\left[r_1(x, \mathbf{A}(n)) - r_2(x, \mathbf{A}(n))\right]$$
$$\dot{n} = n(1-n)f(x) \tag{10.30}$$

where

$$\mathbf{A}(n) = (1-n)\begin{bmatrix} 5 & 1 \\ 3 & 0 \end{bmatrix} + n\begin{bmatrix} 3 & 0 \\ 5 & 1 \end{bmatrix}. \tag{10.31}$$

Using the phase plane as a guide and restricting to the case $S_0 = 1$ and $P_0 = 0$, modulate the difference between R_0 and T_0 so that it is both >0 corresponding to cooperation during depleted times and <0 corresponding to defection during depleted times. Simulate the model and try to find the critical level of difference that enables aversion of the tragedy of the commons, such that the long-term value of $n(t) > 0$. In doing so, address whether or not the answer depends on initial conditions on the interior of the space and justify your answers with phase plane dynamics.

10.8 TECHNICAL APPENDIX

Derivation of the replicator dynamic model for the HD game The dynamics of hawks (x) in the HD game are

$$\dot{x} = r_1(x, \mathbf{A})x - \langle r \rangle(x, \mathbf{A})x. \tag{10.32}$$

As stated in the chapter, the fitness values for the HD game are

$$r_1 = \frac{(G-C)}{2}x_1 + Gx_2 \tag{10.33}$$

$$r_2 = 0 \times x_1 + \frac{G}{2}x_2 = \frac{G}{2}x_2 \tag{10.34}$$

and the average fitness is

$$\langle r \rangle = \frac{(G-C)}{2}x_1^2 + Gx_1x_2 + \frac{G}{2}x_2^2. \tag{10.35}$$

Recalling that $x_1 = x$ and $x_2 = 1-x$, the hawk dynamics can be rewritten as

$$\begin{aligned}
\dot{x} &= \left(\frac{(G-C)}{2}x + G(1-x)\right)x - \left(\frac{(G-C)}{2}x^2 + Gx(1-x) + \frac{G}{2}(1-x)^2\right)x \\
&= x\left[\left(\frac{(G-C)}{2}x + G(1-x)\right) - \left(\frac{(G-C)}{2}x^2 + Gx(1-x) + \frac{G}{2}(1-x)^2\right)\right] \\
&= x\left[\frac{(G-C)}{2}x(1-x) + G(1-x)^2 - \frac{G}{2}(1-x)^2\right]
\end{aligned}$$

$$= x(1-x)\left[\frac{(G-C)}{2}x + G(1-x) - \frac{G}{2}(1-x)\right]$$

$$= x(1-x)\left[\frac{(G-C)}{2}x + \frac{G}{2}(1-x)\right] \tag{10.36}$$

as noted in the chapter text.

Derivation of the replicator dynamic model for the T6SS system Consider a two-player game with red and blue types such that the frequency-dependent fitness functions are

$$r_1(x) = Gx/2 + (1-x)(G-C)/2 \tag{10.37}$$

$$r_2(x) = (G-C)x/2 + (1-x)G/2 \tag{10.38}$$

$$\langle r \rangle = r_1 x + r_2(1-x) \tag{10.39}$$

The dynamics of $\dot{x}$ follow:

$$\dot{x} = x\left[r_1(x) - \langle r \rangle\right], \tag{10.40}$$

which can be written as

$$\dot{x} = x\left[r_1 - r_1 x - r_2(1-x)\right] \tag{10.41}$$

$$= x\left[r_1(1-x) - r_2(1-x)\right] \tag{10.42}$$

$$= x(1-x)(r_1 - r_2) \tag{10.43}$$

The difference between $r_1(x)$ and $r_2(x)$ is

$$r_1 - r_2 = Gx/2 + (1-x)(G-C)/2 - (G-C)x/2 - (1-x)G/2, \tag{10.44}$$

$$= Cx - C/2, \tag{10.45}$$

$$= C\left(x - \frac{1}{2}\right) \tag{10.46}$$

such that

$$\dot{x} = Cx(1-x)(x-1/2) \tag{10.47}$$

as analyzed in the chapter.

Derivation of the game-environment feedback model and local stability Consider the environment-dependent payoff matrix as defined in the chapter:

$$\mathbf{A}(n) = (1-n)\begin{bmatrix} 5 & 1 \\ 3 & 0 \end{bmatrix} + n\begin{bmatrix} 3 & 0 \\ 5 & 1 \end{bmatrix}. \tag{10.48}$$

This linear interpolation can be reduced to

$$\mathbf{A}(n) = \begin{bmatrix} 5-2n & 1-n \\ 3+2n & n \end{bmatrix}. \tag{10.49}$$

As a result, the fitness values for cooperators and defectors are

$$r_1(x, n) = 1 + 4x - n(x + 1) \tag{10.50}$$

$$r_2(x, n) = 3x + n(x + 1) \tag{10.51}$$

such that their difference is

$$\begin{aligned} r_1(x, n) - r_2(x, n) &= 1 + 4x - n(x + 1) - 3x - n(x + 1) \\ &= 1 + x - 2n(1 + x) \\ &= (1 + x)(1 - 2n) \end{aligned} \tag{10.52}$$

which is the basis for the modified replicator dynamics

$$\begin{aligned} \epsilon \dot{x} &= x(1 - x)(1 + x)(1 - 2n) \\ \dot{n} &= n(1 - n)\left[-1 + (1 + \theta)x\right] \end{aligned} \tag{10.53}$$

as analyzed in the chapter.

The local stability of each fixed point can be evaluated via the Jacobian. The Jacobian at each of the corner fixed points is

$$\mathbf{J}(0,0) = \begin{bmatrix} P - S & 0 \\ 0 & -1 \end{bmatrix} \qquad \mathbf{J}(1,0) = \begin{bmatrix} R - T & 0 \\ 0 & \theta \end{bmatrix}$$

$$\mathbf{J}(0,1) = \begin{bmatrix} S - P & 0 \\ 0 & 1 \end{bmatrix} \qquad \mathbf{J}(1,1) = \begin{bmatrix} T - R & 0 \\ 0 & -\theta \end{bmatrix} \tag{10.54}$$

where $R = 3$, $S = 0$, $T = 5$, and $P = 1$. As a result of the fact that $T > R$ and $P > S$, there is one positive eigenvalue at each of the corner fixed points. Hence, each of the four corner fixed points are unstable with respect to perturbations in the interior. Similarly, the Jacobian of the interior fixed point is

$$\mathbf{J} = \begin{bmatrix} 0 & -2x^*(1 - x^*)(1 + x^*) \\ \theta/4 & 0 \end{bmatrix} \tag{10.55}$$

where $0 < x^* = 1/(1 + \theta) < 1$. The eigenvalues for the interior fixed point are

$$\lambda_{interior} = \pm i \sqrt{\frac{\theta x^*(1 - x^*)(1 + x^*)}{2}}.$$

The interior point is neutrally stable such that small deviations are expected to oscillate—this is the case even when analyzing the full model. The full dynamics in Eq. (10.29) can be separated into functions of x or n, which means that it is possible to take the derivative of dx/dn and find a conservation law (analogous to the conservation law in the original Lotka-Volterra model). As a result, these periodic orbits are not limit cycles; instead, each is neutrally stable and linked to the initial conditions (full details in Weitz et al. 2016).

Chapter Eleven

Eco-evolutionary Dynamics

11.1   THE POWER OF EXPONENTIALS

The study of how populations grow and decline is perhaps the oldest of all branches in mathematical biology. In his famous *Essay on the Principle of Population* written in 1798, Thomas Malthus (1970) encapsulated what is now known as the Malthusian principle of population growth:

> I say, that the power of population is indefinitely greater than the power in the earth to produce subsistence for man.

Population, when unchecked, increases in a geometric ratio. That is, two become four, and four become eight, and eight become sixteen, and very soon the numbers are astronomical—far greater than the word *astronomical* could even convey, i.e., greater than the stars in the sky. The power of 2 was of interest well before Malthus. The thirteenth-century Islamic scholar Ibn Khallikān is thought to have proposed the following puzzle as a reward from the sovereign. His requests were modest: he only wanted one grain of rice on the first chessboard, two on the second, four on the third, eight on the fourth, and so on until all 64 squares were filled. The reward of course was not modest at all. By the end, it would take $2^{64} - 1$ grains of rice to fill up the board, or nearly 2 billion billion grains! Exponentials are fast, and indeed, this "power"—in the Malthusian sense—stands in contrast to the finite nature of resources. Malthus (1970) recognized this dichotomy:

> Subsistence increases only in an arithmetical ratio. A slight acquaintance with numbers will shew the immensity of the first power in comparison of the second.

In modern parlance, this means that if populations grow as $N(t+1) = rN(t)$ and the availability of food grows as $R(t+1) = R(t) + a$, then very soon $R/N \ll 1$. The balance of populations and their food, or consumers and resources, is central to studies of ecological and evolutionary dynamics.

This chapter provides an introduction, data, theory, and methods to consider how individuals making up a population interact, how such interactions drive changes in the abundances of populations, and how differential reproduction and survival over time can

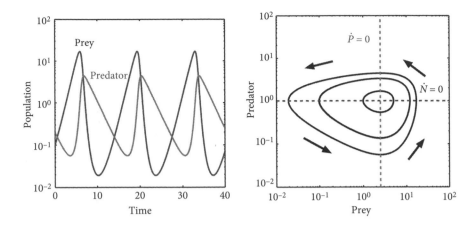

Figure 11.1: Canonical predator-prey dynamics in the original Lotka-Volterra model. (Left) Dynamics as a function of time. (Right) Phase plane view, where time is implicit. In the phase plane, dynamics follow trajectories imprinted by distinct initial conditions, as explained in Section 11.3. The $\dot{N} = 0$ and $\dot{P} = 0$ dashed lines denote nullclines, with a single intersection denoting the coexistence point. The direction of dynamics is indicated by the arrows; it is counterclockwise in the predator-prey plane. The dynamics here are those of Eqs. (11.14)–(11.15), with $r = 1$ hr^{-1}, $c = 10^{-6.5}$ ml/hr, $\epsilon = 0.1$, and $m = 0.1$ hr^{-1}.

reinforce trait differences that drive the evolution of complex and diverse communities. In doing so, it is important to state at the outset what is different in this chapter from what is typically introduced as the canonical view of ecological dynamics (Figure 11.1). The modern origins of population dynamics are attributed to the work of Lotka (1925) and Volterra (1926). As explained in the preface, Lotka and Volterra were concerned with the question of dynamic change of populations. Specifically, were oscillations in population abundances attributable to external drivers, or could they be driven by internal feedback? Volterra was inspired by interyear variations in the yield of Adriatic fisheries. Perhaps oscillations with periods that spanned multiple years resulted from variation in some time-varying environmental property, like temperature, salinity, or mixing, that influenced the annual catch size. However, Volterra reasoned that changes in fishing pressure—a potential significant mortality factor—could be sufficient to drive changes in both predator and prey species as a result of changes in interaction rates between them. The result—a model of coupled predator-prey dynamics to be introduced formally in this chapter—yields oscillations in which predators consume prey, driving prey downward; then as prey decrease, predators decline, and given low levels of predators, the prey recover, leading to resurgence of predators, and the cycle repeats.

Notably, such coupling between an ecological and a human social system has reemerged to become central to understanding modern ecosystems, at both local and global scales. The most visible evidence comes from studies of climate change, an issue to be revisited in Chapter 13. But perhaps just as important are the ways in which human and "natural" systems intersect, e.g., inspiring the formation of national centers, including the National Socio-Environmental Synthesis Center (SESYNC), an NSF-funded research center that was based in Annapolis, MD. Yet if one were to replace fisheries with predators, then we would have struck at the core of ecological dynamics theories. That is, interactions between populations are sufficient to give rise to nonstationary dynamics, including oscillations, even given a "constant" environment. But these oscillations are unlikely to have the imprint of long-ago initial conditions—a feature of a model but not reality.

Indeed, we are not Lotka nor Volterra, and despite our debt to them, with 100 years of work, it would seem appropriate to aim for something more ambitious and, indeed, more realistic. The study of predator-prey dynamics sits squarely in the field of ecology. That is, measurements and models assume that changes due to predation (or consumption more generally) reflect changes due to vital processes—i.e., life and death—between otherwise genetically homogeneous populations. Such an assumption seems reasonable if one thinks that the change in the frequency of genotypes in a population varies over time scales far slower than population rates of change. But what if that assumption is not reasonable? Indeed, perhaps one of the most exciting developments in predator-prey and consumer-resource dynamics more generally is the realization that evolution is rapid and can take place on time scales similar to those of ecological dynamics (Yoshida et al. 2003; Hairston et al. 2005; Duffy and Sivars-Becker 2007; Cortez and Ellner 2010; Hiltunen et al. 2014; Cortez 2018; Reznick et al. 2019). Indeed, if one doesn't arrive to the field with the built-in bias of time scale separation, or if one approaches the questions given a background in microbial dynamics, then it would seem apparent that the same forces that shape whether an organism lives or dies may also shape the frequency of genotypes in a population.

Understanding eco-evolutionary dynamics requires a nested approach. First, this chapter introduces simple models of competition for a common resource and interactions between a dynamic consumer and resource (e.g., as in predator-prey dynamics). These models are, in essence, the Lotka-Volterra models of competition and of predator-prey dynamics, respectively. Yet, to bridge between classic and modern challenges in ecology, it is important to recognize that such classic models have many assumptions, perhaps none more important than that the consequences of interactions among individuals unfold on time scales much faster than changes in the traits of individuals in those populations.

As a key motivating example, Figure 11.2 shows multiple noncanonical phase relationships between predator and prey (Yoshida et al. 2003). This study examined the population and evolutionary dynamics of rotifers and algae. Rotifers are freshwater zooplankton, i.e., small aquatic organisms that eat phytoplankton like algae (which are themselves photo-autotrophs that fix carbon dioxide into organic carbon). These organisms were inoculated in chemostats. Chemostats are continuous culture devices, i.e., flasks with an input and output port. The input port enables new food to be added to the system while the output port allows all the components to be diluted—simulating a background mortality rate as well as providing opportunities for continuous and noninvasive sampling. As is apparent in panels A–D, mixing rotifers and algae together leads to emergent oscillations, even though the chemostat is a continuous culture device and does not include exogenous changes in inflow or outflow. The experiments in panels A–D also reveal oscillations, albeit short ones, in which predators increase in abundance following peaks in prey abundance, leading to decreases in prey abundance, a decline in predators, and then a recovery of prey. In essence, these four panels reveal classic Lotka-Volterra oscillations expected when combining a single predator with a single prey.

Yet panels E–I reveal quite a different story. In these experiments, the total algal population is plotted against the total rotifer population. However, the experiments include

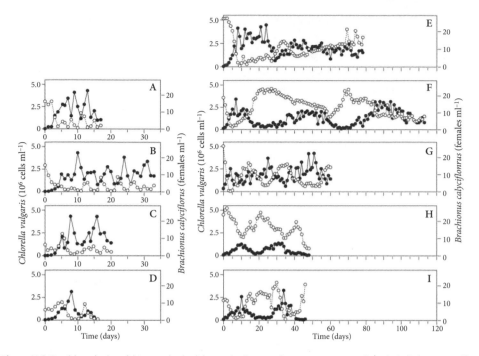

Figure 11.2: Rapid evolution drives ecological dynamics in a predator-prey system. Panels A–D denote rotifer-algal experiments with a single clone, such that the rotifer predator (solid line) has quarter-phase-shifted oscillations relative to the algal prey (dashed line). In contrast, panels E–I show the dynamics of multiclonal systems such that the total algal community is shown, albeit it is actually composed of more than one algal type. In panels E–I, the oscillations tend to be antiphase and not quarter-phase shifted as in classic theory. Reproduced from Yoshida et al. (2003).

multiple algal clones. Hence, rather than a single prey population, the algal population has intrinsic diversity. At the population scale, the dynamics include longer oscillations in which predator and prey populations undergo "antiphase" rather than quarter-phase-shifted oscillations. The term *antiphase* denotes the property that the prey are at their maximum when predators are at their minimum and vice versa. This subtle signature is expected when evolutionary dynamics take place on time scales similar to ecological dynamics. Such antiphase relationships are simply impossible in canonical predator-prey dynamics; not only with the assumptions of Lotka and Volterra but with the typical modifications explored in the decades since. This is not just an example of an unusual exception. To the contrary, this subtle difference in the shape of cycles is in fact common, occurring in many previously published consumer-resource dynamics datasets (Hiltunen et al. 2014). Moreover, the emergence of such antiphase cycles is possible when the frequency of clones changes over the same time scales as the total population dynamics when ecological and evolutionary dynamics are linked (Cortez and Ellner 2010). It was only through a careful combination of mathematical modeling and experimental work that this new feature could supplant a prior paradigm.

Altogether, this chapter addresses a central challenge in the modern study of populations and communities: the link between ecological and evolutionary dynamics. What are the key concepts that drive such changes? How can we start with simple models, extend them (in the right ways), and then connect mechanisms to experiments and observations? How is it that inclusion of rapid evolution modifies the canonical results of Lotka and Volterra to yield qualitatively different signatures of population-level oscillations? To answer these questions, we must first introduce some classic models and their modern extensions, beginning with models preceding those of Lotka and Volterra—corresponding to simple growth in a finite environment.

11.2   CANONICAL MODELS OF POPULATION DYNAMICS

Predator-prey dynamic models utilize templates that can be combined as part of low-diversity systems as well as complex systems involving multiple types of predators and of prey. These templates are also relevant in the quantitative understanding of living systems. This section focuses on two such templates: (i) growth in a finite environment; (ii) competition between two populations of the same trophic level. The former template underpins logistic growth dynamics, an example that is typically used in introductory classes as a means to highlight the saturating nature of growth. Here we use it to make the link between resource conditions and the expected levels of growth and saturation. The latter template is also critical to understanding predator-prey dynamics insofar as the prey population consists of multiple types (e.g., different algal clones as in Figure 11.2). As such, part of the challenge of understanding the ecology of predator interactions with diverse prey is to first assess the potential outcomes of interactions among the prey alone.

11.2.1   Logistic growth dynamics—from explicit to implicit

Populations grow exponentially, until they don't. The impact of the finite limits of space and resources can be accounted for in extensions of exponential growth models. Perhaps the simplest is the classic model of logistic growth:

$$\frac{dN}{dt} = rN\left(1 - N/K\right) \tag{11.1}$$

where N is the abundance of a population, r is the maximum (or Malthusian) growth rate, and K is the carrying capacity of the population. The use of implicit resource dynamics in modeling cellular growth is common. Yet the particular form of logistic growth warrants examination as a limit of a mechanistic model of nutrient uptake and conversion into cell biomass. To explore this limit requires making the implicit explicit. To begin, consider a chemostat model in which a population of consumers, N, takes up a non-reproducing resource with abundance R (Figure 11.3):

$$\frac{dR}{dt} = \omega J_0 - f(R)N - \omega R$$
$$\frac{dN}{dt} = \epsilon f(R)N - \omega N \tag{11.2}$$

Here $f(R)$ is the functional response of the consumer, ϵ is the conversion efficiency of consumers, J_0 is the density of resources that are flowing into the chemostat, and ω is

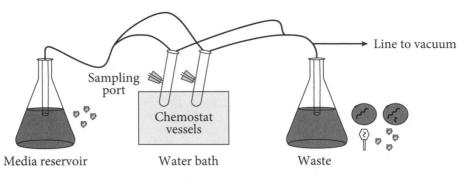

Figure 11.3: Schematic of a chemostat.

the washout rate, i.e., the rate at which both consumers and the resource flow out of the chemostat.

This model of consumer-resource dynamics requires what is termed in the ecological literature a *functional response*. A functional response measures the relationship between resource density and uptake for consumers—whether they are predators eating prey or herbivores consuming plants. To explore one example, consider a consumer that moves "ballistically" through an environment, encountering nutrients in a local sensing zone. Assuming consumers are relatively sparse and each sweeps through a fixed volume (or area) with respect to time and captures a fraction of prey, the functional response should scale like

$$f(R) = cR$$

given a constant c. (In practice, a consumer that does not move may overconsume its local resource, depending on the resource replenishment rate. This would lead to a different form for $f(R)$.) The specification of this simple chemostat model is

$$\frac{dR}{dt} = \omega J_0 - cRN - \omega R$$
$$\frac{dN}{dt} = \epsilon cRN - \omega N$$

(11.3)

This full model can be assessed as a two-dimensional coupled, nonlinear dynamical system. But our aim here is slightly different.

To move from an explicit resource model to an implicit model, we can make the assumption that resource dynamics in this system are much faster than changes in the population of consumers. Operationally, this assumption implies that resource dynamics behave as if N is fixed. In other words, given a current population N, we can apply a fast-slow assumption (as was utilized in the analysis of excitable dynamics in Chapter 6). The equilibrium solution to the fast dynamics of the resource satisfies

$$\omega J_0 - cRN - \omega R = 0$$

(11.4)

such that the resources should rapidly converge to

$$R^q(N) = \frac{\omega J_0}{\omega + cN}$$

(11.5)

where the superscript q denotes the fact that this represents a quasiequilibrium of the consumer-resource system. As is apparent, resource concentration declines with an increasing level of consumers. Now that resource levels are implicitly defined in terms of consumer abundance, it is possible to rewrite the population dynamics strictly in terms of N:

$$\frac{dN}{dt} = \frac{\epsilon \omega c J_0 N}{\omega + cN} - \omega N. \tag{11.6}$$

This assumption holds strictly in the limit that resource dynamics are much faster than consumer population dynamics and otherwise should be recognized as an approximation. In the limit that $N(t) \ll \omega/c$, the dynamics can be further approximated as

$$\frac{dN}{dt} = \overbrace{rN(1 - N/K)}^{\text{logistic growth}} - \omega N \tag{11.7}$$

where $K = \omega/c$ and $r = \epsilon c J_0$ (see the technical appendix on the use of a Taylor approximation for this reduction). A similar approach may be taken with a type II response, in which uptake saturates with increasing resource availability, such that $f(R) = \frac{\gamma R}{Q+R}$. In the limit that $Q \gg R$, then $f(R) = \gamma R/Q = cR$. In that case and in that limit, $K = \omega Q/\gamma$ and $r = \epsilon \gamma J_0/Q$. It is important to keep in mind that the link between the mechanistic and phenomenological parameters only holds strictly in certain limits.

Nonetheless, let us examine the consequences of this modified logistic model. Here this one-dimensional dynamical system has two potential fixed points: $N^* = 0$ and $N^* = K(1 - \omega/r)$. These correspond to the extinction and persistence equilibria, respectively. As should be apparent, the density of the persistence equilibrium is only positive when $r > \omega$. Hence, when maximum growth rates exceed the washout rate (here equivalent to local death), then a small population of consumers can grow to a large, steady state population of consumers. By deriving the model from a resource-explicit uptake model, one can identify the critical condition in this type II functional response model operating at saturation as

$$\left(\frac{\epsilon \gamma J_0}{Q} \right) \times \frac{1}{\omega} > 1. \tag{11.8}$$

In the type I response, in which uptake increases linearly with resources, the condition becomes

$$(\epsilon c J_0) \times \frac{1}{\omega} > 1. \tag{11.9}$$

These may appear difficult to decipher. But consider the following interpretation. Note that $1/\omega$ is the average residence time of the consumer in the chemostat. In both functional responses, the first factor in parentheses corresponds to the division rate per time. Hence, this condition can be restated as follows: a small population of consumers will grow exponentially if the average number of new consumers produced in the lifetime of a consumer within the chemostat is greater than one. Does that resonate? It might. This threshold criterion is equivalent to the threshold condition or $\mathcal{R}_0$ in epidemic outbreaks, yet here the "outbreak" is the spread of a consumer in an otherwise sterile environment. Further discussion of the basic reproduction number can be found in Chapter 12 on outbreaks.

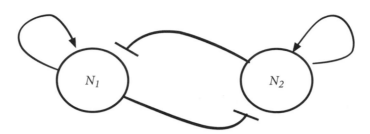

Figure 11.4: Competition between two species, including growth of each population due to intraspecific competition and inhibition of growth due to interspecific competition.

11.2.2 Consumer-resource models

There are two classic consumer-resource models attributed to Lotka and Volterra (Hastings 1997). One of these is the primary subject of this chapter: predator-prey dynamics. But to get there we need to assess the second: a model of competition for resources among two consumers, in which the resource is usually considered to be implicit (Figure 11.4). Understanding the eco-evolutionary dynamics of predators and prey requires integrating both concepts together. In the LV model of competition, consumers are assumed to be at the same *trophic level*, a term that denotes an equivalence in the kinds of foods that consumers may eat and that is usually context dependent when considering food webs (Cohen 1978). Putting aside complications of overlapping levels in food webs (or of the challenge of parasitism in food webs), the model can be assessed using the following dynamics:

$$\frac{dN_1}{dt} = r_1 N_1 \left(1 - \frac{N_1 + a_{12}N_2}{K_1}\right) \tag{11.10}$$

$$\frac{dN_2}{dt} = r_2 N_2 \left(1 - \frac{N_2 + a_{21}N_1}{K_2}\right) \tag{11.11}$$

Here the a_{12} and a_{21} coefficients denote the effect of the densities of each of the two populations on each other. Certain limits are apparent when we consider the case that a_{12} and a_{21} go toward 0. In this event, each population effectively acts independently of the other. Hence, the two populations each can grow to their carrying capacities, i.e., K_1 and K_2, respectively. Moreover, if a_{12} or a_{21} becomes large when the other remains small, then the dominant competitor can outcompete the other. The analysis of this system can be found in standard ecological texts (Hastings 1997) (a brief review is presented in the technical appendix). There are four equilibria in this system:

- Extinction: $(0,0)$
- N_1 dominates: $(K_1, 0)$
- N_2 dominates: $(0, K_2)$
- Coexistence: (N_1^*, N_2^*)

Critically, the extinction point is always unstable as long as $r_1 > 0$ or $r_2 > 0$. Experimentally, one can think of this condition as equivalent to the statement that a flask inoculated with bacteria will enable bacterial growth. Since we expect that both types can grow in the absence of competition, the real ecological question is whether or not the system ends up in a state dominated by a single competitor or in a coexistence state.

As explained in the technical appendix, the condition under which there is coexistence can be summarized as

$$K_1 > a_{12}K_2 \tag{11.12}$$

$$K_2 > a_{21}K_1 \tag{11.13}$$

which implies that intraspecific competition is stronger than interspecific competition. The rationale for Eq. (11.12) is that if species 2 is at its carrying capacity, then the effect of species 2 on species 1 is less than self-limitation of species 1 on itself, and so species 1 can invade a community made up of just species 2. The rationale for Eq. (11.13) is the same with the roles of species 1 and species 2 reversed. More generally, in the event that we assume neither species has a growth advantage, i.e., $K_1 = K = K_2$, then this condition is equivalent to assuming that $\alpha_{12} < 1$ and $\alpha_{21} < 1$. Insofar as there is weak competitions, then both types can persist. This generic finding of coexistence belies the fact that explicit models with a single resource often lead to the exclusion of one competitor. This theory of competitive exclusion (Armstrong and McGehee 1980) posits that there must be as many resources as consumers, at least insofar as the habitat is homogeneous with relatively simple functional forms for uptake. Here weak competition implies that there is, in effect, more than one resource type that enables the coexistence. However, once a predator is added to the system, we will see that competition for resources is not the only way that a consumer can persist. Not dying (i.e, by increasing defenses against predators) is also a key trait and can enable new mechanisms of coexistence.

11.3 PREDATOR-PREY DYNAMICS

Predator-prey systems are, in effect, a kind of consumer-resource system, albeit in which the resource can also reproduce. The following sections build up quantitative models of predator-prey dynamics, beginning with Lotka and Volterra's premise and approaching the types of interactions necessary to integrate ecological and evolutionary dynamics together. Note that the technical appendix includes additional details on the analysis and a summary review of linearization of this class of nonlinear dynamical systems.

11.3.1 Classic models

To begin, consider the following classic model of predator-prey dynamics (Figure 11.5) that was first introduced by Lotka (1925) and Volterra (1926):

$$\frac{dH}{dt} = rH - cHP \tag{11.14}$$

$$\frac{dP}{dt} = \epsilon cHP - mP \tag{11.15}$$

This is a model of predators, P, consuming prey, H, in which predators die exponentially in the absence of prey and prey grow exponentially in the absence of predators. Note that the

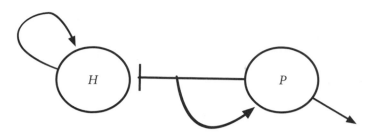

Figure 11.5: Predator-prey dynamics in which the prey (H) grow in the absence of predators (P), predators consume prey, leading to the growth of predators, and predators experience a density-independent mortality.

original model of Lotka and Volterra did not include the saturation of prey growth, e.g., a logistic-like growth term that we explore later in this chapter.

The original LV model of predator-prey dynamics leads to only two possible fixed points: $(0,0)$ and $(m/(\epsilon c), r/c)$. The former denotes the case where there are neither predators nor prey. The latter denotes the coexistence equilibrium. As before, we can ask questions of the long-term dynamics by first investigating the stability properties of the internal equilibria. The technical appendix provides the details on the calculations, which yield the following important result: the eigenvalues of the coexistence equilibrium have zero real parts, i.e., are purely imaginary: $\lambda = \pm i\sqrt{mr}$ where $i = \sqrt{-1}$. Now this is not an imaginary system. Hence, the right (biological) way to interpret this finding is that an initial displacement will neither grow nor decline. Instead, a small deviation from the equilibrium will change like $u(t) \propto \cos(i\omega t)$ where $\omega = \sqrt{mr}$ is the frequency of oscillation. Hence, just like a positive (negative) eigenvalue corresponds to exponential growth (decay), a pair of eigenvalues each with a zero real component and nonzero imaginary components usually corresponds to neutral orbits.

That is precisely what happens. Without a logistic limitation, the dynamics are neutral oscillations such that any initial condition (H_0, P_0) that is not at a boundary of the system nor at the unique interior equilibrium will continue to oscillate. Yet such oscillations are not limit cycles in that they are not isolated periodic orbits. Indeed, there are an infinite number of such orbits, associated with a conserved quantity. For example, by taking the ratio of the prey and predator dynamics, we find

$$\frac{\mathrm{d}H}{\mathrm{d}P} = \frac{H}{(\epsilon c H - m)}\frac{(r - cP)}{P} \tag{11.16}$$

$$\mathrm{d}H\frac{(\epsilon c H - m)}{H} = \mathrm{d}P\frac{(r - cP)}{P} \tag{11.17}$$

such that after integration of both sides and rearranging terms, we find

$$\epsilon c H + cP = r\log P + m\log H + \text{const.} \tag{11.18}$$

What does this constant of motion mean? To physicists, such a constant of motion would be celebrated, as symmetries in systems usually lead to conserved quantities. Not so here.

In practice, the conserved quantity in the original formulation of the predator-prey dynamic model implies that, given an initial value of (H_0, P_0), the system will exhibit

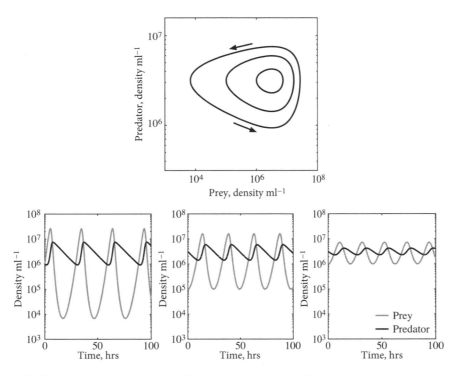

Figure 11.6: Predator-prey dynamics given the Lotka-Volterra model. (Top) Phase plane dynamics where oscillations are counterclockwise, as indicated by the arrows. (Bottom) Dynamics with time, showing the quarter-phase shift of the peaks of predator (black) and prey (gray).

dynamics in a closed loop, on which Eq. (11.18) is satisfied. These are not limit cycles, but rather an infinite number of closed loops that retain their memory of the initial condition, forever. This is not good news for ecology, where the conserved quantity is a fragile feature of the particular choice of functions. One of the problems is that the model is over-simplistic. Indeed, in the absence of predators, this model is unstable. Prey would grow exponentially—forever. That cannot be the case, and relaxing this assumption is precisely the target of Section 11.3.2. Of note, the finding of an infinite number of closed orbits is also consistent with the local stability analysis of the internal coexistence equilibrium.

Yet there is something to be learned from these dynamics, as seen in Figure 11.6 (top). The closed loops have an orientation; they proceed counterclockwise in the predator-prey phase plane. This counterclockwise feature is evident in the time series in the bottom panels of the same figure. For example, beginning at the maximum prey density in any of the closed loops (top panel), the dynamics move "up" in the plane as predators thrive given high levels of prey. Given maximal predator densities, then prey densities decline. With prey densities declining (the leftmost point), then so do predators. With fewer predators (the bottommost point in the loop), the prey recover. Yet with prey recovery, the predators also thrive, and the cycle repeats. Again, this constant of motion is fragile; it only applies when there is no logistic growth. But the relative ordering of the peaks and troughs does have a message that goes beyond this particular choice of functional form.

11.3.2 Predator-prey dynamics with limitations on prey growth

Let us once again revisit predator-prey dynamics by extending the original model to include density limitations on the growth of prey:

$$\frac{dH}{dt} = rH(1 - H/K) - cHP \tag{11.19}$$

$$\frac{dP}{dt} = \epsilon cHP - mP \tag{11.20}$$

Given this new formulation, there are now three fixed points of the system rather than two:

- Extinction: $(0, 0)$
- Prey only: $(K, 0)$
- Coexistence: $\left(\frac{m}{\epsilon c}, \frac{r}{c}\left(1 - \frac{m}{\epsilon cK}\right)\right)$

In this model, coexistence is possible whenever predators can increase in number given that prey are at their maximum, i.e., $\epsilon cK > m$. There is yet another interpretation, in that predators will live on average $1/m$ units of time. In that time, they will produce ϵcK offspring per unit time. Hence, predators will proliferate if their average offspring exceeds one given the addition of a single predator into an otherwise prey-only system. This is termed the *basic reproduction number*, a concept that we revisit in Chapter 12. More details on the stability of this system can be found in the technical appendix. Finally, it is worth mentioning that the dynamics of predator and prey involve endogenous oscillations, whether permanently (as in the model without a carrying capacity) or transiently (as in the model with a carrying capacity).

Figure 11.7 reveals that the inclusion of density limitations fundamentally shifts the nature of the dynamics from persistent neutral orbits to spirals or arcs that converge to a fixed point. The spirals and arcs have the correct handedness, in the sense that the dynamics appear counterclockwise in the phase plane; however, given the parameters in this example, the oscillations are rapidly damped and the system converges to the stable equilibrium. Note that if predators exhibit a type II functional response, then the system can exhibit a limit cycle in which predator and prey oscillate together in a unique orbit, an issue explored next.

11.3.3 Predator-prey dynamics with limitations on prey growth and saturating predation

The prior models both assume that the consumption of prey by predators is limited only by prey availability. In other words, if H increases, then predators can keep eating and eating and eating. Consumption takes time. For example, consider the case where predators can consume at most one prey in a time τ. In that case, the maximum rate of per capita consumption should be on the order of $1/\tau$. Yet, when prey are scarce, the rate of consumption is limited by prey availability and not handling time. These dual limits can be captured in the following model:

$$\frac{dH}{dt} = rH(1 - H/K) - \frac{cH}{1 + H/H_0}P \tag{11.21}$$

$$\frac{dP}{dt} = \frac{\epsilon cH}{1 + H/H_0}P - mP \tag{11.22}$$

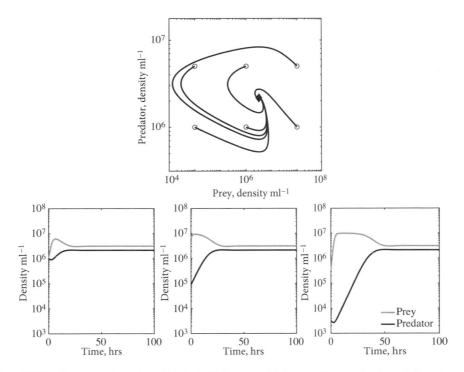

Figure 11.7: Predator-prey dynamics of the Lotka-Volterra model given logistic growth of prey in the absence of predators. (Top) Phase plane dynamics where oscillations are counterclockwise. (Bottom) Time dynamics, showing the quarter-phase shift of the peaks of predator (black) and prey (gray). Parameters: $r = 1$, $c = 10^{-6.5}$, $\epsilon = 0.1$, $m = 0.1$, $K = 10^7$, all in units of cells, mL, and hrs.

This is known as the MacArthur-Rosenzweig model (Rosenzweig and MacArthur 1963), and includes a type II functional response, i.e., a saturating prey consumption function. Note that when $H \gg H_0$, then $cH/(1 + H/H_0) \to cH_0 \equiv 1/\tau$. Hence, another way to write this model is as follows:

$$\frac{dH}{dt} = rH(1 - H/K) - \frac{cH}{1 + c\tau H}P \tag{11.23}$$

$$\frac{dP}{dt} = \frac{\epsilon cH}{1 + c\tau H}P - mP \tag{11.24}$$

where the handling time dependence is now explicit. As a result, there are once again three fixed points:

- Extinction: $(0, 0)$
- Prey only: $(K, 0)$
- Coexistence: (H^*, P^*)

where $H_0 = 1/(c\tau)$, $\epsilon cH^*/m = 1 + H^*/H_0$, or $H^* = \frac{H_0}{H_0\epsilon c/m - 1}$ and $P^* = r/c(1 + H^*/H_0)$ $(1 - H^*/K)$. This model reveals a new phenomenon: stable limit cycles (Figure 11.8). In this regime, the prey and predator continue to oscillate, but unlike in the original LV model, these oscillations do not have an infinite memory of the initial conditions. Instead, the oscillations either increase or dampen and then approach a stable limit cycle (Figure 11.8, top).

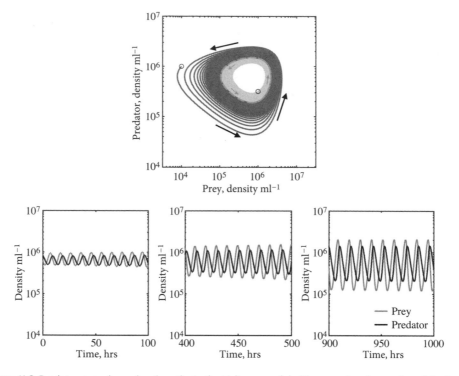

Figure 11.8: Predator-prey dynamics given the Lotka-Volterra model with prey saturation and predator handling of prey. (Top) Phase plane dynamics where oscillations are counterclockwise. Note that two trajectories are shown (dark and light gray) converging to a stable limit cycle (dashed line). (Bottom) Time dynamics, showing the quarter-phase shift of the peaks of predator (black) and prey (gray) approaching a limit cycle. Parameters: $r = 1$, $c = 10^{-6.5}$, $K = 10^7$, $\epsilon = 0.1$, $m = 0.1$, all in units of cells, mL, and hrs.

This implies that efforts to identify the signature of predator-prey interactions in natural systems should focus on phase shifts between the peaks of predators, which should follow those prey. How true this is forms the core point of inquiry of this chapter, and the synthesis of ecological and evolutionary dynamics.

11.4   TOWARD PREDATOR-PREY DYNAMICS WITH RAPID EVOLUTION

The prior sections have included a suite of models spanning the growth of a single population limited by resources, competition between populations of the same trophic level (i.e., interspecific competition), and finally predator-prey dynamics assuming one predator and one prey. These ingredients are precisely what is needed to take the next step and ask what new kinds of ecological phenomena are possible when evolutionary dynamics occur rapidly. By *evolutionary dynamics*, we don't necessarily mean to invoke de novo mutation. Instead, evolutionary dynamics here means that there is a change in the frequency of genotypes in a population. That condition is certainly met if a prey population is made up of multiple clones—and as we shall see, the existence of clones or distinct strains that constitute subpopulations is precisely the driver of a potential fundamental shift in the shape of predator-prey cycles.

11.4.1 Multiple prey and a single predator

There are multiple theories of eco-evolutionary dynamics. Yet perhaps the most straightforward way to think about this link is to revisit the definition of evolution: the heritable change in genotype frequencies from one generation to the next. Hence, envision a prey population consisting of multiple genotypes of prey (i.e., distinct "clones"). It might not be possible to distinguish those clones based on the observational methods commonly used, even if those clones have distinct phenotypes. Moreover, consider a case where two clones have distinct growth and palatabilities such that the dynamics of the entire (unobserved) system is

$$\frac{dH_1}{dt} = r_1 H_1 \left(1 - \frac{H_1 + a_{12}H_2}{K_1}\right) - \frac{c_1 H_1 P}{(1 + c_1 \tau_1 H_1 + c_2 \tau_2 H_2)} \tag{11.25}$$

$$\frac{dH_2}{dt} = r_2 H_2 \left(1 - \frac{H_2 + a_{21}H_1}{K_2}\right) - \frac{c_2 H_2 P}{(1 + c_1 \tau_1 H_1 + c_2 \tau_2 H_2)} \tag{11.26}$$

$$\frac{dP}{dt} = \left(\frac{\epsilon c_1 H_1 + \epsilon c_2 H_2}{1 + c_1 \tau_1 H_1 + c_2 \tau_2 H_2}\right) P - mP \tag{11.27}$$

However, if H_1 and H_2 are indistinguishable, then it is not evident what kind of dynamics might be observable in the P–H phase plane. What can happen is something new: the ratio of genotypes changes on the same time scale as changes in the total population of consumers. Hence, when the three-dimensional system is projected back onto two dimensions, new kinds of dynamics can occur, including antiphase cycles as shown in Figure 11.2.

To understand how such antiphase cycles might arise, consider cases where there is a large cost of prey defense given increases in growth rate (i.e., a growth-defense trade-off). The dynamics of such systems are illustrated in Figure 11.9. This model includes two prey, one well defended (that grows slower) and one that grows faster (but is easier to consume). The dynamics favor the fast-growing prey when predators are rare and the well-defended prey when predators are common. Yet, the trade-off drives a switch in relative densities. When well-defended prey are abundant, this drives down predator densities, which, eventually, leads to the rise and return of fast-growing prey. In turn, the increase of fast-growing prey provides a resource for predators to grow in abundance. What is critical is that the changes in total population densities are occurring *on the same time scale* as are changes in genotype frequencies. The phase plane dynamics (top panel) also reveal another feature: the crossing of phase lines. Because this is a deterministic model, it should not be possible to have phase lines cross in a two-dimensional dynamical system. Instead, the system should move in the same direction whenever it returns to the same point in phase space. Not so here, where the phase line crosses itself—an indicator that the system comprises more than just two components. The phase plane is a projection of the 3D system onto 2D: total prey and predator densities. Here the dynamics seem antiphase; i.e., when the prey are at their

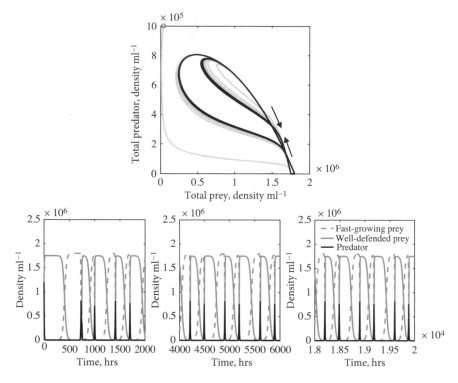

Figure 11.9: Predator-prey dynamics given the Lotka-Volterra model with multiple prey species. (Top) Phase plane dynamics where oscillations are counterclockwise from the first half of the simulation (in gray) and the latter half (in black). The open circle denotes the initial conditions. (Bottom) Time series dynamics, showing the fast-growing prey (dashed gray), well-defended prey (solid gray), and predator (black).

maximum, predators are at their minimum and vice versa. The antiphase dynamics are possible given concurrent changes in genotype composition, evident at the peaks and troughs of prey densities.

11.4.2 Multiple prey and multiple hosts

The dynamics above consider what happens when there are two prey and one predator, and when prey genotype frequencies change on the same time scale as total population dynamics. This rapid evolution can also extend to more than one predator. To see how, consider a model extension to the case of two predators and two prey, in which one prey is relatively well defended against both predators and the other prey grows faster. Likewise, consider the case where one predator is the better consumer of both prey and yet also dies faster in the absence of prey (i.e., has a higher baseline energetic need). As a consequence, this entire set of predators and prey can coexist. Such a result may be hard to identify in theory, but it is not surprising in practice. Complex communities in nature abound. Yet what is surprising is the potential consequence for the qualitative patterns of total prey and predator density given the interaction of multiple clones.

The dynamics in Figure 11.10 illustrate the potential paradox (Cortez and Weitz 2014). When viewed in terms of the total number of prey and predators, it appears that the prey

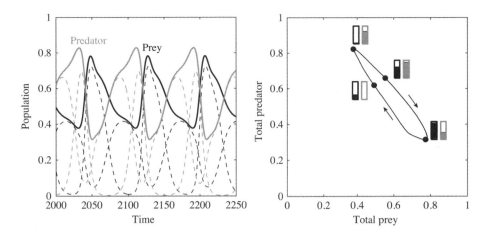

Figure 11.10: Predator-prey dynamics given a two-prey and two-predator model. (Left) Strain-level dynamics and total population dynamics with time. (Right) Phase plane dynamics. The black curve denotes population dynamics, whereas the bars denote the relative frequency of genotypes (black for prey, gray for predator). For prey, the shaded bar denotes low vulnerability and the open bar denotes high vulnerability. For predators, the shaded bar denotes low offense and the open bar denotes high offense. Based on the models described in detail in Cortez and Weitz (2014).

abundances peak *after* those of the predator. In essence, prey appear to eat predators and not the other way around. Such a result is nonsensical, so how else do we interpret these seemingly contradictory findings? The answer lies in the phase plane, where trajectories do in fact move in reverse, such that predator peaks are followed by prey peaks, which are followed by predator troughs, then prey troughs, and finally predator peaks again. Yet, in the process, the relative importance of predator and prey genotypes also changes. These changes are denoted by the shaded bars at multiple points along the clockwise phase plane trajectory.

The dynamics can be explained as follows: when prey abundances are high, this coincides with the state in which prey are not particularly vulnerable and the system has a balance of predator types. However, precisely because predator abundances are low, prey can increase in number, and indeed the prey type that is more vulnerable (but does not pay the cost for defense) can increase. Likewise, in order to attack the limited prey, predators that are better in offense outcompete the other genotype. Therefore, the high-offense predators proliferate on the highly vulnerable prey. As predators increase, the prey do not have advantages when they incur costly defensive strategies (there are simply too many high-offense predators). Instead, the prey switch to low-offense types. As such, predators need not utilize high-offense strategies. This shift in genotypes enables low-vulnerability (and fast-growing) prey to proliferate during a period where predators are dominated by low-offense types. At this point, the population dynamics return to a high-prey, low-predator dynamic, albeit having gone through a *clockwise* cycle in the phase plane. Such dynamics are simply not possible with a conventional predator-prey model (insofar as the functional response is a monotonically increasing function of the prey density). This analysis reveals the critical importance of how rapid evolutionary dynamics can lead to qualitative changes in population dynamics.

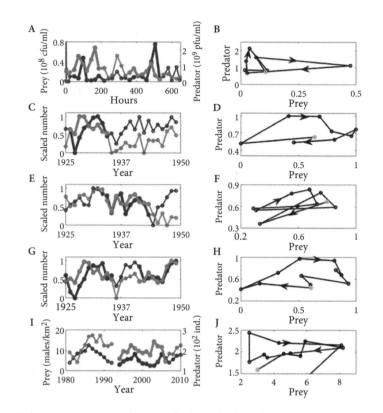

Figure 11.11: Clockwise predator-prey dynamics for bacteriophage (predator) and *Vibrio cholerae* (prey) (A–B), mink (predator) and muskrat (prey) (C–D; E–F; G–H), and gyrfalcon (predator) and rock ptarmigan (prey) (I–J). Reproduced from Cortez and Weitz (2014). In each of the left panels, the prey time series is denoted in black and the predator time series is denoted in gray. In the right panels, the light gray circular point denotes the start of the focal time series section and the arrows denote the direction in time.

Remarkably, such reversed cycles have been found in multiple datasets, including bacteriophage (acting as a predator) and *Vibrio cholerae* (the prey), mink (predator) and muskrat (prey), and gyrfalcon (predator) and rock ptarmigan (prey) systems (Cortez and Weitz 2014). Figure 11.11 shows five such examples in time (left panels) and in the phase plane (right panels). In each, the orientation is clockwise rather than counter-clockwise. Of course, that may suggest that prey eat their predators (they do not), but a different interpretation is that these dynamics are hallmarks of the dynamic variation of subpopulations of predators and prey—projecting down (at least) a four-dimensional dynamical system into a two-dimensional plane. Notably, reverse cycles are not a potential outcome of cases where only the predator or the prey evolves. Instead, the existence of reverse cycles suggests coevolution's impact on ecology. Indeed, in the case of the phage-bacteria system, there is additional evidence to suggest the importance of this multiclonal mechanism.

The experimental study of *V. cholerae* O1 and the bacteriophage JSF4 extended over nearly a month long period with daily sampling (Wei et al. 2011). The time series were considered complex and not necessarily consistent with conventional expectations of predator-prey dynamics. As considered elsewhere (Levin et al. 1977; Weitz 2015), virulent bacteriophage that exclusively infect host cells are expected to induce predator-prey–like dynamics through precisely the same mechanisms as explored in this chapter. Yet, in this case, the experimental team plated out multiple hosts and multiple phage strains from the chemostat experiment. In doing so, they found two key phage variants, what they termed phage T and phage B for "turbid" and "big" plaques, analogous to a less efficient and more efficient predator, given that the turbidity and size of plaques is a metric of efficient replication on the wild-type hosts. Likewise, the team also isolated multiple hosts that they termed T^- and TB^-, denoting largely resistant to the T phage and to both T and B phage, respectively. Rapid coevolution of viruses and bacteria in lab settings is a hallmark of the large population sizes; yet here the impact extends beyond insights into how evolution unfolds in a fixed ecological setting. Together, these findings suggest that evolution and ecology are entangled. This example illustrates the larger point—that reconciling the impact of ecology on evolution requires that we look in both directions.

11.5   TAKE-HOME MESSAGES

- The motivation for this chapter is the observation of antiphase cycles between predator and prey in a rotifer-algae system.
- Antiphase cycles should not be possible in conventional predator-prey dynamics. Instead, predator-prey models, including those with and without density limitation, exhibit quarter-phase-shifted cycles where prey peak and then predators, then prey reach their trough, then predators decline, and finally prey recover and the cycle begins again.
- The particular outcomes of single predator–single prey population dynamics are influenced by functional responses.
- Attracting limit cycles for predator-prey dynamics are possible given density limitation of prey and type II functional responses by the predator.
- In the event of a multiclonal prey population, predator-prey dynamics can exhibit antiphase cycles where one population is at its maximum when the other is at its minimum.
- In the event of a multiclonal prey and multiclonal predator population, reverse cycles can appear—this does not imply that prey eat predators, but rather that frequencies of both genotypes are changing at the same time scale as ecological dynamics.
- Altogether, this chapter provides a nested series of examples showing the impacts of rapid evolution on both ecological and evolutionary dynamics.

11.6   HOMEWORK PROBLEMS

These problems leverage the methodology in the accompanying computational laboratory. This particular laboratory has two distinct paths that function as separate modules. First, the laboratory treats the problem of the origin of functional responses, and how to move from an individual-based model to an effective functional response to use at the population scale. Second, the laboratory covers multiclonal dynamics, including the projection of dynamics from higher- to lower-dimensional systems.

PROBLEM 1. Competition for a Common Resource

A chemostat is a "continuous" culture vessel in which resources, of density R_0, are supplied at a rate ω. Media in the chemostat are then removed at a rate ω—the balance of input and output ensures that the volume is constant. These chemostats can be inoculated with bacterial populations. Assume that such a chemostat has been inoculated with two types of bacteria. These bacteria compete for a single resource; you can assume that their populations obey the following dynamics:

$$\frac{dN_1}{dt} = \gamma_1 \epsilon_1 N_1 R - m_1 N_1$$

$$\frac{dN_2}{dt} = \gamma_2 \epsilon_2 N_2 R - m_2 N_2$$

$$\frac{dR}{dt} = \omega R_0 - \omega R - \gamma_1 N_1 R - \gamma_2 N_2 R$$

a. Find and define the three possible steady states of this model.
b. What are the conditions among bacterial parameters necessary for survival?
c. Develop a program to simulate this model and include the function call in your solution set.
d. In the code, use the values $R_0 = 100$, $\epsilon_{1/2} = 0.5$, and $\omega = 0.1$, and the initial densities $N_1 = 10$, $N_2 = 10$, and $R = 50$. In addition, consider the following table of phenotypes associated with distinct bacterial types:

```
gamma=[0.1 2 10   4   2 1   .3 ]
   m=[1    4 100 10 5 10 1   ]
```

First, evaluate whether or not each type can survive on its own. Next, compete all types that can survive in a pairwise fashion. Graphically display the competition and the statistics of who beats whom.
e. What is the principle, if any, that determines who wins in this competition among distinct microbes for a common resource?

PROBLEM 2. Type I and Type II Functional Responses

This problem is based on the interactive predator-prey simulator available on the book's website:

- master_ibm—the master script for setting parameters of the individual-based model (IBM)
- ibm_predation—dynamic simulator of predation

The chapter text derives an equation for uptake of prey by predators. Modified for a 2D environment, it should be

$$\text{consumption rate} = bHP \qquad (11.28)$$

where the constant approaches $\pi r^2 fk$ for small intervals. Further, when predators take time to handle and consume their prey, then

$$\text{consumption rate} = \frac{bHP}{1 + aH} \qquad (11.29)$$

where b/a is equal to $1/T_h$ and T_h is the handling time. In this problem set, you will use this simulator to test these two ideas. (Hint: When running multiple simulations, disable the visualization altogether.)

a. For the default parameters in the scripts, modify the prey population density from 2 to 40. How well does Eq. (11.27) work? How accurate is your prediction of b? Use graphs and analysis to prove/disprove whether prey consumption is a linear function of prey density.

b. For the default parameters in the scripts, set the handling time to 1 second. By modifying prey density from 2 to 40, evaluate how well Eq. (11.28) works. How good is your prediction of a? Use graphs and analysis to prove/disprove whether prey consumption is a nonlinear function of prey density in the form of Eq. (11.28).

c. Use the software as a method to learn something new—to you—about predator-prey interactions. Include one graph that illustrates what you learned and describe why you find it interesting.

PROBLEM 3. Cycle Time

The theory of predator-prey dynamics predicts an imaginary eigenvalue with magnitude $\lambda = \sqrt{\omega r}$ for the example shown in the text. Parts a–d examine the interpretation and its limitations.

a. Interpret the meaning of the purely imaginary eigenvalue in terms of expected dynamics near the fixed point.

b. Starting with initial conditions near the fixed point in the classic Lotka-Volterra model, find the cycle period of the ensuing oscillations. How do they compare to those predicted by the eigenvalue?

c. Increasing the magnitude of deviations, e.g., from 1% to 10%, find the cycle period and evaluate how the cycle period changes.

d. Now include density limitation in the Lotka-Volterra model using the same parameters as in figure 11.7. Increase $K = 10^7$/ml to 10^9/ml and, using the initial values of 10^6/ml for both predator and prey, evaluate the oscillation periods. To what extent does the classic prediction help inform with respect to the magnitude of the initial oscillations?

PROBLEM 4. Fast-Slow Eco-evolutionary Dynamics

Consider a two-prey and one-predator system:

$$\dot{N}_1 = r_1 N_1 \left(1 - N/K_1\right) - \frac{c_1 N_1 P}{1 + Q} \tag{11.30}$$

$$\dot{N}_2 = r_2 N_2 \left(1 - N/K_2\right) - \frac{c_2 N_2 P}{1 + Q} \tag{11.31}$$

$$\dot{P} = \epsilon P \left(\frac{c_1 N_1}{1 + Q} + \frac{c_2 N_2}{1 + Q}\right) - mP \tag{11.32}$$

where $Q = c_1 \tau_1 N_1 + c_2 \tau_2 N_2$ and $N = N_1 + N_2$. This system has the potential to exhibit anomalous dynamics, including antiphase cycles (i.e., where prey peaks seems to coincide with predator minima and prey minima seem to coincide with predator peaks). The following questions will help find and identify the basis for such behavior. In this problem, you should use the following parameters (typical of a zooplankton interacting with two bacterial types), in units of hrs and mls: $r_1 = 2$, $r_2 = 1$, $K_1 = 2.7 \times 10^7$, $K_2 = 1.5 \times 10^7$, $b_1 = 10^{-6}$, $b_2 = 10^{-7}$, $\epsilon = 0.2$, $c_1 = 2 \times 10^{-7}$, $c_2 = 2 \times 10^{-8}$, $\tau_1 = 1$, $\tau_2 = 1$, and $m = 0.1$. In addition, you may need to run long simulations (around 1000 hrs) to avoid transient effects.

a. Based on the default parameters, describe the functional type of each prey; i.e., is one a defense specialist, growth specialist, or some combination?
b. Simulate the baseline model. Does this system exhibit canonical predator-prey cycles when projecting the system into the $N-V$ plane? Justify your answer using a time domain and/or phase plane view.
c. Can you find a parameter combination that exhibits antiphase cycles? (Hint: Reduce K_1 slowly to decrease the relative maximum capacity of the type I prey.) Again, justify your answer using a time domain and/or phase plane view.
d. Finally, if you found anti phase dynamics, try to reconcile your findings with the fact that a two-dimensional system cannot cross itself (or double back) in phase space. Provide a mechanistic explanation of antiphase dynamics in terms of the full three-dimensional dynamics of the system.

11.7   TECHNICAL APPENDIX

Reducing a consumer-resource model to a logistic model Consider the following equation introduced in the chapter for the changes of consumers in a chemostat with resource introduction density J_0, washout rate ω, uptake rate c, and conversion rate ϵ:

$$\frac{dN}{dt} = \frac{\omega \epsilon c J_0 N}{\omega + cN} - \omega N. \tag{11.33}$$

Note that this equation can be written as

$$\frac{dN}{dt} = \frac{\omega \epsilon c J_0 N}{\omega} \left(1 - cN/\omega\right) - \omega N \qquad (11.34)$$

by approximating $1/(1 + cN/\omega) \approx 1 - cN/\omega$ for $N \ll \omega/c$. Note that the expression $\frac{1}{1+\delta}$ can be approximated as $1 - \delta$ when $\delta \ll 1$. This is a direct result of a Taylor approximation of the expression near $\delta = 0$. By identifying a critical density $K = \omega/c$, this can be rewritten as

$$\frac{dN}{dt} = \frac{\omega \epsilon c J_0 N}{\omega} \left(1 - N/K\right) - \omega N. \qquad (11.35)$$

The dynamics can be reduced to

$$\frac{dN}{dt} = \overbrace{rN\left(1 - N/K\right)}^{\text{logistic growth}} - \omega N \qquad (11.36)$$

where $K = \omega/c$ and $r = \epsilon c J_0$ as described in the chapter. Hence, a logistic model can be thought of as the limiting case of a consumer-resource model in the fast subsystem where resources change far more rapidly than do consumers.

Linearization of nonlinear dynamical systems Recall that the linear stability of fixed points within predator-prey models is determined by the Jacobian, such that if

$$\frac{dH}{dt} = f(H, P) \qquad (11.37)$$

$$\frac{dP}{dt} = g(H, P) \qquad (11.38)$$

then

$$\mathbf{J} = \begin{bmatrix} \frac{\partial f}{\partial H} & \frac{\partial f}{\partial P} \\ \frac{\partial g}{\partial H} & \frac{\partial g}{\partial P} \end{bmatrix}. \qquad (11.39)$$

The Jacobian is evaluated at a specific point corresponding to the equilibrium under investigation: (H^*, P^*). The following deals with the analysis of the model variants discussed in the chapter. The eigenvalues are identified by solving $\mathrm{Det}\,(\mathbf{J} - \lambda \mathbf{I}) = 0$ where $\mathbf{I}$ is the identity matrix.

The rationale is as follows. "Linearizing" around a fixed point is akin to making an approximation of a nonlinear function locally. Just as the local values of a curve can be approximated by taking the current value and then fitting a tangent line at a point of interest, we can use the same technique in higher dimensions. To see how, consider the generalized form of predator-prey models:

$$\frac{dH}{dt} = f(H, P) \qquad (11.40)$$

$$\frac{dP}{dt} = g(H, P) \qquad (11.41)$$

where H denotes the prey and P the predator. A fixed point of this system is denoted as (H^*, P^*) such that $f(H^*, P^*) = 0$ and $g(H^*, P^*) = 0$. Let's now consider points in the state space near the fixed point, i.e., $u = H - H^*$ and $v = P - P^*$. Given a small deviation from the fixed point, these two-dimensional functions can be approximated via a Taylor expansion, i.e.,

$$f(H, P) \approx f(H^*, P^*) + u \left.\frac{\partial f}{\partial H}\right|_{H^*, P^*} + v \left.\frac{\partial f}{\partial P}\right|_{H^*, P^*} + \cdots \tag{11.42}$$

$$g(H, P) \approx g(H^*, P^*) + u \left.\frac{\partial g}{\partial H}\right|_{H^*, P^*} + v \left.\frac{\partial g}{\partial P}\right|_{H^*, P^*} + \cdots \tag{11.43}$$

where the ellipses denote the fact that a full Taylor expansion would include higher-order terms, associated with u^2, uv, and v^2. However, since $u, v \ll 1$, these terms should have magnitudes far smaller than the first-order terms. But the functions—corresponding to the rates of change of H and P, respectively—can be further reduced by recalling that $f(H^*, P^*) = 0$ and $g(H^*, P^*) = 0$:

$$f(H, P) \approx u \left.\frac{\partial f}{\partial H}\right|_{H^*, P^*} + v \left.\frac{\partial f}{\partial P}\right|_{H^*, P^*} \tag{11.44}$$

$$g(H, P) \approx u \left.\frac{\partial g}{\partial H}\right|_{H^*, P^*} + v \left.\frac{\partial g}{\partial P}\right|_{H^*, P^*} \tag{11.45}$$

These approximations represent the 2D equivalent of fitting a tangent line to a curve. Here the equations represent finding the local tangent plane to the functions f and g. This can be written compactly as

$$\begin{bmatrix} f \\ g \end{bmatrix} = \begin{bmatrix} \frac{\partial f}{\partial H} & \frac{\partial f}{\partial P} \\ \frac{\partial g}{\partial H} & \frac{\partial g}{\partial P} \end{bmatrix}_{H^*, P^*} \begin{bmatrix} u \\ v \end{bmatrix} \tag{11.46}$$

where the Jacobian appears as the coefficients in the shape of the tangent plane. Note that because $u = H - H^*$ and $v = V - V^*$, then $du/dt = dH/dt$ and $dv/dt = dP/dt$. Finally, gathering all of this together yields

$$\begin{bmatrix} \frac{du}{dt} \\ \frac{dv}{dt} \end{bmatrix} = \mathbf{J}_{H^*, P^*} \begin{bmatrix} u \\ v \end{bmatrix}. \tag{11.47}$$

Because the matrix $\mathbf{J}$ only has coefficients, it should be apparent that this Taylor expansion leads to a linearized dynamical system—one that we expect to grow exponentially, decay exponentially, or in some marginal cases, have no growth at all.

Consumer-resource Lotka-Volterra model   Consider the following set of equations of the canonical Lotka-Volterra model for competition between two trophic competitors:

$$\frac{dN_1}{dt} = r_1 N_1 \left(1 - \frac{N_1 + a_{21} N_2}{K_1}\right) \tag{11.48}$$

$$\frac{dN_2}{dt} = r_2 N_2 \left(1 - \frac{N_2 + a_{12} N_1}{K_2}\right) \tag{11.49}$$

Here the a_{12} and a_{21} coefficients denote the effect of the densities of each of the two populations on each other. As noted in the chapter there are potentially four equilibria in this system:

- Extinction: $(0,0)$
- N_1 dominates: $(K_1, 0)$
- N_2 dominates: $(0, K_2)$
- Coexistence: (N_1^*, N_2^*)

The coexistence point must satisfy the nullclines of the system, i.e., $\dot{N}_1 = 0$,

$$K_1 = N_1 + \alpha_{12} N_2, \tag{11.50}$$

and $\dot{N}_2 = 0$,

$$K_2 = N_2 + \alpha_{21} N_1. \tag{11.51}$$

Combining both yields the following equations for the coexistence points:

$$N_1^* = \frac{K_1 - \alpha_{12} K_2}{1 - \alpha_{12}\alpha_{21}} \tag{11.52}$$

$$N_2^* = \frac{K_2 - \alpha_{21} K_1}{1 - \alpha_{12}\alpha_{21}} \tag{11.53}$$

This equilibrium point is stable insofar as $K_1 < K_2/\alpha_{21}$ and $K_2 < K_1/\alpha_{12}$; in other words, intraspecific competition is stronger than interspecific competition.

Predator-prey Lotka-Volterra model Consider the following set of equations of the canonical Lotka-Volterra model for predator-prey dynamics:

$$\frac{dH}{dt} = rH - cHP \tag{11.54}$$

$$\frac{dP}{dt} = \epsilon cHP - mP \tag{11.55}$$

For the Lotka-Volterra model, then

$$\mathbf{J} = \begin{bmatrix} r - cP & -cH \\ \epsilon cP & \epsilon cH - m \end{bmatrix}, \tag{11.56}$$

which when evaluated at the equilibrium $(m/(\epsilon c), r/c)$ becomes

$$\mathbf{J} = \begin{bmatrix} 0 & -m/\epsilon \\ \epsilon r & 0 \end{bmatrix}. \tag{11.57}$$

The eigenvalues must satisfy $\mathrm{Det}(\mathbf{J} - \lambda \mathbf{I}) = 0$ where $\mathbf{I}$ is the identity matrix. The matrix $\mathbf{J} - \lambda \mathbf{I}$ can be written as

$$\mathbf{J} - \lambda \mathbf{I} = \begin{bmatrix} -\lambda & -m/\epsilon \\ \epsilon r & -\lambda \end{bmatrix} \tag{11.58}$$

such that the eigenvalues satisfy $\lambda^2 + mr = 0$ or $\lambda = \pm i\sqrt{mr}$ where $i = \sqrt{-1}$. Note that these eigenvalues have zero real components, implying that perturbations do not grow in magnitude. However, perturbations may oscillate. For example, note that $e^{it} = cos(t) + isin(t)$ and $e^{-it} = cos(t) - isin(t)$, such that combinations of the resulting solutions involve only real perturbations $\sim cos(t)$.

Lotka-Volterra model with prey density limitation Consider the Lotka-Volterra model with prey density limitation

$$\frac{dH}{dt} = rH(1 - H/K) - cHP \tag{11.59}$$

$$\frac{dP}{dt} = \epsilon cHP - mP. \tag{11.60}$$

The Jacobian of this system is

$$\mathbf{J} = \begin{bmatrix} r - 2rH/K - cP & -cH \\ \epsilon cP & \epsilon cH - m \end{bmatrix}. \tag{11.61}$$

This Jacobian must be evaluated at three equilibria: $(0,0)$, $(K,0)$, and $(m/(\epsilon c), r/c$ $(1 - m/(\epsilon cK))$, corresponding to the extinction, prey-only, and coexistence cases. The Jacobians and eigenvalues can be evaluated in turn. First, in the case of the extinction equilibrium,

$$\mathbf{J} = \begin{bmatrix} r & 0 \\ 0 & -m \end{bmatrix} \tag{11.62}$$

such that the eigenvalues can be read off the diagonal (always true when one of the off-diagonal terms is zero in a two-by-two case), $\lambda_1 = r$ and $\lambda_2 = -m$. Hence, the $(0,0)$ equilibrium is a saddle point, in which there is one stable direction (with a negative eigenvalue) and one unstable direction (with a positive eigenvalue). Note that these perturbations should correspond to eigenvectors in which the perturbations change exponentially along a single direction. Intuitively, the unstable axis should correspond to adding a small amount of prey (in the absence of predators), i.e., $u_0 > 0$, $v_0 = 0$, such that a small addition of prey only would grow like $u_0 e^{rt}$. Likewise, intuitively, the stable axis should correspond to adding a small amount of predators (in the absence of prey), i.e., $u_0 = 0$ and $v_0 > 0$, such that a small addition of predators only would decay like $u_0 e^{-mt}$. Formally, identifying eigenvectors can provide additional information on the shape of dynamics of the full system.

A digression on eigenvectors Note that the linearization of the predator-prey system with prey density limitation suggests that solutions should be exponential. So consider the ansatz (u_1, v_1) to denote a particular direction of perturbation such that, over time, the system should grow or decay precisely along this direction, i.e., $(u_1 e^{\lambda_1 t}, v_1 e^{\lambda_1 t})$. If so, then the derivative of either perturbation is simply $du/dt = \lambda_1 u_1$ and $dv/dt = \lambda_1 v_1$. The same applies to λ_2. This implies that the linearized system of equations for perturbation along this particular direction—what we term an *eigenvector*—should follow

$$\lambda_1 \begin{bmatrix} u \\ v \end{bmatrix} = \mathbf{J}_{H^*, P^*} \begin{bmatrix} u \\ v \end{bmatrix} \tag{11.63}$$

such that we can reduce this to a system of two equations with two unknowns. The two unknowns are the contributions to the eigenvector. The equations are

$$(\mathbf{J}_{H^*,P^*} - \lambda \mathbf{I}) \begin{bmatrix} u \\ v \end{bmatrix} = 0, \tag{11.64}$$

or more explicitly

$$\begin{bmatrix} J_{11} - \lambda & J_{12} \\ J_{21} & J_{22} - \lambda \end{bmatrix} \begin{bmatrix} u \\ v \end{bmatrix} = 0. \tag{11.65}$$

With this formalism in place, return to the extinction equilibrium. In this case, the two systems of equations can be written in matrix form as

$$\begin{bmatrix} r - \lambda & 0 \\ 0 & -m - \lambda \end{bmatrix} \begin{bmatrix} u \\ v \end{bmatrix} = 0. \tag{11.66}$$

When $\lambda = r$, then the two equations become

$$(r - r)u = 0 \tag{11.67}$$

$$(-m - r)v = 0 \tag{11.68}$$

such that the only way for the second equation to be 0 is for $v_1 = 0$. The first equation is satisfied for any value of u, so we can write the unstable eigenvector as $(1, 0)$, meaning it's a perturbation strictly in the prey direction associated with the eigenvalue $\lambda_1 = r$. The same logic also helps us conclude that the only way to satisfy the two equations in the case of the stable eigenvalue $\lambda_2 = -m$ is for the unstable eigenvector to be $(0, 1)$. Typically, it takes more work to identify eigenvectors.

In the case of the fixed point $(K, 0)$, the Jacobian is

$$\mathbf{J} = \begin{bmatrix} -r & -cK \\ 0 & \epsilon cK - m \end{bmatrix}. \tag{11.69}$$

Because one of the off-diagonal terms is 0, the eigenvalues are $\lambda_1 = -r$ and $\lambda_2 = \epsilon cK - m$. Hence, this prey-only equilibrium is an unstable saddle insofar as $\epsilon cK > m$. When this is the case, there is a single stable and a single unstable direction. Intuitively, we expect that the stable direction corresponds to perturbations along the prey axis only, i.e., $(u_1, 0)$. Note that, in the absence of predators, the rate of decay toward the carrying capacity is controlled by r. In addition, the unstable direction points toward the interior of the phase plane, i.e., (u_2, v_2). We can use the same technique as above to identify the eigenvectors.

The two systems of equations are

$$\begin{bmatrix} -r - \lambda & -cK \\ 0 & \epsilon cK - m - \lambda \end{bmatrix} \begin{bmatrix} u \\ v \end{bmatrix} = 0. \tag{11.70}$$

When $\lambda = -r$, then the two equations become

$$(-r + r)u - cKv = 0 \tag{11.71}$$

$$(\epsilon cK - m + r)v = 0 \tag{11.72}$$

and the only way to satisfy these equations is if $v_1 = 0$. The perturbation associated with the stable eigenvalue $\lambda_1 = -r$ is $(1, 0)$, associated with prey-only perturbations. When $\lambda = \epsilon c K - m$, then the two equations become

$$(-r - \epsilon c K + m)u - c K v = 0 \tag{11.73}$$

$$(\epsilon c K - m - \epsilon c K + m)v = 0 \tag{11.74}$$

such that the first equation defines a relationship (i.e., a line that we denote as the eigenvector) such that

$$\frac{u}{v} = \frac{cK}{-r - \epsilon c K + m}. \tag{11.75}$$

One typically writes down unit eigenvectors. Given that coexistence requires $\epsilon c K > m$ for there to be a coexistence equilibrium, then, irrespective of magnitude, the eigenvector implies that the ratio of the components of the eigenvectors should be negative. In biological terms, the eigenvector points in the direction of decreasing prey and increasing predator density toward the interior!

Returning to the coexistence fixed point Finally, in the event that $\epsilon c K > m$, then there exists an interior fixed point whose Jacobian can be written as

$$\mathbf{J} = \begin{bmatrix} r - 2rH^*/K - r(1 - H^*/K) & -m/\epsilon \\ \epsilon r(1 - H^*/K) & 0 \end{bmatrix} \tag{11.76}$$

or

$$\mathbf{J} = \begin{bmatrix} -rH^*/K & -m/\epsilon \\ \epsilon r(1 - H^*/K) & 0 \end{bmatrix}. \tag{11.77}$$

The eigenvalues are solutions to the equation $\text{Det}(\mathbf{J} - \lambda \mathbf{I}) = 0$ where $\mathbf{I}$ is the identity matrix. In this case, the Jacobian implies a quadratic function of λ. Here the determinant is positive and the trace is negative, which guarantees that the equilibrium is stable. It turns out that the dynamics are that of a spiral sink—refer to a standard linear algebra text or nonlinear dynamics text for the classification of the stability of fixed points in the trace-determinant plane. The chapter text, laboratory, and homework elaborate on the consequences.

Chapter Twelve

Outbreak Dynamics: From Prediction to Control

12.1   MODELING IN THE AGE OF PANDEMICS

On September 26, 2014, the CDC's Ebola Modeling Task Force published an ominous warning: the total number of cases of Ebola virus disease could exceed 1.4 million by January 2015 without large-scale intervention and changes in behavior (Meltzer et al. 2014). Given the high fatality rate, this prediction implied that more than a million individuals could die from Ebola virus disease over a four-month period unless large-scale interventions were initiated. This prediction immediately became news and newsworthy. For the public and policy makers, the number—1.4 million—conveyed a key message the ongoing Ebola epidemic in West Africa was markedly different than every other Ebola epidemic since the first identification of Ebola virus disease in 1975 (Chowell and Nishiura 2014). Already by September 2014, more people had been infected over a wide geographic range and more people were dying than in all prior Ebola epidemics combined (Alexander et al. 2015).

The basis for this original CDC forecast was a dynamic epidemiological model of how Ebola virus disease (EVD) spread through the three most affected countries: Guinea, Liberia, and Sierra Leone. The assumptions of the model, like many other models of infectious outbreaks, were that individuals exposed to the disease eventually become infectious and, only then, could transmit EVD to susceptible individuals (Legrand et al. 2007). Approximately 70% of infected individuals were assumed to subsequently die from the disease. New infections in the model occurred whenever a susceptible individual contracted EVD from an infectious individual through direct contact. A key assumption underlying the CDC model was that individuals within a country had an equally likely chance to interact with anyone else, whether in their village, city, or district or in any other village, city, or district, i.e., the model assumed a "well-mixed" population at the country scale.

The total number of Ebola cases eventually exceeded 30,000 with greater than 11,000 reported deaths. The true number was likely higher given systemic underreporting, and the 2014–2015 EVD outbreak remains the largest documented outbreak (Alexander et al. 2015). Despite its impact, the case counts remained substantially below that of the September projection by the CDC and subsequent projections by the WHO in October 2014 of monthly cases exceeding 50,000 by November and increasing in months to come (World

Health Organization 2014). In reviewing the differences between projections and reported cases, it is important to recall that the CDC included a caveat to the worst-case scenario projection. The CDC modeling report pointed out that, as control measures are "rapidly implemented and sustained, the higher projections presented in this report become very unlikely." As of January 2015, the World Health Organization reported that Liberia's new weekly case counts dropped to nearly zero, and Guinea and Sierra Leone saw substantial improvements from late December 2014. Indeed, the 2014–2015 epidemic soon came to a close, even as new outbreaks (in 2018) and threats of transitions to endemicity remain.

The difference between worst-case scenarios and outcomes should be a reason for some optimism amidst a challenging situation—particularly to the extent that models helped galvanize the response and shape interventions. Indeed, the purpose of models is not simply to predict but also to provide a rationale for why intervention is necessary (Rivers 2014; Lofgren et al. 2014). For example, would we blame the person who called the fire department when seeing a house is on fire, if the house was later saved from ruin? Of course not. Yet some questioned the value of disease forecasts in responding to Ebola, particularly when initial predictions later turned out to have been significant overestimates of the true scope of the outbreak. Just as there are dangers to inaction when faced with a disease outbreak, there are also costs, trade-offs, and dangers to miscalibrating response. If models can be used to support the argument to intervene, then shouldn't we also expect that they should be used to build the appropriate type, scope, and pace of intervention?

These questions remain prescient. In January 2020, news reports began to circulate of case clusters of an influenza-like illness in Wuhan, China. This emerging infectious disease—COVID-19—has transformed the health, movement, and fundamental well-being of individuals across the globe. Given prior experiences with the potential rapid spread and severity of SARS-1, the elevated role of modeling in the EVD response, and increased availability of case data and genomic data (Hadfield et al. 2018), a large number of modeling groups began to respond to this new threat. One group—the Imperial College of London team—provided a key, influential forecast that, like the CDC work of 2014–2015, included no-intervention forecasts compared to projections of the potential utility of scenarios (Ferguson et al. 2020). Panels A and B in Figure 12.1 illustrate the key dilemma. In the absence of aggressive nonpharmaceutical interventions (NPIs), using early case reports from Wuhan to parameterize a compartmental model of spread, the team projected that roughly 80% of a population might become infected. Total fatalities were estimated given an infection fatality rate slightly lower than 1% when taking into account the age dependence of severe disease. As a result, the Imperial College of London team concluded that over 2,000,000 individuals in the United States and over 500,000 individuals in the UK might die as a result of this emerging infection.

Unlike the case of EVD, the unmitigated forecasts for COVID-19 are far closer to the actual outbreak impact. Panel C of Figure 12.1 shows the regional impact of COVID-19 in the United States, including evidence of plateaus and oscillations in fatalities. As of May 2022, there had been more than 1 million documented fatalities in the US alone and over 6 million documented fatalities globally. COVID-19 has become the leading cause of death in many age groups, rivaling even heart disease among the elderly (Woolf et al. 2021). In addition, there is significant uncertainty with respect to the long-term impact of mild or moderate disease (Gandhi et al. 2020). This widespread impact has emerged

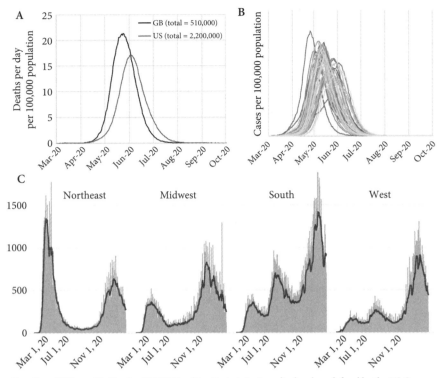

Data from 2020 provided by the COVID Tracking Project. Regions displayed are defined by the US Census.

Figure 12.1: Early COVID-19 modeling scenarios and outcomes. Predictions in the absence of interventions show an excess of 500K fatalities in the UK and more than 2M fatalities in the US. (A) Fatalities/day per 100K. (B) Cases/day per 100K (Reproduced from Ferguson et al. 2020). (C) Realized fatality data from the US reproduced via covidtracking.com broken down into four regions.

despite significant efforts to intervene; guided in part by models that predicted devastating consequences of inaction, as well as the potential impact of NPIs. It is hard to write a textbook chapter about SARS-CoV-2 amidst a response (though other books on virus origins and responses are already in press). Hence, the illustration of impact and contrasting outcomes between EVD and COVID-19 suggest that diseases that, from a modeling perspective, may have similar large-scale impacts—in the absence of intervention and/or behavior change—can have vastly different outcomes when spreading through real populations.

One key difference is that individual and population-level outcomes are not equivalent. For those who get infected, EVD is a horrific disease, with a 70% chance of dying and strikingly few effective treatments—even as new vaccine candidates are being developed. Yet, precisely because of the severity, the risk of asymptomatic spread is relatively low. In contrast, the vast majority of individuals infected with SARS-CoV-2 will survive. Notably, although the infection fatality rate of the disease varies with age, increasing to above 5% for individuals over 75, many individuals—especially younger individuals—can have asymptomatic infections. Hence, for many, an infection with SARS-CoV-2 will not be particularly worse than a cold or flu. Yet, at the population scale, the lack of symptoms is one of the reasons that COVID-19 has become so difficult to stop (Du et al. 2020; Wei et al. 2020; Moghadas et al. 2020; Park, Cornforth, et al. 2020). Asymptomatic and presymptomatic individuals can spread the infection to other more vulnerable individuals, radiating the impact outward through partially silent and then devastating chains.

This chapter is intended to introduce the core concepts of epidemiological models with an eye toward informing the kind of decision making used during an infectious disease outbreak. Whereas many other chapters focus on understanding the core principles of living systems—at whatever scale—here there is an important distinction. The intention of epidemic modeling is to work in Pasteur's quadrant (Stokes 2011), i.e., developing methods to mitigate and control an outbreak rather than predicting a bad outcome accurately. Hence, the purpose of this chapter is to help advance fundamental understanding of disease transmission that could guide and improve critical interventions. The plural is important. In essence, this chapter takes the perspective that models have a crucial role to play in identifying multifaceted, rather than monolithic, responses to epidemic outbreaks. The many facets of such a response may be unconventional, including vaccination, quarantines, and even changes in behavior. The role of each of these responses can be seen by first examining simple dynamic models and then using EVD and SARS-CoV-2 as case studies to put these ideas into practice.

In doing so, this chapter tries to answer the following questions: What are the key drivers that determine whether a disease is likely to spread, or not? What are the priorities for intervention that could forestall the spread of disease or control the spread during an outbreak? How can vaccinations make a difference even if not everyone gets vaccinated? And, finally, how certain can we expect to be about the likely spread and the ideal targets of interventions at the start of an outbreak? The start of an outbreak is precisely when the numbers of infected cases are small, stochasticity is relevant, and behavioral changes in response to disease may be just as important as large-scale interventions.

12.2 THE CORE MODEL OF AN OUTBREAK: THE SIR MODEL

12.2.1 The premise

The spread of infectious disease requires the spread of the causative agent—an infectious pathogen—from one individual to another. Transmission events may occur through various means, e.g., short- or even long-distance transmissions through the air (e.g., sneezing), physical contact (including sexual transmission), indirect transmission (e.g., through the environment via contact with contaminated surfaces), or vector-borne transmission (e.g., via mosquitoes, ticks, flies) (Figure 12.2). Each transmission is a "chance" event; yet, cumulatively, outbreak dynamics can exhibit regular features. Part of the reason for this regularity is that, for large outbreaks, the factors driving individual transmission events lead to predictable changes in the states of individuals with respect to the pathogen. For example, in the case of an infectious disease from which exposure (and survival) leads to lifelong immunity, individuals may be susceptible, infectious, or recovered. The study of epidemiological dynamics aims to connect these microstates of individuals to the macroscopic state of the population, in which there are S, I, and R individuals—those that are either susceptible, infectious, or recovered.

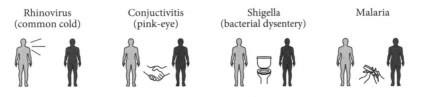

Figure 12.2: Interaction modes for disease transmission, including airborne, contact, environmental, and vector-borne routes.

To begin, consider a population characterized by $(S(t), I(t), R(t))$ such that $S + I + R = N$, where N is the (fixed) number of individuals in the population. This population might represent a town, county, district, state, or country. Now imagine that there is a single infectious individual who has contracted the disease. This infectious individual interacts with c individuals that could potentially lead to disease transmission. Of these individuals, a fraction S/N are susceptible. Hence, the infectious individual has, on average, cS/N potentially infectious contacts. Yet not all contacts lead to transmission. If there is a p probability of transmission for each infectious contact, then one expects the probability of k successful transmission events in one day to be $B(k|c, pS/N)$, where B is the binomial distribution given c trials and pS/N probability of success per trial. Hence, the probability that there are no successful transmission events should be $B(0|c, pS/N)$ or $(1 - pS/N)^c$. Likewise, the probability that there is at least one successful transmission event should be $1 - (1 - pS/N)^c$. Note that when $p \ll 1$ this reduces to cpS/N. We could, in principle, build a stochastic model of disease transmission in this way, tracking the stochastic trajectory of an outbreak and its feedback on the underlying numbers of susceptible, infectious, and (eventually) recovered individuals.

Instead, consider a model that keeps track of very small units of time, where we now interpret the value of c to be a *rate* of contact per unit time. In that way, the probability of a transmission event with one focal infectious individual taking place in a small unit of time is $cp\frac{S}{N}dt$. For I infectious individuals, the probability of a transmission event in a small unit of time is $cp\frac{S}{N}Idt$. Concurrently, infectious individuals can recover at a rate γ. Altogether, a model of disease dynamics would include transmission events and recovery events, which can be represented as a coupled system of nonlinear differential equations:

$$\dot{S} = - \overbrace{\beta\frac{S}{N}I}^{\text{infection}} \tag{12.1}$$

$$\dot{I} = \overbrace{\beta\frac{S}{N}I}^{\text{infection}} - \overbrace{\gamma I}^{\text{recovery}} \tag{12.2}$$

$$\dot{R} = \overbrace{\gamma I}^{\text{recovery}} \tag{12.3}$$

where $\beta \equiv cp$. This is the standard SIR model of infectious disease spread (Figure 12.3). A full derivation beginning with a stochastic model is included in the technical appendix.

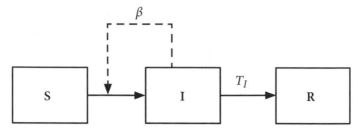

Figure 12.3: The SIR model of disease dynamics.

In practice, the spread of disease in different populations can be compared by rescaling the dynamics in terms of the fraction of individuals in each of the three states. Hence, consider the normalized variables $\tilde{S} = S/N$, $\tilde{I} = I/N$, and $\tilde{R} = R/N$, such that $\tilde{S} + \tilde{I} + \tilde{R} = 1$. In that case, the core model becomes

$$N\frac{d\tilde{S}}{dt} = -N^2 \beta \frac{\tilde{S}}{N} \tilde{I} \tag{12.4}$$

$$N\frac{d\tilde{I}}{dt} = N^2 \beta \frac{\tilde{S}}{N} \tilde{I} - \gamma N\tilde{I} \tag{12.5}$$

$$N\frac{d\tilde{R}}{dt} = \gamma N\tilde{I} \tag{12.6}$$

Factoring out factors of N and removing the ˜ notation, we can write the SIR model in terms of the fraction of the population that is susceptible, infectious, or recovered/removed as

$$\dot{S} = -\beta SI \tag{12.7}$$

$$\dot{I} = \beta SI - \gamma I \tag{12.8}$$

$$\dot{R} = \gamma I \tag{12.9}$$

In this model, the disease-free state corresponds to $(S = 1, I = 0, R = 0)$, although given the fact that $S + I + R = 1$, it is only necessary to focus on two of three state variables. This disease-free state is also an equilibrium of the model in Eqs. (12.7)–(12.9). Precisely because it is an equilibrium, we may want to know: what happens, then, when a small number of infectious individuals are added to the "disease-free" state?

12.2.2   Conditions for disease spread

The fate of a disease can be expressed in multiple ways. One way to examine new disease emergence is to ask: what happens to the fraction of infectious individuals in a population when a very small number of individuals (perhaps only one in N) are infectious? In this model, the dynamics of the infectious fraction are

$$\dot{I} = I(\beta S - \gamma). \tag{12.10}$$

However, when the population is almost, totally susceptible, i.e, $S \approx 1$, then this equation becomes

$$\dot{I} \approx I(\beta - \gamma). \tag{12.11}$$

The fate of the disease depends on the difference between new infections and recovery, or $\beta - \gamma$. This difference is known as the speed of disease spread, $r \equiv \beta - \gamma$. A positive speed implies that an outbreak will occur, with an exponentially increasing number of new cases, at least at first. In contrast, a negative speed implies that the outbreak dissipates, with an exponentially decreasing number of new cases. This difference can also be written in a different way,

$$r = \gamma \left(\mathcal{R}_0 - 1 \right), \tag{12.12}$$

where

$$\mathcal{R}_0 \equiv \frac{\beta}{\gamma} \tag{12.13}$$

is known as the basic reproduction number. The basic reproduction number is "arguably the most important number in the study of epidemiology" (sensu Diekmann et al. 1990 and Diekmann and Heesterbeek 2000). It deserves its own paragraph.

> $\mathcal{R}_0$: The average number of new infections caused by a single infectious individual in an otherwise susceptible population.

It is perhaps even more intuitive to understand $\mathcal{R}_0$ by recognizing that $1/\gamma$ is equal to the average infectious period, T_I. Hence, if β denotes the number of new infections caused by a single individual per unit time, then multiplying by the time of infectiousness yields an average number of new infections over the course of that infection. In essence, $\mathcal{R}_0$ measures disease spread at the individual scale. For example, if $\mathcal{R}_0 = 2$, that implies that a typical infectious individual will generate two new cases before the individual recovers. Those two new cases will each generate two more (or nearly so), which leads to four cases, then eight, sixteen, and so on. At some point, the number of cases may actually deplete the susceptible population and the exponential growth will slow down. Exponentials are not forever. Nonetheless, the strength of the epidemic, as measured in terms of $\mathcal{R}_0$, provides a threshold condition for the spread of the disease. In essence, when $\mathcal{R}_0 > 1$ the disease will spread, because each infectious individual leads to one or more new infectious individuals (exceeding replacement) (Figure 12.4).

Yet you may have a doubt. There is feedback between susceptible and infected individuals. Is it really appropriate to analyze only the $\dot{I}$ equation? The answer, in brief, is yes. Formally, this is because I is the only infected subsystem of the model. But, if you are still not satisfied, then it is worthwhile to take a dynamical systems perspective and ask: what are the conditions in which the disease-free equilibrium is unstable? The local stability of a fixed point can be examined by linearizing the model and exploring the growth or decay of small fluctuations around the fixed point. The process of linearizing a nonlinear model was introduced earlier in this book, in the context of genetic circuits (Chapter 2), neurons (Chapter 6), and predator-prey dynamics (Chapter 11). The same process applies here, as demonstrated in the technical appendix.

There are two eigenvalues for the SIR model given a disease-free equilibrium: $\lambda_1 = 0$ and $\lambda_2 = \beta - \gamma$. Finding a zero eigenvalue implies that there are some perturbations to the fixed point that do not lead to any further change—whether increasing or decreasing—to the state space. It turns out that this corresponds to a particular kind of perturbation, in which a small number of individuals are moved from the S state to the R state, i.e., $(1, 0, 0) \rightarrow (1 - \epsilon, 0, \epsilon)$.

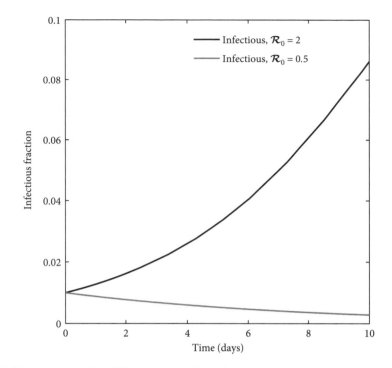

Figure 12.4: Schematic example of differences in outbreak dynamics as a function of the basic reproduction number, $\mathcal{R}_0$. In theory, when $\mathcal{R}_0 > 1$ the disease spreads, whereas when $\mathcal{R}_0 < 1$ the disease dissipates.

The latter state is also an equilibrium of the model. In fact, there is a *whole line* of equilibria, for any combination of S and R that add up to 1! This perturbation also has a biological interpretation: it is what would happen if one vaccinated a fraction ϵ of the population. We return to this concept in a moment to explain precisely why it is possible that vaccinating a fraction of the population can provide "herd immunity" to nearly everyone, even those who are not vaccinated. Hence, although λ_1 does not seem to depend on the etiology or transmissibility of the disease, λ_2 does, and it corresponds precisely to the speed inferred from analyzing only the I equation. Note that, although the analyses coincide, it is preferable to use the linearization approach or even the next-generation method to find robust answers to this question: what are the conditions in which a disease will spread, or not?

12.3   THE SHAPE OF AN OUTBREAK

12.3.1   Outbreak dynamics—the basics

In theory, the spread of a disease begins exponentially. As infectious cases increase, even more "susceptibles" are infected. Then some of the infectious individuals begin to recover, leading to the characteristic increase of recovered and infectious individuals while suscepti-bles decline. Yet the disease does not keep spreading indefinitely. Instead, at a certain point the infectious case count begins to decrease. It does so because of susceptible depletion—the fact that there are fewer and fewer susceptibles available to infect even as the characteristic

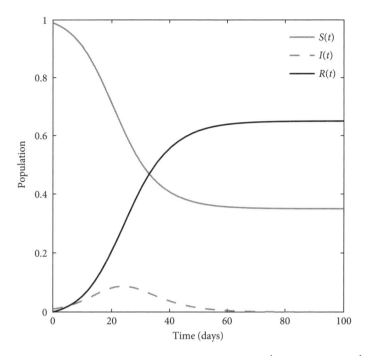

Figure 12.5: Disease outbreak given $\mathcal{R}_0 = 1.6$, $\gamma = 0.25$ days^{-1}, and $\beta = 0.4$ days^{-1}.

recovery rate of the disease remains constant. As is evident in Figure 12.5, there is a critical point at which $\dot{I} = 0$ (corresponding to the peak incidence of the disease), after which the incidence declines. This point must correspond to the condition when

$$\dot{I} = I\left(\beta S - \gamma\right) = 0. \tag{12.14}$$

There are two conditions when this holds. The first is when $I = 0$ (i.e., the disease-free state) and the second is when $S = \frac{\gamma}{\beta} = \frac{1}{\mathcal{R}_0}$. In other words, the disease will begin to die out once the fraction of susceptibles drops below $1/\mathcal{R}_0$. This observation holds the key to multiple facets of epidemiology—from both a prediction and control perspective. Recall that the definition of the basic reproduction number is contextual and assumes that the population is wholly susceptible. Analogously, we can define a state-dependent, dimensionless number termed the *effective reproduction number*:

$\mathcal{R}_{eff} = \frac{\beta S}{\gamma}$: The average number of new infections caused by a single infectious individual in a partially susceptible population, $0 \leq S < 1$.

The connection to the transition between disease spread and disease dissipation is now apparent. The dynamics of $\dot{I}$ can be written as

$$\dot{I} = \gamma I \left(\mathcal{R}_{eff} - 1\right) \tag{12.15}$$

such that the depletion of susceptibles continues to reduce $\mathcal{R}_0$; and because $\mathcal{R}_{eff}$ scales with S, then once S is reduced by a factor of $\mathcal{R}_0$, the effective reproduction number drops below 1.

Once the incident case count starts to decline, it continues to decline until there are no more infectious cases, i.e., the disease has run its course. In the end, not everyone gets sick—not even in theory. For the particular example in Figure 12.5 in which $\mathcal{R}_0 = 1.6$, there is still a substantial fraction of the population, 0.35 to be precise, that was never infected. Hence, the final size of the epidemic was 0.65, in the sense that at one point or another 65% of the population had the disease. Notice that $1 - 1/1.6 = 0.375$, which is not equal to the final number of noninfected individuals because of substantial overshoot after herd immunity is reached. As a reminder, herd immunity is the threshold where cases just replace themselves; this can be a preventative threshold at the outset of an epidemic, but it implies that new cases will continue even after incidence starts to decline. In fact, there is a putative relationship between the strength of the disease (as measured by $\mathcal{R}_0$) and the final size (as measured by S_∞), but that requires more explanation.

12.3.2 Speed, size, and strength

What is the relationship between the strength of the disease and the final size? It would seem intuitive that if infectious individuals spread the disease more readily at the outset this would compound, over time, to yield more cases by the end. Such is the power of exponentials. But diseases spread in finite populations, so the relationship between strength and final size may be more tenuous. To understand the problem, consider what would happen if a single infectious individual entered into a population of size 10^4. As long as $\mathcal{R}_0 > 1$, the disease is expected to spread. But to how many people?

To answer this question, consider a replicate set of populations, e.g., distinct towns, an animal herd, or a crop culture. The strength of a particular pathogen depends both on the etiology of the disease and the connectedness of the population. Hence, the same pathogen could have very different values of $\mathcal{R}_0$ due to the environment in which it spreads. We can determine the final epidemic size by numerically simulating the SIR model. However, there is another way. Note for a moment that the rate of change of infectious and susceptible individuals can be factored, and when taking the ratio of these rates of change yields

$$\frac{\dot{I}}{\dot{S}} = \frac{-\beta SI + \gamma I}{\beta SI}.$$

(12.16)

Because the I variable cancels out, it is possible to integrate these equations, not over time but with respect to the differential change in both infectious and susceptible individuals. The approach is shown in the technical appendix. The result is the following:

$$\mathcal{R}_0 (S_\infty - 1) = \log(S_\infty).$$

(12.17)

Figure 12.6 compares this final size prediction (asterisks) to the results of numerical simulations (solid lines), all beginning with the same initial condition but with increasing $\mathcal{R}_0$. The formula works. However, the lack of a closed form implies that the solution requires finding the crossing of the term on the left-hand side with that on the right. Yet some intuition is possible. Note that the right-hand side of Eq. (12.17) decreases from 0 to $-\infty$ over the interval $0 \leq S_\infty < 1$. The left-hand side denotes a line from 0 to $-\mathcal{R}_0$ over the same interval. These functions will cross at a single point so long as $\mathcal{R}_0 > 1$, ever closer to 0 as $\mathcal{R}_0$ increases. This single crossing point represents the final size of the epidemic.

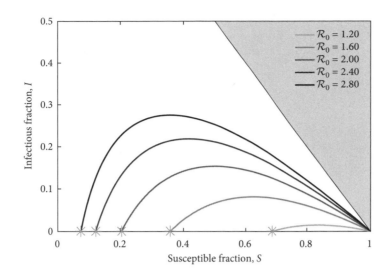

Figure 12.6: Final size dependency on $\mathcal{R}_0$. In each case, trajectories begin near a fully susceptible population albeit with different values of $\mathcal{R}_0$. The asterisks denote the expected final size according to theory.

Some caveats are important when predicting final sizes from the strength of the disease. Indeed, size-strength predictions may have an elegant mathematical underpinning, but they are often violated in practice (Eksin et al. 2019; Scarpino and Petri 2019; Hébert-Dufresne et al. 2020). Critically, models of size-strength coupling assume a well-mixed population with identical risk of exposure, susceptibility, and transmissibility. Yet variation in these levels can lead to fundamental changes in the long-term outcome. For example, given heterogeneity in exposure, it is likely that individuals with more connections may be infected first. However, precisely because these individuals are highly connected, a pathogen is more likely to "waste" interactions with them. The gap between predicted size based on $\mathcal{R}_0$ alone and realized size remains one of the key uncertainties in integrating predictions into practice (Britton et al. 2020; Rose et al. 2021).

12.3.3   From epidemics to endemics

The SIR model is one variant of many representations of epidemic dynamics. The variants often depend on the etiology of the disease, in particular, the extent to which recovered individuals are immune from subsequent infection (as in measles) or susceptible in the near future (as in many common colds caused by rhinoviruses). For example, consider a SIR model in which recovered individuals eventually become susceptible, e.g., at a rate γ. Hence, in this case the standard SIR model becomes

$$\dot{S} = -\beta SI + \alpha R \tag{12.18}$$

$$\dot{I} = \beta SI - \gamma I \tag{12.19}$$

$$\dot{R} = \gamma I - \alpha R \tag{12.20}$$

where α is the rate of loss of immunity. Now, although an outbreak may temporarily deplete susceptibles, these susceptibles will be replenished by the loss of immunity. This suggests

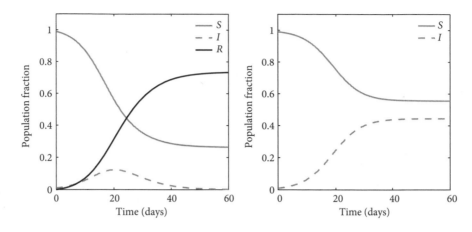

Figure 12.7: From outbreaks in a SIR model (left) to endemics in an SI model (right), both with $\beta = 0.45$ /days and $\gamma = 0.25$ /days.

that there may be a new equilibrium outcome: an endemic disease state. This possibility can be analyzed more readily by taking one particular limit, when $\alpha \gg \gamma$. In that case, the dynamics of the recovered are in a quasi equilibrium, i.e., $\dot{R} = 0$ such that $R^* = \gamma I/\alpha$, and replacing $R(t)$ with this value yields what is termed the SI model:

$$\dot{S} = -\beta SI + \gamma I \tag{12.21}$$

$$\dot{I} = \beta SI - \gamma I \tag{12.22}$$

The SI model denotes a disease that transmits with a rate β and for which infectious individuals recover at a rate γ but are then immediately infectious.

In this model, the conditions for the disease outbreak remain precisely the same as in the SIR model. That is, there is an outbreak when $\mathcal{R}_0 = \beta/\gamma > 1$. However, when there is an outbreak, there is also a different outcome: an endemic disease. Figure 12.7 contrasts the outcomes for a disease with the same values of β and γ; however, the case on the left denotes a disease with permanent immunity and that on the right denotes a disease with no immunity. As such, the dynamics do not lead to a new disease-free equilibrium. Instead, the equilibrium endemic state is

$$S^* = \frac{\gamma}{\beta} = \frac{1}{\mathcal{R}_0} \tag{12.23}$$

$$I^* = 1 - \frac{\gamma}{\beta} = \frac{\mathcal{R}_0 - 1}{\mathcal{R}_0} \tag{12.24}$$

such that increasing strength also corresponds to a larger final size. It is important to note that the final size of the epidemic in the SIR model denotes the number of individuals who ever contracted the disease. In contrast, the SI disease size denotes the disease burden at any given time—it is possible that all will get the disease at one point or another, at least in theory, as long as $\mathcal{R}_0 > 1$. Together, these models suggest that controlling a disease depends critically on steps taken to change the dynamics at the very outset, when there are still a small number of infectious individuals in the population.

12.4 PRINCIPLES OF CONTROL

The reason for focusing so much attention on the strength $\mathcal{R}_0$ is not only because it is a critical threshold condition for understanding when diseases spread, but also because it represents the conceptual basis for understanding how to control and even prevent the spread of an infectious disease. Recall once again the definition of $\mathcal{R}_0$, this time while retaining the individual parameters that underlie its definition:

$$\mathcal{R}_0 = c \cdot p \cdot T_I \cdot \left(\frac{S}{N} \right) \tag{12.25}$$

where c is the number of contacts per unit time, S/N is the fraction of such contacts that are susceptible, and p is the probability that a contact between a susceptible individual leads to transmission. This equation already suggests many of the necessary core strategies, each operating on one or a combination of these contributing factors.

Treatment—T_I Some treatments can reduce symptoms directly by reducing pathogen load and decreasing the time to pathogen clearance, thereby reducing the effective period of infectiousness.

Contact tracing, testing, and targeted isolation—c Contact tracing denotes efforts by public health responders to identify the source of an outbreak. Once an infectious individual is identified, responders will try to work backward to identify a potential transmission chain (i.e., to find out the potential source of the epidemic) as well as forward to try to identify those who may have been exposed. Testing at scale followed by isolation of positive test individuals can be used to reduce the contact rates.

Vaccination—S/N Vaccination reduces the pool of susceptible individuals. As a consequence, a random interaction will less often be between an S and an I individual, and as we will show, when enough potential infection events are wasted, then it is possible to control the disease.

Process engineering or personal protective equipment—p The last line of defense for stopping a disease is process engineering, or the use of PPE (personal protective equipment). For EVD this requires extreme measures (the white hazmat-style suits), whereas for other diseases, like COVID-19, protective measures can include face masks and improved ventilation. Finally, hand washing and other measures are considered standard approaches to reduce p in a preventative sense.

Finally, outcomes matter along with cases. Hospitalization and treatment can have multiple beneficial effects. The most evident effect is the chance to reduce the risk that a case becomes severe (via treatment) or even fatal (via treatment and/or hospitalization). It is important to keep in mind that assessing the impact of an illness must account for both the transmission and impacts on infected individuals.

Rather than work through examples of each intervention, it is worthwhile to highlight a proven approach that has been instrumental in preventing disease outbreaks in the first place: vaccination. Figure 12.8 (Plate 14) shows normalized measles incidence in states across the US before and after the introduction of the measles vaccine in 1963.

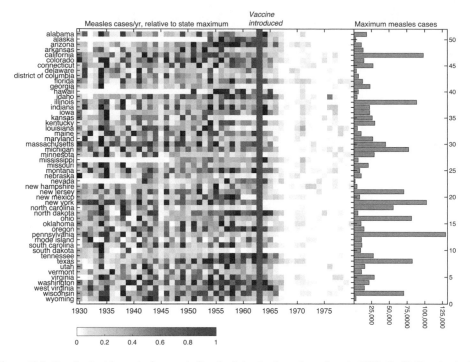

Figure 12.8: Measles incidence before and after the introduction of vaccines in 1963. (Left) Measles incidence per year relative to yearly state maximum (including the District of Columbia). (Right) The maximum number of annual measles cases per region in the time period shown in the left panel. Adapted from a 2015 Project Tycho data visualization. Project Tycho is a global health data initiative based at the University of Pittsburgh: https://www.tycho.pitt.edu.

As is apparent, the introduction of the vaccine led to a rapid decline in case incidences throughout the United States. Prevaccine, there were approximately half a million annual cases, 50,000 hospitalizations, and 500 deaths. Measles is also highly transmissible, with an estimated $\mathcal{R}_0 \approx 15$ (Guerra et al. 2017). How then can vaccines be so effective at stopping it, particularly given that it is unlikely that everyone will receive them?

One step toward an answer starts with an analysis of the SIR model. In practice, a full accounting of a model for measles requires consideration of births, age-dependent susceptibility, transmission, and more. Nonetheless, some insights are possible using the SIR model given the highly contagious nature of the disease. To begin, recall that the initial speed of the disease (corresponding to the positive eigenvalue) is $\beta S_0 - \gamma$. Now consider if a fraction $0 \le f < 1$ of the population is vaccinated and therefore already begins in the R class, so that $S_0 = (1 - f)$. The initial (exponential) speed of the outbreak can be written as

$$r = \gamma \left(\mathcal{R}_0 (1 - f) - 1 \right). \tag{12.26}$$

This speed switches from positive to negative when $\mathcal{R}_0 (1 - f) = 1$, or at the critical vaccination level:

$$f_c = \frac{\mathcal{R}_0 - 1}{\mathcal{R}_0}. \tag{12.27}$$

Figure 12.9 shows that stopping the spread requires that a substantial *fraction* of the population is vaccinated. This fraction increases with increasing strength. The total protection of

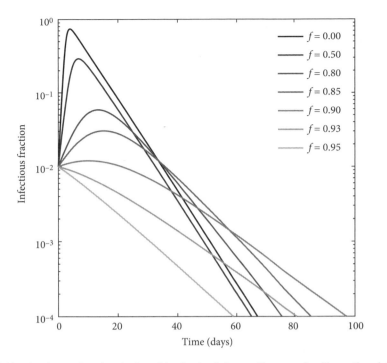

Figure 12.9: Vaccination and outbreaks in epidemic simulations with a measles-like outbreak for which $\mathcal{R}_0 = 14$. Parameters in the model are $\beta = 2$ /days, $\gamma = 0.14$ /days. The vaccination fraction is noted by f.

partial vaccination underlies the concept of herd immunity. Imagine a scenario where an infectious individual A moves to a particular population—will such an introduction subsequently lead to the transmission of the disease to individual C? Imagine further that A interacts with B who interacts with C. If B is vaccinated, then C will be protected, even if C has not been vaccinated. In essence, interactions are "wasted" on vaccinated individuals. The ideal is to have nearly 100% vaccination, but the imperative becomes more important with increasing disease strength.

12.5   EVD: A CASE STUDY IN CONTROL GIVEN UNCERTAINTY

The responses listed in the prior section may seem generic, but they are precisely the kind of responses considered in real outbreaks, particularly when vaccinations or treatments are not available. For example, in the 2014–2015 EVD outbreak in West Africa, an influential response model (Pandey et al. 2014) considered a variety of strategies for containing Ebola. These strategies included the following:

- Transmission precautions for health care workers
- Reduction of general community transmission
- Sanitary burial
- Isolation of Ebola patients
- Contact tracing and quarantine
- General quarantine

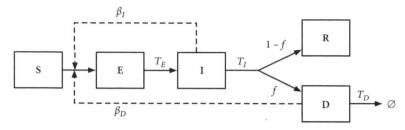

Figure 12.10: SEIRD model of EVD dynamics, including postdeath transmission. Reproduced from Weitz and Dushoff (2015).

Inclusion of these factors requires going beyond the standard SIR model and reflects a suite of actions informed by observations in the midst of an unfolding crisis. Despite the added complexity, these interventions share a common intent: to reduce $\mathcal{R}_{eff}$. How much effort and emphasis to place for each factor depends on robust estimates of the basic reproduction number. Yet such estimates are inherently uncertain, particularly at the start of an outbreak. The extent of this variation sets a lower bound on certainty of the properties of a disease, and should also influence the confidence one has in using a particular control mechanism rather than a set of control mechanisms. Quantifying this stochastic limit to confidence is explained next.

12.5.1   Process noise and stochastic outbreaks

Stochastic variation in the timing of discrete transmission events can generate variation in disease dynamics. Like in the case of population genetic models introduced in Part I of this book, the fact that a disease should grow in case number—if modeled deterministically—does not necessarily mean it will do so. Instead, fluctuations in the number of infectious individuals and the outcomes of transmission at individual scales can translate into meaningful differences at the population scale. For example, rather than growing deterministically, a disease may exhibit periods of stochastic fluctuations before "liftoff"—and only then begin to increase exponentially. Likewise, fluctuations can also induce clustered outbreaks that would otherwise not be expected to increase in number in a fully deterministic model. Even when a disease takes off, stochasticity affects epidemic speed: r_0. The consequences of such stochasticity are critical for inference and control. Because the speed of a disease can be directly measured, variation in estimating speed influences model data fits and the inference of the strength, $\mathcal{R}_0$, including limiting the ability to confidently estimate strength based on parameter variation alone (King et al. 2016).

To illustrate these points, consider the stochastic dynamics of EVD using an extended version of the SIR model. This extended version is termed an SEIRD model, because it includes the dynamics of susceptible (S), asymptomatically exposed (E), infectious (I), recovered (R), or dead but not yet buried (D) individuals (Figure 12.10). The dynamics can be written as

$$\frac{dS}{dt} = -\beta_I SI/N - \beta_D SD/N \tag{12.28}$$

$$\frac{dE}{dt} = \beta_I SI/N + \beta_D SD/N - E/T_E \tag{12.29}$$

Table 12.1: Relationship between epidemic process mechanism and rate

Epidemic process	Reaction	Reaction rate
Predeath infection	$S + I \rightarrow E + I$	$r_1 = \beta_I S \frac{I}{N}$
Postdeath infection	$S + D \rightarrow E + D$	$r_2 = \beta_D I \frac{S}{N}$
Onset of infectiousness	$E \rightarrow I$	$r_3 = \sigma = 1/T_E$
End of infectiousness (survival)	$I \rightarrow R$	$r_4 = (1-f)\gamma = (1-f)/T_I$
End of infectiousness (death)	$I \rightarrow D$	$r_5 = f\gamma = f/T_I$
Burial	$D \rightarrow B$	$r_6 = \rho = 1/T_D$

$$\frac{dI}{dt} = E/T_E - I/T_I \tag{12.30}$$

$$\frac{dR}{dt} = (1-f)I/T_I \tag{12.31}$$

$$\frac{dD}{dt} = fI/T_I - D/T_D \tag{12.32}$$

where the inclusion of transmission from the dead class reflects the fact that postdeath transmission was a critical factor in EVD transmission (Weitz and Dushoff 2015) (e.g., in an extreme case, the burial of an Imam in Mali led to nearly 100 documented cases). The basic reproduction number $\mathcal{R}_0$ in this case is the contribution from the I and D classes, i.e., $\mathcal{R}_0 = \beta_I T_I + f\beta_D T_D$. In this particular example, the transmission rates are set such that a fraction $\rho_D = \frac{\mathcal{R}_0(\text{dead})}{\mathcal{R}_0} = 0.25$ of transmission is attributable to interactions with dead individuals.

The SEIRD model can be simulated stochastically using the Gillespie framework introduced in Chapter 3, in which process rates determine both when the next event occurs and which event occurs. The epidemic process, equivalent epidemiological "reaction," and associated rates are shown in Table 12.1. The reactions denote how the interaction of individuals on the left-hand side of the reaction arrow lead to a different type of individual(s) on the right-hand side of the reaction arrow. In each simulation, the same underlying parameters are chosen and epidemics are initiated with one infectious individual in an otherwise susceptible population. Mathematically, the initial state is $\mathbf{y} = (N_0 - 1, 0, 1, 0, 0)$ for (S, E, I, R, D). In practice, the outbreak proceeds by calculating the time until the next event and then choosing which single event took place. The total rate of outbreak-associated events is $r_{tot} = \sum_{i=1}^{6} r_i$. The time until the next event is exponentially distributed with a mean time $1/r_{tot}$. At this point, the probability that the event corresponds to a particular process is the ratio of the process rate to the sum of rates: r_i/r_{tot}. Each event leads to a corresponding update in the number of individuals in each state. Given the updated profile, reaction rates are recalculated and the process continues—one event at a time. Trajectories are complete when there are no longer any exposed or infectious individuals—the end of the epidemic is an absorbing state of the model.

Figure 12.11A shows five examples of resulting simulations using the Gillespie algorithm. As is apparent, despite having the same underlying rates and initial condition, the trajectories take a different time to "liftoff" (here defined as the time to reach 50 active cases) as well as have different speeds as measured in the time period after liftoff. The ensemble of dynamics after liftoff is shown in Figure 12.11B, making it evident that there can be

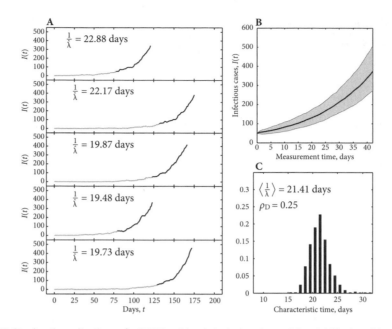

Figure 12.11: Stochastic realizations of a SEIRD epidemic include substantial variability in epidemic growth rate (reproduced from Taylor et al. 2016). (A) Each panel denotes a randomly chosen trajectory for which $I(t)$ exceeds 50. The gray period denotes those times for which $I(t) < 50$, and the black period denotes a 42-day measurement period after the first day for which $I(t) \geq 50$. The estimated epidemic growth rate for each trajectory over the fitting period is denoted in the upper left of each panel. (B) Variation in infectious time series once a threshold of 50 cases is reached. The solid line denotes the average number of cases, and the shaded region denotes a 95% confidence interval. (C) Variation in the estimated characteristic time, τ_c, of the epidemic. In all cases, simulations correspond to stochastic simulations of a SEIRD model in which $N = 10^6$, $\beta_I = 0.25$ /days, $\beta_D = 0.2$ /days, $\sigma = 1/11$ /days, $\gamma = 1/6$ /days, $f = 0.7$, and $\rho_D = 0.25$. The theoretically expected characteristic time is $\tau_c = 21$ days.

substantial differences in apparent speed driven by process noise. Finally, Figure 12.11C shows how the measurement of the characteristic speed of the epidemic and associated characteristic time can vary substantially over this time horizon. Variation in observations of speed are compounded by uncertainties with respect to the generation interval distribution. The result is that confidence in estimating the basic reproduction number and the efficacy of particular controls may be far more limited, when taking into account process noise, than is often assumed by deterministic model fits alone. The fact that uncertainty in the speed of an outbreak directly translates into uncertainty of estimates of $\mathcal{R}_0$ is a theme present in analysis of SARS-CoV-2 as well (Park, Bolker, et al. 2020).

12.5.2 Control strategies given uncertainty

The previous section presented a rationale for why estimates of the strength of a disease are likely to be uncertain—more uncertain than when using deterministic models. Process noise is just one of many factors influencing the uncertainty in the estimate of disease strength. For diseases with multiple transmission routes, this uncertainy can also cloud action taking with respect to optimal control mechanisms, especially those in which interventions require time to implement (Morris et al. 2021). For EVD, part of the uncertainty with respect to estimating strength was fundamental: how much of the transmission occurred due to contact with infectious individuals versus how much occurred after death as burial-associated transmission. In response to the potential for postdeath transmission, safe burial teams were deployed to the region, a process that could be implemented far more rapidly than Ebola treatment units, which took three or more months to deploy at scale and within hard-to-access regions (Nielsen et al. 2015). What is the rationale for this choice? In the case of the SIR model, there is a straightforward relationship between the observed speed r and the inferred strength $\mathcal{R}_0$, i.e., $\mathcal{R}_0 = 1 + rT_I$ where T_I is the infectious period. For EVD, such estimates are more challenging. It is possible to use information on the generation interval distribution as a means to estimate $\mathcal{R}_0$ for a more complex model—like the SEIRD model—from the observed growth rate, r (Wallinga and Lipsitch 2007). Theoretically, this link is

$$\mathcal{R}_0 = \frac{1}{M(-r)} \tag{12.33}$$

where

$$M(z) = \int_0^\infty e^{za} g(a)\, \mathrm{d}a \tag{12.34}$$

with a being the age of infection, $g(a)$ the generational interval distribution, and $M(z)$ a moment-generating function (see the technical appendix for more details). In practice, postdeath transmission extends the duration of the generation interval distribution, stretching out the age between infector and infectee to the period beyond death, due to the potential for transmission via contact with bodily fluids. This also means that eliminating postdeath transmission could represent up to half of the needed reduction in disease strength to eliminate the outbreak altogether (Weitz and Dushoff 2015). In practice, the inclusion of safe burials became a key to ending the EVD outbreak.

The initial CDC model of Ebola virus disease spread introduced in Section 12.1 had two central components: a model of infectious transmissions in the absence of interventions and a model for the effect of interventions. In the absence of control, infectious individuals were assumed to cause approximately 1.5 to 2 new infected cases. Such transmission could be reduced by interventions. The two primary types of intervention considered were hospitalization and home/community isolation. Both home isolation and hospitalization were assumed to reduce spread by more than 90%. As a result, the central intervention prioritized by state-level and international relief agencies was the construction, staffing, and use of Ebola treatment units. What was missed initially was the possibility that other interventions, especially safe burials, could be faster to introduce and nearly as effective. Ultimately, directing the focus of interventions depends on estimates of the potential reduction in spread, a reduction driven by model inference and limited by many types of uncertainty. Recognizing the limits to confidence early can reduce the generation of overly confident predictions as well as keep more intervention options on the table.

12.6 ON THE ONGOING CONTROL OF SARS-COV-2

Nonpharmaceutical interventions and vaccination programs have had an enormous impact on the ongoing SARS-CoV-2 pandemic. Compared to earlier outbreaks, epidemic models have played a far larger role in structuring interventions to SARS-CoV-2, though their impact is debated. Part of that debate may be warranted, but part of it resembles the same debate that was triggered about the use of models in the 2014–2015 response to EVD (Rivers 2014). It is critical to distinguish between models that are meant to highlight the risk of inaction, evaluate the potential benefits of interventions, and perform short-term forecasts (often in the absence of a mechanistic representation of the transmission of disease). Each of these model types has been used and continues to be used to respond to SARS-CoV-2 (Vespignani et al. 2020). Nonetheless, one of the key gaps in all of these approaches has been the challenge of integrating epidemic spread with the response of individuals and societies in changing behavior.

The link between behavior and epidemics remains one of the most important conceptual challenges for the field (Funk et al. 2010). Yet, in practice, the earliest models used by the Imperial College of London team (and the bulk of models used in forecasting and assessing the value of interventions) combined variants of SEIR models with strictly exogenous changes in transmission as a result of policy changes. The policy changes varied, but included a suite of options including generalized lockdowns, restrictions on travel, school closures, and mask wearing. Yet, near the outset, the assumptions in such models were that the compliance with these policy changes and the extent to which individuals would comply with travel restrictions and/or mask wearing was set by the timing of the imposition and release of policies. The reality is more complex—and not only because populations involve spatially distributed networks of connections, but precisely because individuals do not have a static suite of reactions to an unfolding pandemic.

In practice, the global awareness of the unfolding SARS-CoV-2 pandemic has elevated the importance of assessing, integrating, and potentially leveraging the feedback between awareness, behavior, and disease spread. In the absence of such feedback, then, SARS-CoV-2 transmission should have led to a single peak corresponding to reaching herd immunity and then a rapid decline. Yet, with NPIs, the appearance of peaks in scenario models were connected to the timing of policies, such that a second wave would reflect the fact that early changes in policy led to decreases in transmission in the absence of susceptible depletion. In turn, when policies relaxed, then the disease would rebound. Figure 12.12 reveals that exogenous changes in transmission are not the only route to generate plateaus and oscillations. Instead, such patterns may be characteristic of feedback between disease and awareness.

In this SEIR model, the transmission rate decreases as a function of the awareness of the severity. Consider the extended SEIR model:

$$\dot{S} = -\frac{\beta S I}{\left[1 + (\delta/\delta_c)^k\right]} \tag{12.35}$$

$$\dot{E} = \frac{\beta S I}{\left[1 + (\delta/\delta_c)^k\right]} - \mu E \tag{12.36}$$

$$\dot{I} = \mu E - \gamma I \tag{12.37}$$

$$\dot{R} = (1 - f_D)\gamma I \tag{12.38}$$

$$\dot{H} = f_D \gamma I - \gamma_H H \tag{12.39}$$

$$\dot{D} = \gamma_H H \tag{12.40}$$

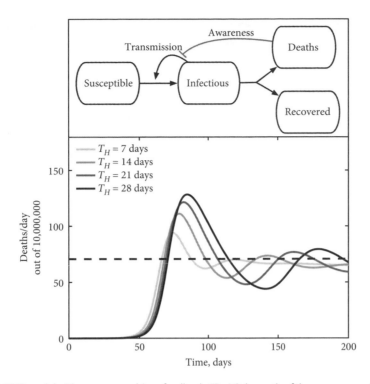

Figure 12.12: SEIR model with awareness-driven feedback. (Top) Schematic of the awareness-driven model. (Bottom) SEIR model with awareness can lead to emergent oscillations given parameters associated with the transmission of SARS-CoV-2. Parameters are $\beta = 0.5$ /days, $\mu = 0.5$ /days, $\gamma = 1/6$ /days, $f = 0.01$, $\mathcal{R}_0 = 3$, $\delta_c = 50$ /days, $k = 2$, and $N = 10^7$. Reproduced from Weitz et al. (2020).

where β is the transmission rate, f_D is the infection fatality rate, $T_H = 1/\gamma_H$ defines the average time in a hospital stay before a fatality, μ is the exposure to infectiousness rate, γ is the average infectious period, β is a transmission rate, and $\delta \equiv \dot{D}$ denotes the level of new fatalities, which scales behavior given a critical level δ_c and a shape exponent k. This model variant includes time spent in the hospital in the case of severe illness. As a result, the feedback between disease severity and awareness implies the possibility of oscillatory dynamics. When severity is high, individuals change behavior, but because of time lags between behavior change and the onset of cases that eventually become severe and lead to fatalities, there can be lags before behaviors change again. These cycles drive the system toward a persistent plateau, which represents a quasi-stationary state in the disease with $\mathcal{R}_{eff} \approx 1$ (see Figure 12.12). Although such models do exhibit qualitative (and even quantitative) features of joint disease spread and behavior change, it is evident there remain significant gaps to a full integration. For example, these models assume deterministic responses and do not include the impacts of stochastic dynamics that can lead to similar, emergent plateaus (Berestycki et al. 2021; Tkachenko et al. 2021). The absence of integrating behavior and epidemics comes with peril: neglecting endogenous behavioral changes may lead to missed opportunities to improve the efficacy of planning, control, and intervention efforts. Indeed, an ongoing challenge in public health modeling is to develop effective and realizable approaches to increase awareness of disease risk and modify behaviors to protect the health of individuals and communities.

12.7   TAKE-HOME MESSAGES

- Mathematical models of epidemic spread are increasingly utilized in responding to infectious disease outbreaks.
- Core epidemic models are driven by interactions that lead to changes in the disease status of individuals.
- The criterion for spread when the bulk of the population is susceptible is that a single infectious individual infects one (or more) new individuals on average.
- Transmission modulates both the strength and speed of disease.
- The final size of epidemics should be linked to the initial strength. However, caution is needed given the many assumptions used in size-strength formulations that are mathematically elegant but often incorrect in practice.
- Application of epidemic models to the case of EVD suggested the need for interventions; however, the absence of certain categories (like postdeath transmission) also limited the nature of appropriate recommendations.
- An ongoing challenge arising from SARS-CoV-2 is to integrate behavior change into improved scenario, forecasting, and intervention modeling.

12.8   HOMEWORK PROBLEMS

Problem 1. Nuanced Etiology

Extend the SIR model of emerging infectious disease to an SEIR model:

$$\dot{S} = -\frac{\beta SI}{N} \tag{12.41}$$

$$\dot{E} = \frac{\beta SI}{N} - \eta E \tag{12.42}$$

$$\dot{I} = \eta E - \gamma I \tag{12.43}$$

$$\dot{R} = \gamma I \tag{12.44}$$

Given the parameters $\beta = 1.25$/week, $\gamma = 1$/week calculate the basic reproduction number for the SIR version of this model (i.e., without an exposed state). Then systematically modulate the value of η from 4/week to 0.25/week. Using simulations and other methods, estimate the growth rate of the epidemic and compare it to that expected in the limit where individuals are immediately infectious. Next, assume a mismatch between the real value of η and the prior assumption of its value, e.g., $\eta_0 = 1$/week. Infer the estimate value of $\mathcal{R}_0$ from growth data for each pair of true versus prior values of η and discuss how mismatches between the period of infectiousness and actual periods of infectiousness can impact estimates of the strength of an outbreak in terms of $\mathcal{R}_0$.

Problem 2. Stochastic Liftoff

Extend the SEIR model presented in Problem 1 to a stochastic framework using the techniques developed in the computational lab guide associated with this chapter. Given these parameter sets, simulate 100 outbreaks in a population of size $N = 1000$ beginning with a single individual in the exposed state. What is the final size distribution of these outbreaks? Next, compare and contrast these final size distributions assuming that there are 10 initially exposed individuals. If you have time, use intermediate values of the number of initially exposed individuals to assess outcomes, including the potential for stochastic liftoff.

Problem 3. Cruise Ship Outbreaks

Consider a cruise ship with $N = 2200$ individuals in which there is an outbreak of Norwalk disease (equivalent to norovirus causing gastroenteritis), with the following features:

- Infections are highly contagious (approximately 5% chance of infection given an interaction with an infected person)
- Recovery time is typically five days. Recovered individuals become immune for six months.
- A typical cruise ship passenger interacts with a small percentage of other passengers in a day (usually 1%).

a. Propose a model for the dynamics, including estimates of all relevant parameters.
b. The trip is meant to last three weeks. One passenger is carrying Norwalk virus and becomes ill on the first day; show the expected dynamics of the disease. Do you think the disease dies out or does it spread? Why or why not?
c. Using the deterministic SIR model and assumptions, determine how many people got sick, how many people never got sick, and the maximum number of sick people at any given time.
d. Revisit parts (b) and (c) above in a stochastic framework: does the disease die out or does it spread? Why or why not? And, in the event that the disease does spread, characterize the features of the outbreak.
e. Your grandparents invite you to go on a three-week Alaskan cruise. Sounds fun. You have the option of choosing cruise ships with 1400 people, 800 people, or 200 people. You have heard about Norwalk and are concerned about being on a cruise ship with an epidemic. Assuming that one person is infected on any of these cruise ships, and using the parameter values in part (a), identify which of these ships is susceptible to an outbreak. Show graphs and provide explanations to support your results.

Problem 4. Outbreaks and Behavior

(Caveat: This problem was first utilized as a weekly assignment in November 2019—prior to the outbreak of SARS-CoV-2.) Thus far we have assumed that interactions remain constant throughout the course of the epidemic. This problem addresses the consequences of feedback between disease spread and behavior, given the following dynamics:

$$\dot{S} = -\frac{\beta(I, R)SI}{N} \tag{12.45}$$

$$\dot{E} = \frac{\beta(I, R)SI}{N} - \eta E \tag{12.46}$$

$$\dot{I} = \eta E - \gamma I \tag{12.47}$$

$$\dot{R} = \gamma I \tag{12.48}$$

where the infection rate β is equal to the product of the average number of contacts per unit time and the probability of infection spread per contact; these may depend on the state of the system. In this example, consider dynamics associated with an airborne-transmitted disease (like SARS) in a city of size $N = 5 \times 10^4$ in which individuals usually encounter ≈ 50 individuals per day (under baseline conditions), with an average incubation duration of five days and average infectious period of five days.

a. Simulate the baseline dynamics of the system with a 1% probability of infection. Given the introduction of a single infectious individual in an otherwise susceptible population, what is the peak time of the epidemic, how many total individuals are infected over the course of the epidemic, and what is the speed of the disease spread? Use visualizations of the epidemic to support your answer.

b. Given public health campaigns, individuals with symptoms (i.e., I indivdiuals) start to wear masks, which reduce the spread of disease per contact to virtually 0%. However, the compliance with mask wearing scales with disease incidence, such that the fraction of individuals wearing masks scales like $p = \epsilon \frac{I+R}{N}$ where $\epsilon = 0.4$, i.e., representing the maximum fraction of mask wearing. Simulate the expected spread of disease and estimate the strength, speed, and final size of the epidemic. Which of these differs significantly from the baseline case, and why?

c. Modify the previous model assuming that the campaign targets all individuals, including S, E, I, and R individuals. Is there any difference in outcome? Why or why not?

d. If you had to design a public health policy surrounding mask wearing, what level of preexisting compliance (i.e., what level of preexisting p) would be necessary to prevent an outbreak at the very outset?

Problem 5. Heterogeneity and Epidemics

Not only can individuals behave differently, they may have fundamentally different levels of susceptibility and infectiousness. As a first step toward examining the relationship between heterogeneity, spread, and outcome, consider a standard SIR model albeit with a number of susceptibility classes m (see Rose et al. 2021), such that the model becomes

$$\dot{S}_i = -\beta \epsilon_i S_i I$$

$$\dot{I} = \sum_{i=1}^{m} \beta \epsilon_i S_i I - \gamma I$$

$$\dot{R} = \gamma I$$

where ϵ_i can be drawn from a suitable distribution. Assume that ϵ_i is drawn from an exponential distribution with mean 1 and that $\beta = 2$ and $\gamma = 1$.

    a. Using $m = 2, 3, 5$, and 10 classes of susceptibility, simulate an epidemic outbreak beginning with a small fraction of infectious individuals. Show the dynamics of prevalence across each class (as well as the total prevalence).

    b. Compare and contrast the outcome to that of the SIR model prediction, in terms of initial speed, maximum prevalence, and final size.

    c. Find the final size of the epidemic using the strength-size relationship. How does the final size in heterogeneous models to expectations?

    d. Evaluate the change in average susceptibility among the remaining susceptibles over time in the population. How does $\bar{\epsilon}(t)$ change during the pandemic?

    e. Bonus for self-exploration: Based on your findings, provide a hypothesis as to how heterogeneity may impact the speed, strength, and final size of epidemics.

12.9   TECHNICAL APPENDIX

Derivation of the mean field SIR model   SIR model equations represent the dynamics of the expected numbers of susceptible, infectious, and recovered individuals. In essence, the SIR are mean field equations, neglecting fluctuations and correlated changes in population states. To derive these equations, consider the population state in terms of three integers, (S, I, R), such that $S + I + R = N$. A stochastic equation for the change in values of these three states at time step Δt in the future is

$$S(t + \Delta t) = S(t) - \chi(S, I) \tag{12.49}$$

$$I(t + \Delta t) = I(t) + \chi(S, I) - \psi(I) \tag{12.50}$$

$$R(t + \Delta t) = R(t) + \psi(I) \tag{12.51}$$

where χ and ψ are binomial distributed variables, such that

$$\chi \sim B(S, \beta \frac{I}{N} \Delta t) \tag{12.52}$$

and

$$\psi \sim B(I, \gamma \Delta t). \tag{12.53}$$

There can be at most S new infections and I recovery events in a time period Δt. However, when $\Delta t \to 0$, that becomes unlikely; indeed, there will be, at most, a single event of either kind per small unit of time. The equations above describe a stochastic process. We are interested in the expected value, i.e.,

$$\mathbb{E}\left(S(t+\Delta t)\right) = \mathbb{E}\left(S(t) - \chi(S, I)\right) \tag{12.54}$$

$$\mathbb{E}\left(I(t+\Delta t)\right) = \mathbb{E}\left(I(t) + \chi(S, I) - \psi(I)\right) \tag{12.55}$$

$$\mathbb{E}\left(R(t+\Delta t)\right) = \mathbb{E}\left(R(t) + \psi(I)\right) \tag{12.56}$$

We will overload notation and use the same variables S, I, and R for the expected values, such that we can rewrite these equations as

$$S(t + \Delta t) = S(t) - \beta S \frac{I}{N} \Delta t \tag{12.57}$$

$$I(t + \Delta t) = I(t) + \beta S \frac{I}{N} \Delta t - \gamma I \Delta t \tag{12.58}$$

$$R(t + \Delta t) = R(t) + \gamma I \Delta t \tag{12.59}$$

Finally, noting that $\frac{dx}{dt} = \lim_{\Delta t \to 0} \frac{x(t+\Delta) - x(t)}{\Delta t}$, we derive the SIR model:

$$\frac{dS}{dt} = -\beta S \frac{I}{N} \tag{12.60}$$

$$\frac{dI}{dt} = \beta S \frac{I}{N} - \gamma I \tag{12.61}$$

$$\frac{dR}{dt} = \gamma I \tag{12.62}$$

Local stability analysis of the SIR model Consider the normalized SIR model and the fixed point $(1, 0)$, i.e., $S^* = 1$ and $I^* = 0$ and, by definition, $R^* = 0$ given that $S + I + R = 1$. If $\dot{S} = f(S, I)$ and $\dot{I} = g(S, I)$, then the Jacobian is

$$\mathbf{J} = \begin{bmatrix} \frac{\partial f}{\partial x} & \frac{\partial f}{\partial y} \\ \frac{\partial g}{\partial x} & \frac{\partial g}{\partial y} \end{bmatrix} \tag{12.63}$$

where for this model

$$\mathbf{J} = \begin{bmatrix} -\beta I & -\beta S \\ \beta I & \beta S - \gamma \end{bmatrix}, \tag{12.64}$$

which evaluated at the fixed point $(1, 0)$ becomes

$$\mathbf{J} = \begin{bmatrix} 0 & -\beta \\ 0 & \beta - \gamma \end{bmatrix}. \tag{12.65}$$

The eigenvalues can be solved by finding the solution to $\mathrm{Det}\,(\mathbf{J} - \lambda \mathbb{I})$ where $\mathbb{I}$ is the identity matrix $\begin{bmatrix} 1 & 0 \\ 0 & 1 \end{bmatrix}$. For 2×2 matrices with a zero off-diagonal term, the eigenvalues can be read off the diagonal of the $\mathbf{J}$ matrix. To see this explicitly, note that $\mathrm{Det}\,(\mathbf{J} - \lambda \mathbb{I}) = -\lambda\,(\lambda - (\beta - \gamma))$. Hence, $\lambda_1 = 0$ and $\lambda_2 = \beta - \gamma$. The interpretation of these eigenvalues is discussed in the chapter.

Strength–final size relationship In the SIR model, the final size of the disease R_∞ is related to the strength of the outbreak, $\mathcal{R}_0$. The final size denotes the total number of individuals who were infected by the disease. Because all individuals eventually recover (or are removed), then $R_\infty = 1 - S_\infty$ represents the quantity of interest. The final size relationship in the chapter text can be derived as follows. Recall that in the SIR model the fraction of susceptible and infectious individuals changes according to

$$\dot{S} = -\beta S I \tag{12.66}$$

$$\dot{I} = \beta S I - \gamma I \tag{12.67}$$

such that the ratio of the derivatives, $\dot{I}/\dot{S}$, can be written as

$$\frac{dI}{dS} = \frac{-\beta S I + \gamma I}{\beta S I}$$

$$\frac{dI}{dS} = \frac{-\beta S + \gamma}{\beta S} = -1 + \gamma/(\beta S) \tag{12.68}$$

The term on the right-hand side is a function of S only, which means that this equation is separable and can be solved by multiplying both sides by dS and integrating, i.e.,

$$\int dI = \int dS \left(-1 + \frac{\gamma}{\beta S} \right), \tag{12.69}$$

yielding

$$I = -S + \frac{\gamma}{\beta} \log(S) + C \tag{12.70}$$

where C is an integration constant. This equation can be solved by recalling that at the start of the epidemic $S = 1$ and $I = 0$ such that $C = 1$. The critical insight to solving this final size problem is to recognize that the relationship between I and S also applies at the end of the epidemic where $S = S_\infty$ and $I_\infty = 0$. Hence, this leads to the following relationship:

$$S_\infty - 1 = \frac{\log(S_\infty)}{\mathcal{R}_0} \tag{12.71}$$

or

$$\mathcal{R}_0\,(S_\infty - 1) = \log(S_\infty). \tag{12.72}$$

Note that there is a slightly different size-strength relationship when the population is initially partially susceptible, i.e., $S(t=0) = S_0$. In that case, the relationship

$$I = -S + \frac{\gamma}{\beta} \log(S) + C \tag{12.73}$$

implies that

$$C = S_0 - \frac{\gamma}{\beta} \log S_0. \tag{12.74}$$

Recall that $\mathcal{R}_{eff} = \mathcal{R}_0 S_0$ where S_0 is the initial fraction susceptible. As such, we can rewrite the integration constant as

$$C = S_0 \left(1 - \frac{\log S_0}{\mathcal{R}_{eff}}\right). \tag{12.75}$$

Therefore, the final size must satisfy

$$C = S_\infty - \frac{\gamma}{\beta} \log S_\infty \tag{12.76}$$

or

$$\left(1 - \frac{\log S_0}{\mathcal{R}_{eff}}\right) = \frac{S_\infty}{S_0} - \frac{\log S_\infty}{\mathcal{R}_{eff}} \tag{12.77}$$

or finally

$$\mathcal{R}_{eff}\left(1 - \frac{S_\infty}{S_0}\right) = \log \frac{S_0}{S_\infty}. \tag{12.78}$$

Local stability analysis of the SI model Consider the normalized SI model and the fixed point $(1, 0)$, i.e., $S^* = 1$ and $I^* = 0$. If $\dot{S} = f(S, I)$ and $\dot{I} = g(S, I)$, then the Jacobian is

$$\mathbf{J} = \begin{bmatrix} \frac{\partial f}{\partial x} & \frac{\partial f}{\partial y} \\ \frac{\partial g}{\partial x} & \frac{\partial g}{\partial y} \end{bmatrix} \tag{12.79}$$

where for this model

$$\mathbf{J} = \begin{bmatrix} -\beta I & -\beta S + \gamma \\ \beta I & \beta S - \gamma \end{bmatrix}, \tag{12.80}$$

which evaluated at the fixed point $(1, 0)$ becomes

$$\mathbf{J} = \begin{bmatrix} 0 & 0 \\ 0 & \beta - \gamma \end{bmatrix}. \tag{12.81}$$

Using the methods described above, $\lambda_1 = 0$ and $\lambda_2 = \beta - \gamma$. The interpretation of these eigenvalues is discussed in the chapter.

Renewal equation and inferring strength from speed Consider a SIR model in which the infectious dynamics follow

$$\dot{I} = i(t) - \gamma I \tag{12.82}$$

where $i(t)$ is the number of new cases per unit time, i.e., the incidence βSI. Initially, both prevalence $I(t)$ and incidence $i(t)$ should grow exponentially $\sim e^{rt}$ where r is the speed. Incidence depends on prior infections:

$$i(t) = \int_0^\infty \mathrm{d}a\, i(t-a)n(a) \qquad (12.83)$$

where $n(a)$ is the number of new infections per unit time caused by an individual infected a units of time ago. This new infection rate can be written as $n(a) = l(a)m(a)$ where $l(a)$ is the probability of being infected at time a after infection and $m(a)$ is the number of new infections per unit time for infections of time a. Precisely because of the exponential feature, then $i(t) = i(t-a)e^{ra}$ such that

$$i(t) = \int_0^\infty \mathrm{d}a\, i(t)e^{-ra}n(a) \qquad (12.84)$$

$$1 = \int_0^\infty \mathrm{d}a\, e^{-ra}n(a). \qquad (12.85)$$

Recall that the basic reproduction number is defined as the average number of new infections caused by a single infectious individual in an otherwise susceptible population over a typical infectious period life span, or

$$\mathcal{R}_0 = \int_0^\infty \mathrm{d}a\, n(a). \qquad (12.86)$$

Using this definition, the distribution of new infections with the age of infection can be normalized, i.e., $g(a) = n(a)/\mathcal{R}_0$ such that $\int \mathrm{d}a\, g(a) = 1$. The term $g(a)$ is the generation interval distribution; it denotes the distribution of times between the infector and the infectee. We can now replace $n(a)$ with $\mathcal{R}_0 g(a)$, yielding

$$\frac{1}{\mathcal{R}_0} = \int_0^\infty \mathrm{d}a\, e^{-ra}g(a) \equiv M(-r) \qquad (12.87)$$

where

$$M(z) = \int_0^\infty \mathrm{d}a\, e^{za}g(a) \qquad (12.88)$$

is the moment-generating function.

This renewal equation formalism can be applied to infer strength $\mathcal{R}_0$ from the measured speed r. In the case of EVD, we consider a gamma-distributed exposed period, with $T_E = 11$ days and shape parameters $n_E = 6$ and $b_E = n_E/T_E$, whose generating function is

$$M_E(-\lambda) = \left(\frac{b_E}{b_E + \lambda}\right)^{n_E}. \qquad (12.89)$$

We also consider two extremal conditions for the infectious periods of the I and D classes: exponentially distributed and delta-distributed periods. The former is conventionally used while the latter provides an upper bound on the possible inferred values of $\mathcal{R}_0$. The generating functions are, in the first scenario,

$$M_I(-\lambda) = \frac{\gamma}{\gamma + \lambda} \tag{12.90}$$

$$M_D(-\lambda) = \frac{\rho}{\rho + \lambda} \tag{12.91}$$

where $\gamma = 1/T_I$ and $\rho = 1/T_D$, and in the second scenario,

$$M_I(-\lambda) = e^{-\lambda T_I} \tag{12.92}$$

$$M_D(-\lambda) = e^{-\lambda T_D} \tag{12.93}$$

Therefore, for the SEIRD model, it is possible to estimate $\mathcal{R}_0$ using the generating function method given observations of an epidemic growth rate and suitable information-on epidemiological models and parameters. Full details on the use of moment-generating functions can be found in Wallinga and Lipsitch (2007), with application to EVD in Weitz and Dushoff (2015), and using the gamma function approximation in Park et al. (2019).

Part IV

The Future of Ecosystems

Chapter Thirteen

Ecosystems: Chaos, Tipping Points, and Catastrophes

13.1 ECOSYSTEMS—THE INTEGRATED FRONTIER

Saharan grasslands, Arctic tundra, and deep sea hydrothermal vents along with a multitude of ecosystems across our planet harbor a diversity of species. The interactions among diverse species impact population dynamics, the organization of complex communities, and the flow of nutrients and matter. The variation among ecosystems may seem to defy efforts to identify governing principles. Yet identifying robust principles could support the very efforts needed to protect and sustain ecosystems vital to human and environmental health. Like other complex systems, ecosystem dynamics arise from the interactions of heterogeneous agents across scales from molecules to organisms, a combination of biotic and abiotic factors, and often act under the influence of endogenous and exogenous stochasticity. Such a combination of factors would tend to suggest the need to build larger and larger models with ever more interactions, feedback, and parameters. But features of ecosystem dynamics can often transcend the particular details.

One clarion example comes from decades of monitoring and experimental studies of the relationship between nutrient input and ecosystem status in shallow lakes (Scheffer et al. 1993; Carpenter et al. 2011). Lake systems are ideal models to evaluate the link between abiotic factors and ecosystem structure. Lakes receive variable levels of nutrient input, both over seasonal cycles and on longer time scales. Variation in nutrient input can trigger changes in the relative balance of algal growth and consumption by zooplankton in the water column. Disruption of this balance can have dramatic consequences. In a healthy lake, nutrient input from sediment influx drives the growth of cyanobacteria that become food for microscopic grazers, which themselves are prey for a spectrum of larger copepods that eventually provide food for top predators (in the Midwest, these would include common sport fish like smallmouth bass). Due to the relative efficiency of algal consumption by grazers, a lake can remain in an oligotrophic regime—where *oligotrophic* denotes a system with relatively low standing availability of essential nutrients, including nitrogen and phosphorus, and often clear blue water. Oligotrophic (and mesotrophic) lake ecosystems are increasingly threatened by the accumulation of nitrogen and phosphorus input from human activities—especially agricultural runoff. The increase in nutrient input can lead to a boon for cyanobacteria and other algae. These algae proliferate in numbers that can outstrip the ability of grazers to control their population. Increasing in abundance, the denser populations of algae shade the lower levels of a lake such that photosynthesis-driven CO_2 consumption and oxygen production is reduced. This then compounds the problem, leading to oxygen depletion and, eventually, fish kills—the absence of fish can then cascade across the food web, leading to a new stable state involving a turbid lake with few large fish and an overabundance of (potentially harmful) algae (Scheffer et al. 2001).

Subtle changes in nutrient sinks and sources can lead to long-term changes in ecosystem state, durable well after elevated levels of nutrient input are removed (Scheffer et al. 1993). For example, an oligotrophic lake could remain in its state even given transient addition of nutrient inputs. This is an example of one type of resilience—the ability of a system to maintain a macroscopic state feature despite perturbations (Holling 1973). However, resilience is not without limits. Beyond a critical point, additional nutrient input can lead to a switch from an oligotrophic to a eutrophic (i.e., high-nutrient and turbid) condition. Even worse, because the new state is also resilient, it can often take more aggregate removal of nutrients than were added to shift the system back to the healthy state. This asymmetry between the input required to drive a system from one state to another and back again is termed *hysteresis*. Figure 13.1 (Plate 15) provides a visual demonstration of the phenomenon of alternative stable states in lake systems at two scales. The top panel shows satellite images from lakes in Iron County, Wisconsin. The variation in lake conditions is readily apparent from the variation in water clarity as measured using satellite imaging. The variation in relative opacity spans relatively clear and relatively turbid lake states. Lake turbidity can vary even within local regions. In fact, stochasticity and feedback within the lake ecosystem can lead to durable switches in water clarity, fish levels, and ecosystem productivity.

The bottom panel of Figure 13.1 provides a zoomed-in view of such alternative stable states. The two lakes are located in Michigan's Upper Peninsula. Peter (the larger lake) and Paul (the smaller lake) have been the subject of a long-term ecological monitoring effort to understand the relationship between nutrient input, ecological interactions, and trophic cascades (i.e., the ways in which changes in one level of a food web change outcomes at other levels) (Elser et al. 1986). At various points in the study, these two lakes diverged significantly in state, often with relatively little indication that a change was imminent. In light of the apparent unpredictability of state changes, a group of ecologists led by Stephen Carpenter initiated an ecosystem-wide manipulation of the lake's food web. They did so by adding larger fish predators—largemouth bass—to Peter Lake, which had a viable population of smallmouth bass and other fishes whose diet included water fleas (Carpenter et al. 2011). The addition of a larger predator drove the smaller fish into refuges as a means to avoid predation. Given the reduced activity of small fish, water fleas proliferated. In turn, the higher abundance of water fleas triggered larger fluctuations in their resource—algae— that could sometimes lead to rapid state transitions in the ecosystem as a whole. Notably, by using a continuous monitoring framework, the researchers identified early indicators of critical state transitions—including increasing fluctuations in both biotic and abiotic components. The existence of early indicators of critical transitions has a practical interpretation for management: continuous monitoring may be able to trigger interventions to redirect an ecosystem away from a shift from a healthy to an unhealthy or even collapsed state (Wilkinson et al. 2018).

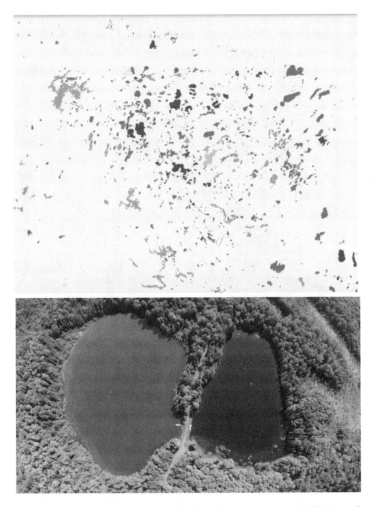

Figure 13.1: Alternative stable states in shallow lakes in Wisconsin (top) and Michigan (bottom). (Top) Lake water clarity derived from Landsat images of Iron County, Wisconsin, from July 15, 2004. Source: https://lakesat.org/LakesTSI.php at the University of Wisconsin-Madison (credit: Sam Batzli). Darker blue corresponds to more transparent conditions and lighter green corresponds to more turbid conditions (see Plate 15). (Bottom) Photograph of Peter (left) and Paul (right) Lakes. Peter Lake was subject to an experimental manipulation resulting in a eutrophic, turbid state, while Paul Lake was left undisturbed and served as the control of an oligotrophic, clear water state. Photo credit: Steve Carpenter/University of Wisconsin–Madison Center for Limnology.

Studies of Midwest lakes have shed light on the principles governing alternative stable states, resilience, rapid transitions, and even the possibility of tragedy and renewal. These ecosystem features can transcend the particular species and interactions in a focal site, but precisely because of system complexity it is usually challenging to identify salient drivers. What then is the right balance to strike in an effort to identify a salient series of problems (and models) to form the basis for a principled study of ecosystems science? If made too simplistic, models become no more than metaphors or, worse, objects meant to be admired, not falsified. If made too complex, it becomes harder to identify the key mechanisms that could be tested and used to both understand and predict how ecosystems will behave as they are increasingly pushed outside of their "natural" limits. This book has attempted to strike a balance between simplicity and complexity by contextualizing and motivating the biology (and physics, when necessary) of model choices so as to connect dynamical models with mechanisms. In doing so, each chapter has been organized as a case study, focused on a specific problem, whether the Luria-Delbrück experiments, robustness in *E. coli* chemotaxis, the excitation of cardiac and neuronal cells, flocking, or even recent Ebola and COVID-19 outbreaks. These case studies reveal how dynamic feedback mechanisms give rise to systemwide behavior that can serve as a bridge between models and data. Nonetheless, this kind of work, when adopted at ecosystem scales, poses certain challenges.

As already noted, ecosystems have many components, including diverse populations, intraspecies variants, nutrient inputs, patches/habitats for colonization, and other physical drivers. Yet what has given life to ecosystem models are a few simple ideas (Levin 1992; Beisner et al. 2003; Scheffer et al. 2012; Allesina and Tang 2012): , for example, (i) small changes in input can lead to large changes in outcomes; (ii) states—when entered—can persist for far longer and under more extreme conditions than it took to enter them; (iii) certain types of feedback can lead to runaway, positive feedback events and even to catastrophe; (iv) scale matters when trying to understand the link between pattern and mechanism.

Hence, rather than utilizing a single case study, this chapter uses a shorter series of examples to help identify the dynamical ingredients that lead to the emergence of chaos, catastrophes, alternative stable states, and even the excitability of diverse communities and complex ecosystems. Because this chapter focuses on ingredients at the end of a book (and for some of you, a long semester), each of the model sections is "small." The state-of-the-art approaches to each of these systems often includes many more pieces, particularly true of global climate models (whether developed by academic research groups or the Intergovernmental Panel on Climate Change). Nonetheless, this chapter tries to generate some needed interest in looking for minimally reducible models that help engender a quantitative mindset to guide real-world interventions.

13.2   CHAOS IN COMMUNITIES

Populations can change across growing seasons or discrete periods such that, in practice, it is both practical and reflective of the underlying biology to describe population dynamics in terms of discrete dynamical systems. Indeed, the Wright-Fisher model is one such example where dynamics are assumed to occur across discrete generations. The WF model focuses on changes in the relative frequencies of genotypes in a population. In contrast, in the mid-1970s, Robert May (1979) reviewed the population dynamics associated with the discrete logistic equation. In the discrete logistic model, the population x_t changes according to the following rule:

$$x_{t+1} = r x_t \left(1 - x_t\right). \tag{13.1}$$

This model acts as a "map" over time—where time may represent a generation, season, or epoch. For low values of x_t, the population at the next time, x_{t+1}, will also be low due to the limited reproductive ability of individuals. However, for high values of x_t, the population at the next time, x_{t+1}, will also be low, albeit for a different reason: competition, crowding, and limitation that leads to a relative collapse. This map is shown in Figure 13.2. Given this model and prior experience with the dynamics of the *continuous* logistic model, it would be

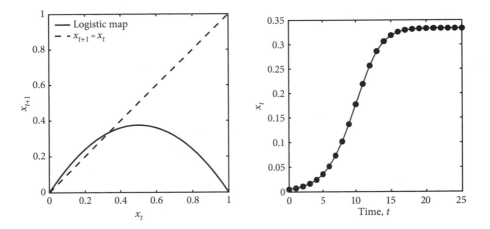

Figure 13.2: Discrete logistic model and dynamics. (Left) The map, $x_{t+1} = rx_t(1-x_t)$ (solid), with the line $x_{t+1} = x_t$ as a reference (dashed). (Right) Discrete dynamics given $r = 1.5$.

natural to presume that the dynamics of the *discrete* logistic model should follow a similar course; that is, increasing toward an equilibrium and remaining there for all time. Indeed, for certain values of r, that is precisely what happens (right panel of Figure 13.2). This kind of exponential increase followed by saturation is the only class of nontrivial dynamics for the continuous logistic model. However, the discrete logistic model has far more to teach us about how short-term predictions can fail over the long term and how small models can lead to big, complex results.

To begin, it is evident that the discrete model does not converge to 1. Indeed, if the system were to be initialized at its maximum capacity, $x_t = 1$, then the next time the system would collapse to $x^* = 0$. However, there is another fixed point in the system: the intersection shown in the left panel of Figure 13.2. This intersection point corresponds to the value x^* such that

$$x^* = f(x^*) = rx^*(1-x^*),\tag{13.2}$$

which has a solution of $x^* = 1 - 1/r = \frac{r-1}{r}$. The other solution is $x^* = 0$. Hence, the discrete logistic model has two fixed points: either the extinction case of $x^* = 0$ or the persistence case of $x^* = \frac{r-1}{r}$. Do these fixed points also represent the steady state of the system? Answering this question requires characterizing the (local) stability of fixed points as a function of r.

The local stability of a discrete dynamical system can be assessed by evaluating how small perturbations near a fixed point grow or decay. Assume that $x_t = x^* + u_t$ where u_t denotes the size of the perturbation. In that case, the discrete logistic model can be rewritten as

$$x_{t+1} = f(x_t)\tag{13.3}$$

$$= f(x^* + u_t)\tag{13.4}$$

$$= f(x^*) + u_t \left.\frac{\partial f}{\partial x_t}\right|_{x_t = x^*} + \mathcal{O}\left(u_t^2\right)\tag{13.5}$$

by writing out terms in a Taylor expansion. This map can be simplified by recalling that $f(x^*) = x^*$, which is the condition of a fixed point that maps back onto itself. This means the dynamics can be written to first order in u_t as

$$x^* + u_{t+1} = x^* + u_t \left. \frac{\partial f}{\partial x_t} \right|_{x_t = x^*} \tag{13.6}$$

$$u_{t+1} = u_t \left. \frac{\partial f}{\partial x_t} \right|_{x_t = x^*}. \tag{13.7}$$

Hence, perturbations follow discrete exponential growth dynamics characterized by the discrete growth number $R \equiv \left. \frac{\partial f}{\partial x_t} \right|_{x_t = x^*}$. When $|R| > 1$, then the deviation grows and the fixed point is locally unstable. In contrast, when $|R| < 1$, then the deviation shrinks and the fixed point is locally stable. The use of absolute magnitudes is critical, as if $R < -1$, then the magnitude of the deviation will grow, in absolute terms, even as the relative value will oscillate back and forth from less than to greater than zero. This general form can then be applied to the specific case of $f(x_t) = rx_t(1 - x_t)$.

In the case of the logistic model, the discrete growth number is

$$R = r - 2rx_t|_{x^*} \tag{13.8}$$

$$= r - 2r\frac{r-1}{r} \tag{13.9}$$

$$= 2 - r. \tag{13.10}$$

Hence, when $|2 - r| < 1$, the system is stable. This implies that for $1 \leq r \leq 3$ the system is stable and converges to $x^* = \frac{r-1}{r}$. However, for $r > 3$, small perturbations increase despite the fact that there are no other fixed points (besides $x^* = 0$, which is also unstable). Instead, something remarkable happens. The system initiates a series of period-doubling bifurcations, such that for a suitable choice of $r = 3 + \epsilon$, the dynamics oscillate such that $x_{t+2} = x_t$. This finding implies that

$$x_t = f(f(x_t)) \tag{13.11}$$

or

$$x_t = r^2 x_t (1 - x_t) \left(1 - rx_t(1 - x_t)\right). \tag{13.12}$$

In practice, this new oscillatory state is stable, at least for a range of r. But this is far from the only steady state.

Figure 13.3 shows how trajectories change as a function of r below and above $r = 3$. For $1 < r < 3$, initially divergent trajectories will converge to the globally stable fixed point; whereas, for $r > 3 + \epsilon$, the trajectories will also converge, albeit to an oscillatory state. Increasing the value of r leads to another period doubling, and then another, and then another, all in a finite region of r. In seminal work that has shaped the development of chaos theory, Mitchell Feigenbaum (1978, 1984) showed that there are an infinite number of such period doublings, each occurring ever more closely to the previous doubling. Remarkably, the gap between period doublings follows a universal curve, depending on the nonlinearity of the original map, such that there exists a critical point beyond which the system no longer oscillates in a period-predictable fashion. Instead, there is aperiodic behavior that exhibits extreme sensitivity to small differences in initial conditions. Notably, a hallmark of chaos can be seen in the bottom row of Figure 13.3. This row shows what happens in two

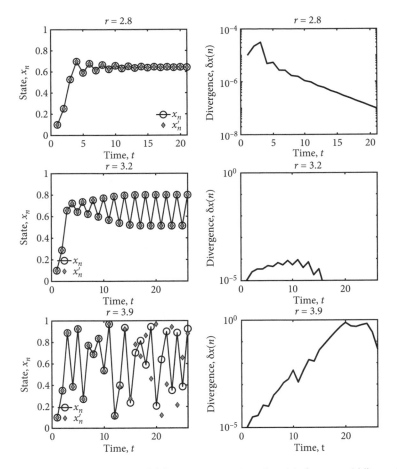

Figure 13.3: Dynamics of the logistic model for $r = 2.8$, $r = 3.2$, and $r = 3.9$, for top, middle, and bottom, respectively. The left panels denote two trajectories with an initial deviation of 10^{-5}. The right panels denote the divergence between these trajectories. As is evident, the divergence decreases in the top scenarios and increases in the bottom one.

instances of the discrete logistic model that share the same value of $r = 3.9$, albeit differing only by an infinitesimal value in their initial conditions. One might expect that, since the model is deterministic, the trajectories should remain close to one another. To the contrary, trajectories do not remain close to one another in chaotic regimes. In practice, it is possible to quantify the divergence of two trajectories that started from nearby initial conditions. Consider two trajectories, x_n and x_n', which are initialized at the initial conditions x_0 and $x_0 \pm \delta_0$. Formally, the divergence after n steps can be measured as the absolute value of the difference between states:

$$\delta x(n) \equiv |x_n - x_n'|. \tag{13.13}$$

In the case of a chaotic system, these divergences will grow exponentially, such that even nearly identically prepared systems will soon become unpredictably different. This is precisely what is seen for sufficiently large values of r, where $\delta x(n) \sim e^{an}$, i.e., growing exponentially large with the number of iterations.

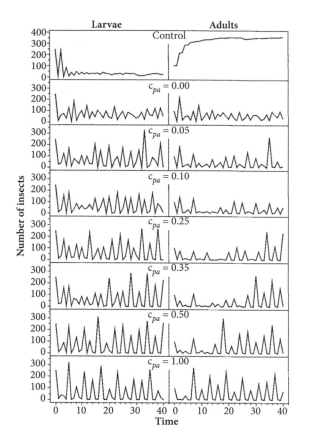

Figure 13.4: Chaos in a flour beetle population, reprinted from Constantino et al. (1997). The panels denote a manipulation of the pupa survival probability in the presence of A adults, i.e., $e^{-c_{pa}A}$. The top panel is the control, with increasing adult-induced mortality of pupae shown in the subsequent panels. The resulting dynamics are aperiodic, with time units measured in two-week intervals. The Lyapunov exponents for the underlying deterministic model of the state-structured population are predicted to be positive for the values $c_{pa} = 0.25$, 0.35, and 0.5. Full details in Constantino et al. (1997).

The discrete logistic model may seem like a small model far removed from practical relevance. Yet the same notion of deterministic chaos that is part of the discrete logistic model is also the basis for chaos in atmospheric systems—what is commonly known as the *butterfly effect* (Lorenz 1963). The butterfly effect suggests that the flapping of a butterfly's wings in one of two otherwise identical systems could eventually lead to divergences in atmospheric flow, limiting the ability to make reliable long-term predictions even in a fully deterministic system. And observations of chaos are also features of real-world population dynamics—whether for plankton (Huisman and Weissing 1999) or flour beetles (Constantino et al. 1997). Figure 13.4 provides one example of a transition to chaos in a flour beetle population—including examples in which there are oscillatory population dynamics and dynamics with the hallmarks of chaos, i.e., a positive Lyapunov exponent. The meaning, and caution, is that nonlinear feedback in ecological systems can sometimes put fundamental limits on our ability to predict systems even when we know the rules. Indeed, it may turn out to be that chaos is far more prevalent in ecological systems than had previously been recognized (Munch et al. 2022).

13.3 CONDORCET AND CATASTROPHES

Populations have a limited capacity to grow. The limitations are myriad: space, nutrients, stress, and other populations. Yet populations also have the ability to reshape their environment. In certain realms of ecology, such ideas are often framed as niche construction. The term *niche* refers to the idea, popularized by G. Evelyn Hutchinson (1957), that there are certain combinations of biotic and abiotic factors that enable an organism to persist. These fundamental constraints may be greater than the realized constraints due to ecological and evolutionary feedbacks (Holt 2009). Niche construction suggests that organismal behavior, when compounded at the population scale, can lead to feedback with the environment, reshaping it so that the environment is in fact more conducive to population growth (Odling-Smee et al. 1996). In layperson's terms, one can think of a beaver dam—in which the beaver restructures the flow of water and local habitat to make it more conducive to its survival. In a similar sense, agriculture is also a form of transforming an environment to expand the viability of a particular environment that has another consequence: increasing the potential for increases in population density. But for many microbes, their interactions and transformation of carbon and nutrients can transform the Earth system. One could argue that the oxygenation of Earth approximately 2.3 billion years ago by cyanobacteria was the greatest niche construction event of all time. Photosynthetic cyanobacteria take in carbon dioxide and release oxygen. Over a few hundred million years, the growth and replication of cyanobacteria in the global oceans led to the massive release of oxygen that displaced methane and transformed the atmosphere and oceans into an environment in which organisms that used aerobic metabolism could thrive.

Building a full model of the great oxidation event would take more than a small model. Hence, this section takes a simplified view of niche construction by exploring what happens when a population is limited by its environment (as in the logistic growth model), but with a twist. In the model explored next, population growth can reshape the environment, potentially increasing its own capacity for growth. As the model shows, there can be too much of a good thing, and such feedback can have unexpected and even catastrophic outcomes. To begin, recall that the continuous logistic model represents the impact of *negative* density dependence on the growth of a particular population. But what if the density were to have a stimulating effect on growth, rather than an inhibitory effect? Such a model can be considered by generalizing the logistic model as follows:

$$\frac{dN(t)}{dt} = \rho N(t) \left(K(t) - N(t) \right) \tag{13.14}$$

$$\frac{dK(t)}{dt} = c \frac{dN(t)}{dt} \tag{13.15}$$

where $N(t)$ is the population, $K(t)$ is the carrying capacity, and ρ is a scaled growth rate. In the spirit of Joel Cohen's (1995) proposal, we term this the Malthus-Condorcet model. The key concept in the Malthus-Condorcet model is that the carrying capacity increases at a rate equal to that of the rate of change of the population modified by the Condorcet parameter, c. It is also important to note that in this formulation the Malthus parameter ρ has a slightly different interpretation. For example, when $N \ll K$, then the growth rate is exponential with a rate of ρK. This model can be mapped to the logistic model when $c = 0$. Three regimes are relevant for some form of positive feedback $c > 0$: (i) $c < 1$; (ii) $c = 1$; (iii) $c > 1$. The value of c indicates whether or not the rate of expanding resources/capacity of a system scales more slowly than, equal to, or faster than the population.

Figure 13.5 illustrates the potential of this model to do unusual things. These three models all start with the same initial conditions. If K is constant, then the maximum growth rate would be $\rho K_0 = 0.05$, setting a time scale of approximately 20. But the carrying capacity is not constant. Instead, in each case the carrying capacity increases, leading to a qualitative change in dynamics. One can think of this increase in carrying capacity as a result of joint changes in technology and education (in the case of human populations) that make it possible for the carrying capacity to increase as the population increases. The units are arbitrary and are simply meant to illustrate the dynamics. For the case $c = 0.5$ (top), the resulting

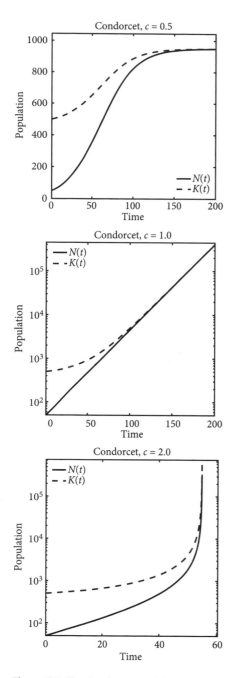

Figure 13.5: The Condorcet model with sublinear feedback, threshold feedback, and superlinear feedback. Each model uses the parameters and initial conditions $\rho = 10^{-4}$, $K_0 = 500$, and $N_0 = 50$, albeit given $c = 0.5$, $c = 1$, and $c = 2$, from top to bottom, respectively.

population dynamics appear to be logistic, albeit with deviations from standard logistic growth. The carrying capacity nearly doubles from 500 to ≈ 950. As the population approaches this saturating level, the difference between K and N decreases, and because $c < 1$, the carrying capacity increases less than does the population. Over time, the population equals that of the carrying capacity, and both $\dot{N}$ and $\dot{K}$ equal 0.

Next, for the case $c = 1$ (middle), the system grows exponentially after a transient period—note the use of log-scaled population axes in this plot. The key idea is that every bit of increase in population leads to a concomitant increase in the carrying capacity. Hence, the population can never reach it, and in fact, the population remains as far from the carrying capacity as it was at the very start. Note that for the case at long times the dynamics converge to $N(t) \sim Ae^{\bar{r}t}$, where A is a constant. Because $c = 1$, then N and K must have the same exponential growth rates, $\bar{r}$ but could differ by a factor C, such that $K(t) = Ae^{\bar{r}t} + C$. In this case, the factor is $C \equiv (K(0) - N(0))$ and the effective exponential growth rate is $\rho(K(0) - N(0))$.

Finally, for the case of $c = 2$ (bottom), the system does something seemingly unexpected. It grows faster than exponentially! The faster-than-exponential growth can be identified by the positive curvature of $N(t)$ on log-scaled population axes. What is striking is that, in contrast to the exponential growth model in which populations go to infinity—rapidly to be sure, but in an infinite amount of time—the Malthus-Condorcet model reaches infinity in a finite time when $c > 1$! This seems impossible, and of course in the real world the system would not reach an infinite population. But it would spike up in a seeming singularity, increasing superexponentially, to a time in the not-so-distant future—a "doomsday" scenario caused by the ability of the system to continually expand its capacity for growth (of note, related ideas were proposed in a model to explain rises in methane at the end of the Permian era (Rothman et al. 2014)). According to von Foerster and colleagues, the doomsday when fitting to human population growth data in the late 1950s corresponds to Friday, November 13, 2026 (Von Foerster et al. 1960). The early fits of population growth were in fact prescient, until the population growth rate began to slow down in the 1970s (Ehrlich 1968). In reality, the date is unlikely to be "doomsday." Yet, in many ways, governments and populations continue to ignore the caution that ever-accelerating growth is not in our best interests—perhaps at our peril.

13.4   THRESHOLDS IN ECOSYSTEMS AND THE EARTH SYSTEM

This chapter began with a synopsis of factors that can give rise to alternative stable states in ecosystems, with a focus on shallow lakes and their switch between eutrophic and oligotrophic conditions. A eutrophic lake is one characterized by high nutrient levels (typically measured in terms of phosphorus or nitrogen) and turbid conditions, with few if any fish, and algal blooms. In contrast, an oligotrophic lake is one characterized by lower nutrient levels and clear conditions in the surface waters, with fish comprising an essential part of the food web. This phenomenon of alternative stable states resembles that of the mechanisms underlying bistability in gene regulation introduced in Chapter 3. The premise is that changes in phosphorus input can drive critical changes in the steady state concentration of phosphorus. A full ecosystem model of lake processes would include nutrient dynamics and biotic interactions, as well as hydrodynamic models that span the water column and the sediment. Yet the essential nature of the switch between states can be captured in simplified models.

Stephen Carpenter and colleagues proposed the following simplified model of the impacts of nutrient input on lakes:

$$\frac{\mathrm{d}P}{\mathrm{d}t} = I - sP + \frac{rP^q}{m^q + P^q} \tag{13.16}$$

where P is the phosphorus concentration in algae and I is the phosphorus input per unit time (Carpenter et al. 1999). In this model, s is the loss rate (e.g., due to sedimentation, outflow, or uptake by consumers), r is the maximum recycling rate (typically from sediments), m is the half-saturation constant, and q is a nonlinearity exponent that denotes how the presence of P stimulates recycling. This model does not represent the ecosystem in terms of its explicit biological components. Instead, it presumes that the increase in algae leads to greater recycling of phosphorus due to the anoxic conditions—a potentially self-reinforcing effect.

Figure 13.6 provides a summary of simulation results that illustrate the potential for the phosphorus model in Eq. (13.16) to generate alternative stable states, resilience, rapid transitions, and hysteresis. The top panel shows the steady states, P^*, as a function of the phosphorus input level—the units are arbitrary and are meant to illustrate the qualitative features of the dynamics. For low levels of phosphorus input, the sources and sinks only equal each other at a single value (Figure 13.6, bottom left). This low steady state concentration of phosphorus corresponds to an oligotrophic regime. Yet, as the phosphorus input increases, the system undergoes a bifurcation. Instead of a single steady state, there are now three, two of which are stable and one of which is unstable (Figure 13.6, bottom middle). The two stable states have large differences in the value of P^*, i.e., representing an oligotrophic state with low P^* and a eutrophic state with high P^*. In between these two states is an unstable fixed point. Small perturbations below this unstable fixed point would be driven toward the oligotrophic regime, while small perturbations above this unstable fixed point would be driven toward the eutrophic regime. In this sense, each of the alternative stable states is resilient to perturbations—within limits. But when a sufficiently large perturbation is made to the system, then it is possible for a rapid transition to take place toward the

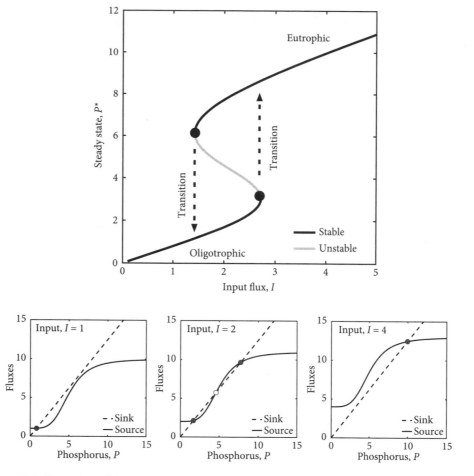

Figure 13.6: Alternative stable states and hysteresis in a simplified model of oligotrophic to eutrophic conditions in shallow lakes. The panels utilize Eq. (13.16) with the parameters $s = 1.25$, $r = 9$, $m = 5$, and $q = 4$, with I varying. (Top) Steady state outcomes as a function of input flux, I, reveals a transition from a single, low-phosphorus state to alternative stable states and then a single, high-phosphorus state. (Bottom) Comparison of sources and sinks as a function of the phosphorus concentration in lake algae. The sources include input and recycling terms. The sink is the per capita loss of algae. The intersection of sources and sinks denotes a fixed point, with the stability marked as stable (closed circle) and unstable (open circle). For $I = 1$ (left panel), there is only a single, stable fixed points, corresponding to an oligotrophic, low-phosphorus state. For $I = 2$ (middle panel), there are three fixed points, including two stable states—an oligotrophic and a eutrophic state. For $I = 4$ (right panel), there is only a single, stable fixed point, corresponding to a eutrophic, high-phosphorus state.

alternative steady state. Finally, when phosphorus input is sufficiently high, then there is yet again only a single stable state, corresponding to a eutrophic system (Figure 13.6, bottom right).

The bifurcation diagram in Figure 13.6 (top) summarizes the potential steady states as a function of phosphorus input. But rather than think about dynamics strictly in terms of a fixed input level, consider what would happen if I were to slowly increase—such that the system effectively reached its equilibrium before the next incremental change in I. Beginning with low levels of nutrient input, the system would remain in the oligotrophic state, denoted by the lower part of the curve in Figure 13.6 (top). The dynamics change when $I \approx 2.71$. At this point, further increases in I would lead to a rapid transition of the system to the upper curve and a eutrophic system. Hence, a very small change in input would lead to a very large change in output. Increases in I beyond that would lead to a nearly linear increase in P^*. Now consider that the system in a eutrophic state starts to experience less phosphorus input. Hence, rather than continuing to increase I, envision what would happen to the system when decreasing I. In this case, the system would follow the upper curve in Figure 13.6 (top) until $I \approx 1.43$. At this point, further decreases in I would lead to a rapid transition of the system to the lower curve and an oligotrophic system. This asymmetry is an example of a hysteresis—in this case, the phosphorus input must be lowered to nearly half that of the point at which the oligotrophic to eutrophic transition was made in order to restore an oligotrophic environment. Hysteresis also provides a caution: restoring a healthy state can take more work than it took to disrupt it in the first place. This message may also have relevance for the Earth system as a whole. Full-scale models of the Earth system are far beyond what can be treated with small-scale dynamical systems models. Earth system models include three-dimensional coupling of hydrodynamics, chemistry, biology, and the integration of processes across massive spatiotemporal scales (e.g., Follows et al. 2007). There are good reasons to develop and contribute to such collaborative efforts. Indeed, these models can make a scope of predictions that a simple model cannot. Yet a simple model has another advantage: it has the potential to provide insight and perhaps even a narrative, to understand how perturbations in a nonlinear system can, given certain feedback, drive a system from a seemingly long-term resilient state to somewhere very different.

There is precedence for such departures. The Earth system has experienced mass extinction events in which life was fundamentally altered. With time, other species proliferated, but in some cases, entire groups (e.g., the dinosaurs) ceased to exist. These major extinction events are often associated with disruptions of the carbon cycle (Rothman 2001). Observational data indicates that disruptions of the carbon cycle in the past exhibit a characteristic flux—measured in terms of the injection of carbon divided by the time of the change (an example of this data is shown in Figure 13.7 (Plate 16)). This critical value could be reached if a large amount of carbon is injected in a relatively longer time period or a somewhat smaller amount of carbon is injected in a relatively shorter time period. This flux could be caused by volcanism (as in the past) or anthropogenic release of carbon into the atmosphere (as in the present). One interpretation of this observational data is that if the Earth system is forced at a rate beyond the capacity of its abiotic and biotic processes to sequester the influx, then it is possible the system could initiate an excitable excursion from its resident state, similar to the excursion of a neuron from rest given a sufficiently large perturbation.

A recent model has proposed a way to link carbon cycles and mass extinction by considering how dissolved forms of inorganic carbon influence carbon fluxes within the ocean (Rothman 2017, 2019). The model details are explored in the computational laboratory associated with this chapter. The model focuses on the nonlinear feedback between dissolved inorganic carbon and carbonate ions in the upper lit layer of the ocean, mediated by the action of photosynthetic organisms—predominantly phytoplankton. The role of phytoplankton is embedded into the following feedback mechanism: high levels of CO_2 lower carbonate levels, which in turn impairs the growth of shelled planktonic organisms

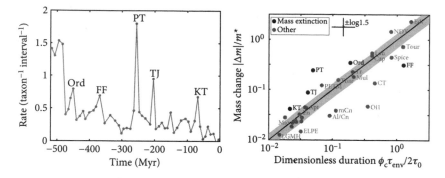

Figure 13.7: Carbon cycle disruptions and mass extinction. (Left) The loss of taxa over time (in millions of years) where five mass extinctions are labeled, including the end-Cretacious (KT), end-Triassic (TJ), end-Permian (PT), end-Ordovician (Ord), and Frasnian-Fammenian (FF). Reproduced from Arnscheidt and Rothman (2022) using data from Alroy (2014). (Right) Influx of carbon into the ocean as a function of a dimensionless duration for 31 different disruptions of the global carbon cycle over the past 540 million years. Reproduced from Rothman (2017).

(that require calcium carbonate for their shells). When they die, these shelled organisms are an essential part of the flux of carbon from the surface to deep oceans and into ocean sediments. Hence, when carbonate is low, there is less growth of shelled organisms and less export to the deep, so more of the organic carbon in the upper lit layers of the ocean is respired back into CO_2. This further accelerates the process of increasing CO_2 levels and decreasing carbonate levels.

This feedback mechanism suggests that the Earth system may be a type of excitable system with long excursions from a rest state given sufficiently strong forcing. Excitable systems exhibit such characteristic behavior too—because they are essentially remnants of limit cycles. Limit cycles are intrinsic properties of nonlinear systems—as we have seen in many circumstances in this book. Viewed in this light, the mass extinction events of the past may reflect self-amplification with the carbon cycle that reaches a common, critical rate rather than some external perturbation that would be unlikely to share a common magnitude. The relevance of this view of the Earth system as an excitable system is particularly relevant in the case of injections of CO_2 over short time scales, e.g., anthropogenic emissions. The flux rates are high, but the time has been relatively short, and the risk is that their product may soon cross a tipping point. The consequences, however, are radically different than many conventional perspectives, which focus on the challenges of moving CO_2 back below 450 ppm. Instead, a model of excitability suggests that feedback processes beyond a critical, forcing rate could lead to very long excursions in the Earth system's carbon cycle that lead us far from equilibrium, including through dynamic regimes that could be devastating to existing life. And that is a large lesson to learn from a small model.

13.5   THE CHALLENGE CONTINUES

This book has stretched from molecules to organisms to populations and now, finally, to ecosystems. Yet the work is far from done. As originally conceived, there is no final exam for the class; instead, students are encouraged to identify a paper written in the past decade that integrates quantitative models as part of the study of living systems at any scale of life. The aim is to facilitate student curiosity, but also to ensure that students begin to embed these lessons into their practice. There is another reason—in that there are pressing challenges at all scales of life where quantitative principles are needed to advance both foundational understanding and translation that can help improve human and environmental health. Real change will take a new generation of scientists willing to combine rigorous theory, data-focused computation, and a genuine sense of understanding of the culture and problems of biological sciences. If not now, when?

REFERENCES

Adler, F. R. (2012). *Modeling the Dynamics of Life: Calculus and Probability for Life Scientists*. Brooks/Cole Cengage Learning, Boston.

Aguilar, J., Zhang, T., Qian, F., Kingsbury, M., McInroe, B., Mazouchova, N., Li, C., Maladen, R., Gong, C., Travers, M., et al. (2016). A review on locomotion robophysics: The study of movement at the intersection of robotics, soft matter and dynamical systems. *Reports on Progress in Physics*, 79(11):110001.

Alexander, K. A., Sanderson, C. E., Marathe, M., Lewis, B. L., Rivers, C. M., Shaman, J., Drake, J. M., Lofgren, E., Dato, V. M., Eisenberg, M. C., et al., (2015). What factors might have led to the emergence of Ebola in West Africa? *PLoS Neglected Tropical Diseases*, 9(6):e0003652.

Allen, R. C., Popat, R., Diggle, S. P., and Brown, S. P. (2014). Targeting virulence: Can we make evolution-proof drugs? *Nature Reviews Microbiology*, 12(4):300–308.

Allen, S. M., and Cahn, J. W. (1972). Ground state structures in ordered binary alloys with second neighbor interactions. *Acta Metallurgica*, 20(3):423–433.

Allesina, S., and Tang, S. (2012). Stability criteria for complex ecosystems. *Nature*, 483:205–208.

Alon, U. (2006). *An Introduction to Systems Biology: Design Principles of Biological Circuits*. Chapman and Hall/CRC, Boca Raton, FL.

Alon, U., Surette, M. G., Barkai, N., and Leibler, S. (1999). Robustness in bacterial chemotaxis. *Nature*, 397(6715):168–171.

Alroy, J. (2014). Accurate and precise estimates of origination and extinction rates. *Paleobiology*, 40(3):374–397.

Anderson, P. W. (1972). More is different. *Science*, 177(4047):393–396.

Aranson, I. S., and Tsimring, L. S. (2006). Patterns and collective behavior in granular media: Theoretical concepts. *Reviews of Modern Physics*, 78(2):641.

Aristotle (ca.350). *De Motu Animalium*.

Armstrong, R. A., and McGehee, R. (1980). Competitive exclusion. *American Naturalist*, 115:151–170.

Arnold, L., Jones, C., Mischaikow, K., and Raugel, G. (1995). Geometric singular perturbation theory. In *Dynamical Systems*, 1609:44–118. Springer, Berlin.

Arnscheidt, C. W., and Rothman, D. H. (2022). The balance of nature: A global marine perspective. *Annual Review of Marine Science*, 14:49–73.

Astley, H., Abbott, E., Azizi, E., Marsh, R., and Roberts, T. (2013). Chasing maximal performance: A cautionary tale from the celebrated jumping frogs of Calaveras County. *Journal of Experimental Biology*, 216:3947–3953.

Astley, H., and Roberts, T. (2012). Evidence for a vertebrate catapult: Elastic energy storage in the plantaris tendon during frog jumping. *Biology Letters*, 8:386–389.

Astley, H. C., Mendelson, J. R., III, Dai, J., Gong, C., Chong, B., Rieser, J. M., Schiebel, P. E., Sharpe, S. S., Hatton, R. L., Choset, H., et al., (2020). Surprising simplicities and syntheses in limbless self-propulsion in sand. *Journal of Experimental Biology*, 223(5):jeb103564.

Balázsi, G., Van Oudenaarden, A., and Collins, J. J. (2011). Cellular decision making and biological noise: From microbes to mammals. *Cell*, 144(6):910–925.

Ballerini, M., Cabibbo, N., Candelier, R., Cavagna, A., Cisbani, E., Giardina, I., Lecomte, V., Orlandi, A., Parisi, G., Procaccini, A., et al. (2008). Interaction ruling animal collective behavior depends on topological rather than metric distance: Evidence from a field study. *Proceedings of the National Academy of Sciences*, 105(4):1232–1237.

Barenblatt, G. I. (2003). *Scaling*. Cambridge University Press, Cambridge, UK.

Barkai, N., and Leibler, S. (1997). Robustness in simple biochemical networks. *Nature*, 387(6636):913–917.

Barrangou, R., Fremaux, C., Deveau, H., Richards, M., Boyval, P., Moineau, S., Romero, D. A., and Horvath, P. (2007). CRISPR provides acquired resistance against viruses in prokaryotes. *Science*, 315:1709–1712.

Basler, M., and Mekalanos, J. (2012). Type 6 secretion dynamics within and between bacterial cells. *Science*, 337(6096):815.

Bauch, C. T., and Earn, D. J. (2004). Vaccination and the theory of games. *Proceedings of the National Academy of Sciences*, 101(36):13391–13394.

Beisner, B. E., Haydon, D. T., and Cuddington, K. (2003). Alternative stable states in ecology. *Frontiers in Ecology and the Environment*, 1(7):376–382.

Benedek, G. B., and Villars, F. M. (2000a). *Physics with Illustrative Examples from Medicine and Biology: Electricity and Magnetism*. Springer-Verlag, New York.

Benedek, G. B., and Villars, F. M. (2000b). *Physics with Illustrative Examples from Medicine and Biology: Mechanics*. Springer-Verlag, New York.

Benedek, G. B., and Villars, F. M. (2000c). *Physics with Illustrative Examples from Medicine and Biology: Statistical Physics*. Springer-Verlag, New York.

Berestycki, H., Desjardins, B., Heintz, B., and Oury, J.-M. (2021). Plateaus, rebounds and the effects of individual behaviours in epidemics. *Scientific Reports*, 11:18339.

Berg, H. C. (1993). *Random Walks in Biology*. Princeton University Press, Princeton, NJ.

Berg, H. C. (2006). Marvels of bacterial behavior. *Proceedings of the American Philosophical Society*, 150: 428–442.

Berg, H. C., and Anderson, R. A. (1973). Bacteria swim by rotating their flagellar filaments. *Nature*, 245(5425):380–382.

Berg, H. C., and Brown, D. A. (1972). Chemotaxis in *Eschericia coli* analyzed by 3-dimensional tracking. *Nature*, 239:500–504.

Berg, H. C., and Purcell, E. M. (1977). Physics of chemoreception. *Biophysical Journal*, 20:193–219.

Bialek, W., Cavagna, A., Giardina, I., Mora, T., Silvestri, E., Viale, M., and Walczak, A. M. (2012). Statistical mechanics for natural flocks of birds. *Proceedings of the National Academy of Sciences*, 109(13):4786–4791.

Blickhan, R. (1989). The spring-mass model for running and hopping. *Journal of Biomechanics*, 22(11-12):1217–1227.

Blickhan, R. and Full, R. (1993). Similarity in multilegged locomotion: Bouncing like a monopode. *Journal of Comparative Physiology A*, 173(5):509–517.

Blount, Z. D., Barrick, J. E., Davidson, C. J., and Lenski, R. E. (2012). Genomic analysis of a key innovation in an experimental *Escherichia coli* population. *Nature*, 489(7417):513–518.

Blount, Z. D., Borland, C. Z., and Lenski, R. E. (2008). Historical contingency and the evolution of a key innovation in an experimental population of *Escherichia coli*. *Proceedings of the National Academy of Sciences*, 105(23):7899–7906.

Blount, Z. D., Lenski, R. E., and Losos, J. B. (2018). Contingency and determinism in evolution: Replaying life's tape. *Science*, 362(6415).

Bodine, E. N., Lenhart, S., and Gross, L. J. (2014). *Mathematics for the Life Sciences*. Princeton University Press, Princeton, NJ.

Borisyuk, A., Ermentrout, G. B., Friedman, A., and Terman, D. H. (2005). *Tutorials in Mathematical Biosciences I: Mathematical Neuroscience*. Springer, New York.

Box, G. E. (1976). Science and statistics. *Journal of the American Statistical Association*, 71(356):791–799.

Bray, D. (2002). Bacterial chemotaxis and the question of gain. *Proceedings of the National Academy of Sciences*, 99:7–9.

Breitbart, M., Bonnain, C., Malki, K., and Sawaya, N. A. (2018). Phage puppet masters of the marine microbial realm. *Nature Microbiology*, 3(7):754–766.

Britton, T., Ball, F., and Trapman, P. (2020). A mathematical model reveals the influence of population heterogeneity on herd immunity to SARS-CoV-2. *Science*, 369(6505):846–849.

Brockmann, H. J. and Barnard, C. (1979). Kleptoparasitism in birds. *Animal Behaviour*, 27:487–514.

Brown, J. H., Gillooly, J. F., Allen, A. P., Savage, V. M., and West, G. B. (2004). Toward a metabolic theory of ecology. *Ecology*, 85(7):1771–1789.

Buhl, J., Sumpter, D. J. T., Couzin, I. D., Hale, J. J., Despland, E., Miller, E. R., and Simpson, S. J. (2006). From disorder to order in marching locusts. *Science*, 312(5778):1402–1406.

Burmeister, A. R., Hansen, E., Cunningham, J. J., Rego, E. H., Turner, P. E., Weitz, J. S., and Hochberg, M. E. (2021). Fighting microbial pathogens by integrating host ecosystem interactions and evolution. *BioEssays*, 43(3):2000272.

Cahn, J. W. (1961). On spinodal decomposition. *Acta Metallurgica*, 9(9):795–801.

Cahn, J. W., and Hilliard, J. E. (1958). Free energy of a nonuniform system. I. Interfacial free energy. *Journal of Chemical Physics*, 28(2):258–267.

Cai, L., Friedman, N., and Xie, X. S. (2006). Stochastic protein expression in individual cells at the single molecule level. *Nature*, 440(7082):358–362.

Canton, B., Labno, A., and Endy, D. (2008). Refinement and standardization of synthetic biological parts and devices. *Nature Biotechnology*, 26(7):787–793.

Carpenter, S. R., Cole, J. J., Pace, M. L., Batt, R., Brock, W. A., Cline, T., Coloso, J., Hodgson, J. R., Kitchell, J. F., Seekell, D. A., et al. (2011). Early warnings of regime shifts: A whole-ecosystem experiment. *Science*, 332(6033):1079–1082.

Carpenter, S. R., Ludwig, D., and Brock, W. A. (1999). Management of eutrophication for lakes subject to potentially irreversible change. *Ecological Applications*, 9(3):751–771.

Cassidy, C. K., Himes, B. A., Sun, D., Ma, J., Zhao, G., Parkinson, J. S., Stansfeld, P. J., Luthey-Schulten, Z., and Zhang, P. (2020). Structure and dynamics of the *E. coli* chemotaxis core signaling complex by cryo-electron tomography and molecular simulations. *Communications Biology*, 3(1):1–10.

Cavagna, A., Cimarelli, A., Giardina, I., Parisi, G., Santagati, R., Stefanini, F., and Viale, M. (2010). Scale-free correlations in starling flocks. *Proceedings of the National Academy of Sciences*, 107(26):11865–11870.

Cavagna, A., Giardina, I., and Grigera, T. S. (2018). The physics of flocking: Correlation as a compass from experiments to theory. *Physics Reports*, 728:1–62.

Changeux, J.-P., and Edelstein, S. J. (2005). Allosteric mechanisms of signal transduction. *Science*, 308(5727):1424–1428.

Chen, D., and Arkin, A. P. (2012). Sequestration-based bistability enables tuning of the switching boundaries and design of a latch. *Molecular Systems Biology*, 8:620.

Cherry, E. M., and Fenton, F. H. (2008). Visualization of spiral and scroll waves in simulated and experimental cardiac tissue. *New Journal of Physics*, 10(12):125016.

Cherry, J. L., and Adler, F. R. (2000). How to make a biological switch. *Journal of Theoretical Biology*, 203:117–133.

Chisholm, S. W. (2017). *Prochlorococcus*. *Current Biology*, 27(11):R447–R448.

Chisholm, S. W., Olson, R. J., Zettler, E. R., Goericke, R., Waterbury, J. B., and Welschmeyer, N. A. (1988). A novel free-living prochlorophyte abundant in the oceanic euphotic zone. *Nature*, 334(6180):340–343.

Chou, T., Mallick, K., and Zia, R. K. (2011). Non-equilibrium statistical mechanics: From a paradigmatic model to biological transport. *Reports on Progress in Physics*, 74(11):116601.

Chowell, G., and Nishiura, H. (2014). Transmission dynamics and control of Ebola virus disease (EVD): A review. *BMC Medicine*, 12(1):1–17.

Clutton-Brock, T. H., and Albon, S. D. (1979). The roaring of red deer and the evolution of honest advertisement. *Behaviour*, 69(3-4):145–170.

Cohen, J. E. (1978). *Food Webs and Niche Space*. Princeton University Press, Princeton, NJ.

Cohen, J. E. (1995). Population growth and earth's human carrying capacity. *Science*, 269(5222):341–346.

Cole, K. S., and Curtis, H. J. (1939). Electric impedance of the squid giant axon during activity. *Journal of General Physiology*, 22(5):649–670.

Collins, S., Ruina, A., Tedrake, R., and Wisse, M. (2005). Efficient bipedal robots based on passive-dynamic walkers. *Science*, 307(5712):1082–1085.

Constantino, R. F., Desharnais, R. A., Cushing, J. M., and Dennis, B. (1997). Chaotic dynamics in an insect population. *Science*, 275(5298):389–391.

Cortez, M. H. (2018). Genetic variation determines which feedbacks drive and alter predator–prey eco-evolutionary cycles. *Ecological Monographs*, 88(3):353–371.

Cortez, M. H., and Ellner, S. P. (2010). Understanding rapid evolution in predator-prey interactions using the theory of fast-slow dynamical systems. *American Naturalist*, 176:E109–E127.

Cortez, M. H., and Weitz, J. S. (2014). Coevolution can reverse predator-prey cycles. *Proceedings of the National Academy of Sciences*, 111(20):7486–7491.

Cull, P. (2007). The mathematical biophysics of Nicolas Rashevsky. *BioSystems*, 88(3):178–184.

Darnton, N. C., Turner, L., Rojevsky, S., and Berg, H. C. (2007). On torque and tumbling in swimming *Escherichia coli*. *Journal of Bacteriology*, 189(5):1756–1764.

Darwin, C. (1859). *On the Origin of Species*. John Murray, London.

Davies, N. G., Abbott, S., Barnard, R. C., Jarvis, C. I., Kucharski, A. J., Munday, J. D., Pearson, C. A. B., Russell, T. W., Tully, D. C., Washburne, A. D., et al. (2021). Estimated transmissibility and impact of SARS-CoV-2 lineage B.1.1.7 in England. *Science*, 372(6538): eabg3055.

Desroches, M., and Jeffrey, M. R. (2011). Canards and curvature: The "smallness of ϵ" in slow–fast dynamics. *Proceedings of the Royal Society A: Mathematical, Physical and Engineering Sciences*, 467(2132):2404–2421.

d'Herelle, F. (1917). Sur un microbe invisible antagoniste des bacilles dysentèriques. *Comptes rendus de l'Académie des Sciences Paris*, 165.

Díaz-Muñoz, S. L., Sanjuán, R., and West, S. (2017). Sociovirology: Conflict, cooperation, and communication among viruses. *Cell Host and Microbe*, 22(4):437–441.

Diekmann, O., and Heesterbeek, J. A. P. (2000). *Mathematical Epidemiology of Infectious Diseases: Model Building, Analysis and Interpretation*, volume 5. Wiley, Chichester, UK.

Diekmann, O., Heesterbeek, J. A. P., and Metz, J. A. (1990). On the definition and the computation of the basic reproduction ratio R_0 in models for infectious diseases in heterogeneous populations. *Journal of Mathematical Biology*, 28(4):365–382.

Dill, L. M. (1974a). The escape response of the zebra danio (*Brachydanio rerio*) I. The stimulus for escape. *Animal Behaviour*, 22(3):711–722.

Dill, L. M. (1974b). The escape response of the zebra danio (*Brachydanio rerio*) II. The effect of experience. *Animal Behaviour*, 22(3):723–730.

Dill, L. M., and Fraser, A. H. G. (1997). The worm re-turns: Hiding behavior of a tube-dwelling marine polychaete, *Serpula vermicularis*. *Behavioral Ecology*, 8(2):186–193.

Dobzhansky, T. (1973). Nothing in biology makes sense except in light of evolution. *American Biology Teacher*, 35:125–129.

Dodds, P. S., Rothman, D. H., and Weitz, J. S. (2001). Re-examination of the "3/4-law" of metabolism. *Journal of Theoretical Biology*, 209(1):9–27.

Doebeli, M., and Hauert, C. (2005). Models of cooperation based on the prisoner's dilemma and the snowdrift game. *Ecology Letters*, 8(7):748–766.

Du, Z., Xu, X., Wu, Y., Wang, L., Cowling, B. J., and Meyers, L. A. (2020). Serial interval of COVID-19 among publicly reported confirmed cases. *Emerging Infectious Diseases*, 26(6):1341.

Duffy, M. A., and Sivars-Becker, L. (2007). Rapid evolution and ecological host-parasite dynamics. *Ecology Letters*, 10:44–53.

Ehrlich, P. R. (1968). *The Population Bomb*. Ballantine Books, New York.

Einstein, A. (1956). *Investigations on the Theory of the Brownian Movement*. Dover, Mineola, NY.

Eksin, C., Paarporn, K., and Weitz, J. S. (2019). Systematic biases in disease forecasting: The role of behavior change. *Epidemics*, 27:96–105.

Elena, S. F., Cooper, V. S., and Lenski, R. E. (1996). Punctuated evolution caused by selection of rare beneficial mutations. *Science*, 272(5269):1802–1804.

Elena, S. F., and Lenski, R. E. (2003). Microbial genetics: Evolution experiments with microorganisms: The dynamics and genetic bases of adaptation. *Nature Reviews Genetics*, 4(6):457.

Eling, N., Morgan, M. D., and Marioni, J. C. (2019). Challenges in measuring and understanding biological noise. *Nature Reviews Genetics*, 20(9):536–548.

Elowitz, M. B., Levine, A. J., Siggia, E. D., and Swain, P. S. (2002). Stochastic gene expression in a single cell. *Science*, 297:1183–1186.

Elser, M. M., Elser, J. J., and Carpenter, S. R. (1986). Paul and Peter Lakes: A liming experiment revisited. *American Midland Naturalist*, 116(2):282–295.

Evans, M. R. (2000). Phase transitions in one-dimensional nonequilibrium systems. *Brazilian Journal of Physics*, 30(1):42–57.

Evans, M. R., and Hanney, T. (2005). Nonequilibrium statistical mechanics of the zero-range process and related models. *Journal of Physics A: Mathematical and General*, 38(19):R195.

Farley, C. T., Glasheen, J., and McMahon, T. A. (1993). Running springs: Speed and animal size. *Journal of Experimental Biology*, 185(1):71–86.

Feigenbaum, M. J. (1978). Quantitative universality for a class of nonlinear transformations. *Journal of Statistical Physics*, 19(1):25–52.

Feigenbaum, M. J. (1984). Universal behavior in nonlinear systems. In *Universality in Chaos*, pages 49–84. Adam Hilger, Bristol, UK.

Feller, W. (1971). *An Introduction to Probability Theory and Its Applications*, volume 2. Wiley, New York, 2nd edition.

Fenton, F., and Karma, A. (1998). Vortex dynamics in three-dimensional continuous myocardium with fiber rotation: Filament instability and fibrillation. *Chaos: Interdisciplinary Journal of Nonlinear Science*, 8(1):20–47.

Fenton, F. H., Cherry, E. M., Hastings, H. M., and Evans, S. J. (2002). Multiple mechanisms of spiral wave breakup in a model of cardiac electrical activity. *Chaos: Interdisciplinary Journal of Nonlinear Science*, 12(3):852–892.

Ferguson, N., Laydon, D., Nedjati Gilani, G., Imai, N., Ainslie, K., Baguelin, M., Bhatia, S., Boonyasiri, A., Cucunuba Perez, Z., Cuomo-Dannenburg, G., et al. (2020). Report 9: Impact of non-pharmaceutical interventions (NPIs) to reduce COVID-19 mortality and healthcare demand. *Imperial College London Report*, March 16, 2020.

Fidler, I. J., and Kripke, M. L. (1977). Metastasis results from preexisting variant cells within a malignant tumor. *Science*, 197(4306):893–895.

Fisher, R. A. (1930). *The Genetical Theory of Natural Selection*. Oxford University Press, Oxford, UK.

FitzHugh, R. (1961). Impulses and physiological states in theoretical models of nerve membrane. *Biophysical Journal*, 1(6):445–466.

Follows, M. J., Dutkiewicz, S., Grant, S., and Chisholm, S. W. (2007). Emergent biogeography of microbial communities in a model ocean. *Science*, 315:1843–1846.

Fox, J. W., and Lenski, R. E. (2015). From here to eternity—the theory and practice of a really long experiment. *PLoS Biology*, 13(6):e1002185.

Friedman, N., Cai, L., and Xie, X. S. (2006). Linking stochastic dynamics to population distribution: An analytical framework of gene expression. *Physical review letters*, 97(16):168302.

Full, R. J., and Koditschek, D. E. (1999). Templates and anchors: Neuromechanical hypotheses of legged locomotion on land. *Journal of Experimental Biology*, 202(23):3325–3332.

Funk, S., Salathé, M., and Jansen, V. A. (2010). Modelling the influence of human behaviour on the spread of infectious diseases: A review. *Journal of the Royal Society Interface*, 7(50):1247–1256.

Gandhi, R. T., Lynch, J. B., and Del Rio, C. (2020). Mild or moderate COVID-19. *New England Journal of Medicine*, 383(18):1757–1766.

Gardner, T. S., Cantor, C. R., and Collins, J. J. (2000). Construction of a genetic toggle switch in *Escherichia coli*. *Nature*, 403:339–342.

Gell-Mann, M. (1994a). Complex adaptive systems. In *Complexity: Metaphors, Models, and Reality*, pages 17–45. Addison-Wesley, Reading, MA.

Gell-Mann, M. (1994b). *The Quark and the Jaguar: Adventures in the Simple and the Complex*. W. H. Freeman, New York.

Gillespie, D. T. (1977). Exact stochastic simulation of coupled chemical reactions. *Journal of Physical Chemistry*, 82(25):2340–2361.

Gillooly, J. F., Brown, J. H., West, G. B., Savage, V. M., and Charnov, E. L. (2001). Effects of size and temperature on metabolic rate. *Science*, 293(5538):2248–2251.

Goldberger, A. L., Amaral, L. A. N., Hausdorff, J. M., Ivanov, P. C., Peng, C. K., and Stanley, H. E. (2002). Fractal dynamics in physiology: Alterations with disease and aging. *Proceedings of the National Academy of Sciences*, 99(3, suppl 1):2466–2472.

Golding, I., and Cox, E. C. (2004). RNA dynamics in live *Escherichia coli* cells. *Proceedings of the National Academy of Sciences*, 101(31):11310–11315.

Golding, I., Paulsson, J., Zawilski, S. M., and Cox, E. C. (2005). Real-time kinetics of gene activity in individual bacteria. *Cell*, 123(6):1025–1036.

Goldstein, R. E. (2018). Point of view: Are theoretical results "results"? *Elife*, 7:e40018.

Good, B. H., McDonald, M. J., Barrick, J. E., Lenski, R. E., and Desai, M. M. (2017). The dynamics of molecular evolution over 60,000 generations. *Nature*, 551(7678):45.

Gould, S., and Eldredge, N. (1993). Punctuated equilibrium comes of age. *Nature*, 366(6452):223–227.

Graham, M. D. (2018). *Microhydrodynamics, Brownian Motion, and Complex Fluids*. Cambridge University Press, Oxford, UK.

Grant, P. R., and Grant, B. R. (2014). *40 Years of Evolution: Darwin's Finches on Daphne Major Island*. Princeton University Press, Princeton, NJ.

Gray, J., and Hancock, G. (1955). The propulsion of sea-urchin spermatozoa. *Journal of Experimental Biology*, 32(4):802–814.

Großkinsky, S., Schütz, G. M., and Spohn, H. (2003). Condensation in the zero range process: Stationary and dynamical properties. *Journal of Statistical Physics*, 113(3):389–410.

Guckenheimer, J. (1980). Dynamics of the van der Pol equation. *IEEE Transactions on Circuits and Systems*, 27(11):983–989.

Guerra, F. M., Bolotin, S., Lim, G., Heffernan, J., Deeks, S. L., Li, Y., and Crowcroft, N. S. (2017). The basic reproduction number (R_0) of measles: A systematic review. *Lancet Infectious Diseases*, 17(12):e420–e428.

Hadfield, J., Megill, C., Bell, S. M., Huddleston, J., Potter, B., Callender, C., Sagulenko, P., Bedford, T., and Neher, R. A. (2018). Nextstrain: Real-time tracking of pathogen evolution. *Bioinformatics*, 34(23):4121–4123.

Hairston, N. G., Ellner, S. P., Geber, M. A., Yoshida, T., and Fox, J. A. (2005). Rapid evolution and the convergence of ecological and evolutionary time. *Ecology Letters*, 8:1114–1127.

Hastings, A. (1997). *Population Biology: Concepts and Models*. Springer-Verlag, New York.

Hatton, R. L., and Choset, H. (2013). Geometric Swimming at Low and High Reynolds Numbers. *IEEE Transactions on Robotics*, 29(3):615–624.

Hauert, C., and Doebeli, M. (2004). Spatial structure often inhibits the evolution of cooperation in the snowdrift game. *Nature*, 428(6983):643–646.

Hébert-Dufresne, L., Althouse, B. M., Scarpino, S. V., and Allard, A. (2020). Beyond R_0: Heterogeneity in secondary infections and probabilistic epidemic forecasting. *Journal of the Royal Society Interface*, 17(172):20200393.

Helbing, D., Szolnoki, A., Perc, M., and Szabó, G. (2010). Punish, but not too hard: How costly punishment spreads in the spatial public goods game. *New Journal of Physics*, 12(8):083005.

Henrich, J., Boyd, R., Bowles, S., Camerer, C., Fehr, E., Gintis, H., and McElreath, R. (2001). In search of *Homo economicus*: Behavioral experiments in 15 small-scale societies. *American Economic Review*, 91(2): 73–78.

Hiltunen, T., Hairston, N. G., Hooker, G., Jones, L. E., and Ellner, S. P. (2014). A newly discovered role of evolution in previously published consumer-resource dynamics. *Ecology Letters*, 17:915–23.

Hodgkin, A. L., and Huxley, A. F. (1952). A quantitative description of membrane current and its application to conduction and excitation in nerve. *Journal of Physiology*, 117:500–544.

Hofbauer, J., and Sigmund, K. (1998). *Evolutionary Games and Population Dynamics*. Cambridge University Press, Cambridge, UK.

Holland, J. H. (1992). Complex adaptive systems. *Daedalus*, 121(1):17–30.

Holling, C. S. (1973). Resilience and stability of ecological systems. *Annual Review of Ecology and Systematics*, 4(1):1–23.

Holt, R. D. (2009). Bringing the Hutchinsonian niche into the 21st century: Ecological and evolutionary perspectives. *Proceedings of the National Academy of Sciences*, 106(Supplement 2):19659–19665.

Huang, A. S. (1973). Defective interfering viruses. *Annual Reviews in Microbiology*, 27(1):101–118.

Hublin, J.-J., Ben-Ncer, A., Bailey, S. E., Freidline, S. E., Neubauer, S., Skinner, M. M., Bergmann, I., Le Cabec, A., Benazzi, S., Harvati, K., et al., (2017). New fossils from Jebel Irhoud, Morocco and the pan-African origin of *Homo sapiens*. *Nature*, 546(7657):289–292.

Huisman, J., and Weissing, F. J. (1999). Biodiversity of plankton by species oscillations and chaos. *Nature*, 402(6760):407–410.

Hutchinson, G. E. (1957). Concluding remarks. In *Population Studies: Animal Ecology and Demography*, volume 22 of *Cold Spring Harbor Symposia on Quantitative Biology*, pages 415–427.

Hutchinson, G. E. (1959). Homage to Santa Rosalia, or why are there so many kinds of animals? *American Naturalist*, 93:254–259.

Ijspeert, A. J., Nakanishi, J., and Schaal, S. (2002). Movement imitation with nonlinear dynamical systems in humanoid robots. *Proceedings of the 2002 IEEE International Conference on Robotics and Automation*, 2:1398–1403.

Israel, G., and Gasca, A. (2002). *The Biology of Numbers: The Correspondence of Vito Volterra on Mathematical Biology*. Birkhäuser, Basel.

Izhikevich, E. M. (2007). *Dynamical Systems in Neuroscience*. MIT Press, Cambridge, MA.

Judson, H. F. (1979). *The Eighth Day of Creation: Makers of the Revolution in Biology*. Simon & Schuster, New York.

Karma, A. (1994). Electrical alternans and spiral wave breakup in cardiac tissue. *Chaos: Interdisciplinary Journal of Nonlinear Science*, 4(3):461–472.

Kawecki, T. J., Lenski, R. E., Ebert, D., Hollis, B., Olivieri, I., and Whitlock, M. C. (2012). Experimental evolution. *Trends in Ecology & Evolution*, 27(10):547–560.

Kessler, D. A., and Levine, H. (2013). Large population solution of the stochastic Luria-Delbrück evolution model. *Proceedings of the National Academy of Sciences*, 110(29):11682–11687.

Kim, K., Spieler, P., Lupu, E.-S., Ramezani, A., and Chung, S.-J. (2021). A bipedal walking robot that can fly, slackline, and skateboard. *Science Robotics*, 6(59):eabf8136.

Kim, S., Laschi, C., and Trimmer, B. (2013). Soft robotics: A bioinspired evolution in robotics. *Trends in Biotechnology*, 31(5):287–294.

King, A. A., Nguyen, D., and Ionides, E. L. (2016). Statistical inference for partially observed Markov processes via the R package pomp. *Journal of Statistical Software*, 69(12):1–43.

Klausmeier, C. A. (1999). Regular and irregular patterns in semiarid vegetation. *Science*, 284(5421):1826–1828.

Kleiber, M. (1932). Body size and metabolism. *Hilgardia*, 6:315–353.

Kleiber, M. (1947). Body size and metabolic rate. *Physiological Reviews*, 27:511–541.

Kleiber, M. (1961). *The Fire of Life. An Introduction to Animal Energetics*. Wiley, New York.

Koonin, E. V., and Wolf, Y. I. (2009). Is evolution Darwinian or/and Lamarckian? *Biology Direct*, 4(1):42.

Kouzy, R., Abi Jaoude, J., Kraitem, A., El Alam, M. B., Karam, B., Adib, E., Zarka, J., Traboulsi, C., Akl, E. W., and Baddour, K. (2020). Coronavirus goes viral: Quantifying the COVID-19 misinformation epidemic on Twitter. *Cureus*, 12(3):e7255.

Krotkov, E., Hackett, D., Jackel, L., Perschbacher, M., Pippine, J., Strauss, J., Pratt, G., and Orlowski, C. (2017). The DARPA robotics challenge finals: Results and perspectives. *Journal of Field Robotics*, 34(2):229–240.

Labrie, S. J., Samson, J. E., and Moineau, S. (2010). Bacteriophage resistance mechanisms. *Nature Reviews Microbiology*, 8:317–327.

Lang, G. I., Rice, D. P., Hickman, M. J., Sodergren, E., Weinstock, G. M., Botstein, D., and Desai, M. M. (2013). Pervasive genetic hitchhiking and clonal interference in forty evolving yeast populations. *Nature*, 500(7464):571.

Lauga, E., and Powers, T. R. (2009). The hydrodynamics of swimming microorganisms. *Reports on Progress in Physics*, 72(9):096601.

Lea, D. E., and Coulson, C. A. (1949). The distribution of the numbers of mutants in bacterial populations. *Journal of Genetics*, 49(3):264–285.

Lederberg, J., and Lederberg, E. M. (1952). Replica plating and indirect selection of bacterial mutants. *Journal of Bacteriology*, 63(3):399–406.

Legrand, J., Grais, R. F., Boelle, P.-Y., Valleron, A.-J., and Flahault, A. (2007). Understanding the dynamics of Ebola epidemics. *Epidemiology & Infection*, 135(4):610–621.

Lenski, R. E., Rose, M. R., Simpson, S. C., and Tadler, S. C. (1991). Long-term experimental evolution in *Escherichia coli*. I. Adaptation and divergence during 2,000 generations. *American Naturalist*, 138(6):1315–1341.

Lenski, R. E., and Travisano, M. (1994). Dynamics of adaptation and diversification: A 10,000-generation experiment with bacterial populations. *Proceedings of the National Academy of Sciences*, 91(15):6808–6814.

Levin, B. R., Stewart, F. M., and Chao, L. (1977). Resource-limited growth, competition, and predation: A model and experimental studies with bacteria and bacteriophage. *American Naturalist*, 111:3–24.

Levin, S. A. (1992). The problem of pattern and scale in ecology. *Ecology*, 73:1943–1967.

Levin, S. A. (2003). Complex adaptive systems: Exploring the known, the unknown, and the unknowable. *Bulletin of the American Mathematical Society*, 40(1):3–19.

Lin, Y.-H., and Weitz, J. S. (2019). Spatial interactions and oscillatory tragedies of the commons. *Physical Review Letters*, 122(14):148102.

Lofgren, E. T., Halloran, M. E., Rivers, C. M., Drake, J. M., Porco, T. C., Lewis, B., Yang, W., Vespignani, A., Shaman, J., Eisenberg, J. N., et al. (2014). Opinion: Mathematical models: A key tool for outbreak response. *Proceedings of the National Academy of Sciences*, 111(51):18095–18096.

Lorenz, E. N. (1963). Deterministic nonperiodic flow. *Journal of Atmospheric Sciences*, 20(2):130–141.

Lotka, A. (1925). *Elements of Physical Biology*. Dover, New York, reprinted 1956 edition.

Lubchenco, J. (1980). Algal zonation in the New England rocky intertidal community: An experimental analysis. *Ecology*, 61(2):333–344.

Luo, L. (2015). *Principles of Neurobiology*. Garland Science, New York.

Luria, S. E. (1984). *A Slot Machine, a Broken Test Tube: An Autobiography*. Alfred P. Sloan Foundation series. Harper & Row, New York, 1st edition.

Luria, S. E., and Delbrück, M. (1943). Mutations of bacteria from virus sensitivity to virus resistance. *Genetics*, 28:491–511.

Luther, S., Fenton, F. H., Kornreich, B. G., Squires, A., Bittihn, P., Hornung, D., Zabel, M., Flanders, J., Gladuli, A., Campoy, L., et al. (2011). Low-energy control of electrical turbulence in the heart. *Nature*, 475(7355):235–239.

Madigan, M., Martinko, J. M., Dunlap, P. V., and Clark, D. V. (2009). *Brock Biology of Microorganisms*. Pearson Benjamin Cummings, San Francisco.

Makarova, K. S., Haft, D. H., Barrangou, R., Brouns, S. J., Charpentier, E., Horvath, P., Moineau, S., Mojica, F. J., Wolf, Y. I., Yakunin, A. F., et al. (2011). Evolution and classification of the CRISPR–Cas systems. *Nature Reviews Microbiology*, 9(6):467–477.

Maladen, R. D., Ding, Y., Li, C., and Goldman, D. I. (2009). Undulatory swimming in sand: Subsurface locomotion of the sandfish lizard. *Science*, 325(5938):314–318.

Maladen, R. D., Ding, Y., Umbanhowar, P. B., Kamor, A., and Goldman, D. I. (2011). Mechanical models of sandfish locomotion reveal principles of high performance subsurface sand-swimming. *Journal of the Royal Society Interface*, 8(62):1332–1345.

Malthus, T. R. (1970). *An Essay on the Principle of Population; and a Summary View of the Principle of Population*. A. Flew, ed. Penguin Books, London.

May, R. M. (1979). Simple mathematical models with complicated dynamics. *Nature*, 261(10):3.

McGeer, T. (1990). Passive dynamic walking. *International Journal of Robotics Research*, 9(2):62–82.

McMahon, T. A. (1973). Size and shape in biology. *Science*, 179:1201–1204.

McNally, L., Bernardy, E., Thomas, J., Kalziqi, A., Pentz, J., Brown, S. P., Hammer, B. K., Yunker, P. J., and Ratcliff, W. C. (2017). Killing by type VI secretion drives genetic phase separation and correlates with increased cooperation. *Nature Communications*, 8(1):14371.

Meltzer, M. I., Atkins, C. Y., Santibanez, S., Knust, B., Petersen, B. W., Ervin, E. D., Nichol, S. T., Damon, I. K., and Washington, M. L. (2014). Estimating the future number of cases in the Ebola epidemic—Liberia and Sierra Leone, 2014–2015. *Morbidity and Mortality Weekly Report*, 63(3):1–14.

Mézard, M., Parisi, G., and Virasoro, M. A. (1987). *Spin Glass Theory and Beyond: An Introduction to the Replica Method and Its Applications*, volume 9. World Scientific, Singapore.

Milo, R., Jorgensen, P., Moran, U., Weber, G., and Springer, M. (2010). BioNumbers–the database of key numbers in molecular and cell biology. *Nucleic Acids Research*, 38:D750–D753.

Milo, R., Shen-Orr, S., Itzkovitz, S., Kashtan, N., Chklovskii, D., and Alon, U. (2002). Network motifs: Simple building blocks of complex networks. *Science*, 298:824–827.

Moghadas, S. M., Fitzpatrick, M. C., Sah, P., Pandey, A., Shoukat, A., Singer, B. H., and Galvani, A. P. (2020). The implications of silent transmission for the control of COVID-19 outbreaks. *Proceedings of the National Academy of Sciences*, 117(30):17513–17515.

Moore, L. R., Rocap, G., and Chisholm, S. W. (1998). Physiology and molecular phylogeny of coexisting *Prochlorococcus* ecotypes. *Nature*, 393(6684):464–467.

Morris, D. H., Rossine, F. W., Plotkin, J. B., and Levin, S. A. (2021). Optimal, near-optimal, and robust epidemic control. *Communications Physics*, 4(1):1–8.

Munch, S. B., Rogers, T. L., Johnson, B. J., Bhat, U., and Tsai, C.-H. (2022). Rethinking the prevalence and relevance of chaos in ecology. *Annual Review of Ecology, Evolution, and Systematics*, 53(1):227–249.

Nadell, C. D., Xavier, J. B., and Foster, K. R. (2008). The sociobiology of biofilms. *FEMS Microbiology Reviews*, 33(1):206–224.

Nagumo, J., Arimoto, S., and Yoshizawa, S. (1962). An active pulse transmission line simulating nerve axon. *Proceedings of the IRE*, 50(10):2061–2070.

Nash, J. F. (1950). Equilibrium points in *n*-person games. *Proceedings of the National Academy of Sciences*, 36(1): 48–49.

Nelson, G., Saunders, A., and Playter, R. (2019). The PETMAN and Atlas robots at Boston Dynamics. *Humanoid Robotics: A Reference*, 169:186.

Nelson, P. (2021). *Physical Models of Living Systems*. Chiliagon Science, Philadelphia.

Neuhauser, C. (2011). *Calculus for Biology and Medicine*. Prentice Hall, Upper Saddle River, NJ.

Nevozhay, D., Adams, R. M., Van Itallie, E., Bennett, M. R., and Balázsi, G. (2012). Mapping the environmental fitness landscape of a synthetic gene circuit. *PLoS Computational Biology*, 8(4):e1002480–e1002480.

Nielsen, C. F., Kidd, S., Sillah, A. R., Davis, E., Mermin, J., and Kilmarx, P. H. (2015). Improving burial practices and cemetery management during an Ebola virus disease epidemic—Sierra Leone, 2014. *Morbidity and Mortality Weekly Report*, 64(1):20.

Niklas, K. J. (1994). Size-dependent variations in plant-growth rates and the 3/4-power rules. *American Journal of Botany*, 81:134–144.

Noble, D. (1962). A modification of the Hodgkin-Huxley equations applicable to Purkinje fibre action and pacemaker potentials. *Journal of Physiology*, 160(2):317–352.

Novick-Cohen, A., and Segel, L. A. (1984). Nonlinear aspects of the Cahn-Hilliard equation. *Physica D: Nonlinear Phenomena*, 10(3):277–298.

Nowak, M. A. (2006). *Evolutionary Dynamics: Exploring the Equations of Life*. Belknap Press of Harvard University Press, Cambridge, MA.

Nowak, M. A., and May, R. M. (1992). Evolutionary games and spatial chaos. *Nature*, 359(6398):826–829.

Odling-Smee, F. J., Laland, K. N., and Feldman, M. W. (1996). Niche construction. *American Naturalist*, 147(4):641–648.

Ozbudak, E. M., Thattai, M., Kurtser, I., Grossman, A. D., and Van Oudenaarden, A. (2002). Regulation of noise in the expression of a single gene. *Nature Genetics*, 31(1):69–73.

Ozbudak, E. M., Thattai, M., Lim, H. N., Shraiman, B. I., and Van Oudenaarden, A. (2004). Multistability in the lactose utilization network of *Escherichia coli*. *Nature*, 427(6976):737–740.

Paine, R. T. (1974). Intertidal community structure. *Oecologia*, 15(2):93–120.

Paine, R. T., and Levin, S. A. (1981). Intertidal landscapes: Disturbance and the dynamics of pattern. *Ecological Monographs*, 51(2):145–178.

Pandey, A., Atkins, K. E., Medlock, J., Wenzel, N., Townsend, J. P., Childs, J. E., Nyenswah, T. G., Ndeffo-Mbah, M. L., and Galvani, A. P. (2014). Strategies for containing Ebola in West Africa. *Science*, 346(6212): 991–995.

Parisi, G. (1988). *Statistical Field Theory*. Addison-Wesley, Redwood City, CA.

Park, S. W., Bolker, B. M., Champredon, D., Earn, D. J., Li, M., Weitz, J. S., Grenfell, B. T., and Dushoff, J. (2020). Reconciling early-outbreak estimates of the basic reproductive number and its uncertainty: Framework and applications to the novel coronavirus (SARS-CoV-2) outbreak. *Journal of the Royal Society Interface*, 17(168):20200144.

Park, S. W., Champredon, D., Weitz, J. S., and Dushoff, J. (2019). A practical generation-interval-based approach to inferring the strength of epidemics from their speed. *Epidemics*, 27:12–18.

Park, S. W., Cornforth, D. M., Dushoff, J., and Weitz, J. S. (2020). The time scale of asymptomatic transmission affects estimates of epidemic potential in the covid-19 outbreak. *Epidemics*, 31:100392.

Parkinson, J. S., Hazelbauer, G. L., and Falke, J. J. (2015). Signaling and sensory adaptation in *Escherichia coli* chemoreceptors: 2015 update. *Trends in Microbiology*, 23(5):257–266.

Paulsson, J. (2004). Summing up the noise in gene networks. *Nature*, 427(6973):415–418.

Peebles, J. (2019). Nobel Prize telephone interview with Prof. James Peebles, December 2019. https://www.nobelprize.org/prizes/physics/2019/peebles/interview/.

Peters, R. H. (1983). *The Ecological Implications of Body Size*. Cambridge University Press, Cambridge, UK.

Phillips, R., Kondev, J., and Theriot, J. (2009). *Physical Biology of the Cell*. Garland Science, New York.

Platt, J. R. (1964). Strong inference: Certain systematic methods of scientific thinking may produce much more rapid progress than others. *Science*, 146(3642):347–353.

Price, C. A., Weitz, J. S., Savage, V. M., Stegen, J., Clarke, A., Coomes, D. A., Dodds, P. S., Etienne, R. S., Kerkhoff, A. J., McCulloh, K., et al. (2012). Testing the metabolic theory of ecology. *Ecology Letters*, 15(12):1465–1474.

Ptashne, M. (2004). *A Genetic Switch: Phage Lambda Revisited*. Cold Spring Harbor Laboratory Press, Cold Spring Harbor, NY, 3rd edition.

Purcell, E. M. (1977). Life at low Reynolds number. *American Journal of Physics*, 45(1):3–11.

Raina, J.-B., Fernandez, V., Lambert, B., Stocker, R., and Seymour, J. R. (2019). The role of microbial motility and chemotaxis in symbiosis. *Nature Reviews Microbiology*, 17(5):284–294.

Ratcliff, W. C., Denison, R. F., Borrello, M., and Travisano, M. (2012). Experimental evolution of multicellularity. *Proceedings of the National Academy of Sciences*, 109(5):1595–1600.

Reznick, D. N., Losos, J., and Travis, J. (2019). From low to high gear: There has been a paradigm shift in our understanding of evolution. *Ecology Letters*, 22(2):233–244.

Rietkerk, M., Dekker, S. C., De Ruiter, P. C., and van de Koppel, J. (2004). Self-organized patchiness and catastrophic shifts in ecosystems. *Science*, 305(5692):1926–1929.

Rivers, C. (2014). Ebola: Models do more than forecast. *Nature*, 515(7528):492.

Roca, C. P., Cuesta, J. A., and Sánchez, A. (2009). Evolutionary game theory: Temporal and spatial effects beyond replicator dynamics. *Physics of Life Reviews*, 6(4):208–249.

Rolls, E. T., and Treves, A. (1997). *Neural Networks and Brain Function*, volume 572. Oxford University Press, Oxford, UK.

Rose, C., Medford, A. J., Goldsmith, C. F., Vegge, T., Weitz, J. S., and Peterson, A. A. (2021). Heterogeneity in susceptibility dictates the order of epidemic models. *Journal of Theoretical Biology*, 528:110839.

Rosenzweig, M. L., and MacArthur, R. H. (1963). Graphical representation and stability conditions of predator-prey interactions. *American Naturalist*, 97:209–223.

Rothman, D. H. (2001). Global biodiversity and the ancient carbon cycle. *Proceedings of the National Academy of Sciences*, 98:4305–4310.

Rothman, D. H. (2017). Thresholds of catastrophe in the Earth system. *Science Advances*, 3(9):e1700906.

Rothman, D. H. (2019). Characteristic disruptions of an excitable carbon cycle. *Proceedings of the National Academy of Sciences*, 116(30):14813–14822.

Rothman, D. H., Fournier, G. P., French, K. L., Alm, E. J., Boyle, E. A., Cao, C., and Summons, R. E. (2014). Methanogenic burst in the end-Permian carbon cycle. *Proceedings of the National Academy of Sciences*, 111(15):5462–5467.

Sanjuán, R., Nebot, M. R., Chirico, N., Mansky, L. M., and Belshaw, R. (2010). Viral mutation rates. *Journal of Virology*, 84(19):9733–9748.

Savage, V. M., Gillooly, J. F., Woodruff, W. H., West, G. B., Allen, A. P., Enquist, B. J., and Brown, J. H. (2004). The predominance of quarter-power scaling in biology. *Functional Ecology*, 18(2):257–282.

Scarpino, S. V., and Petri, G. (2019). On the predictability of infectious disease outbreaks. *Nature Communications*, 10(1):1–8.

Schaechter, E. (2014). ASM blog: Esther Lederberg, pioneer of bacterial genetics. *ASM Blog*, July 28, 2014.

Scheffer, M., Carpenter, S., Foley, J. A., Folke, C., and Walker, B. (2001). Catastrophic shifts in ecosystems. *Nature*, 413(6856):591–596.

Scheffer, M., Carpenter, S. R., Lenton, T. M., Bascompte, J., Brock, W., Dakos, V., van de Koppel, J., Van de Leemput, I. A., Levin, S. A., Van Nes, E. H., et al. (2012). Anticipating critical transitions. *Science*, 338(6105):344–348.

Scheffer, M., Hosper, S. H., Meijer, M. L., Moss, B., and Jeppesen, E. (1993). Alternative equilibria in shallow lakes. *Trends in Ecology & Evolution*, 8(8):275–279.

Schmidt-Nielsen, K. (1984). *Scaling: Why Is Animal Size So Important?* Cambridge University Press, Cambridge, UK.

Schuster, P. (2011). Mathematical modeling of evolution: Solved and open problems. *Theory in Biosciences*, 130(1):71–89.

Schwiening, C. J. (2012). A brief historical perspective: Hodgkin and Huxley. *Journal of Physiology*, 590(Pt 11):2571.

Scott, S. H. (2004). Optimal feedback control and the neural basis of volitional motor control. *Nature Reviews Neuroscience*, 5(7):532–545.

Segel, L. A. (1984). *Modeling Dynamic Phenomena in Molecular and Cellular Biology*. Cambridge University Press, Cambridge, UK.

Sender, R., Fuchs, S., and Milo, R. (2016). Revised estimates for the number of human and bacteria cells in the body. *PLoS Biology*, 14(8):e1002533.

Shamir, M., Bar-On, Y., Phillips, R., and Milo, R. (2016). SnapShot: Timescales in cell biology. *Cell*, 164(6):1302.

Shapere, A., and Wilczek, F. (1987). Self-propulsion at low Reynolds number. *Physical Review Letters*, 58(20):2051.

Shetty, R. P., Endy, D., and Knight, T. F. (2008). Engineering BioBrick vectors from BioBrick parts. *Journal of Biological Engineering*, 2(1):5.

Shubin, N. (2008). *Your Inner Fish: A Journey into the 3.5-Billion-Year History of the Human Body*. Vintage Books, New York.

Skotheim, J. M., and Mahadevan, L. (2005). Physical limits and design principles for plant and fungal movements. *Science*, 308(5726):1308–1310.

Smith, J. M. (1973). The stability of predator-prey systems. *Ecology*, 54:384–91.

Smolke, C. D. (2009). Building outside of the box: iGEM and the BioBricks Foundation. *Nature Biotechnology*, 27(12):1099–1102.

Sniegowski, P. D., Gerrish, P. J., and Lenski, R. E. (1997). Evolution of high mutation rates in experimental populations of *E. coli. Nature*, 387(6634):703.

Sourjik, V., and Wingreen, N. S. (2012). Responding to chemical gradients: Bacterial chemotaxis. *Current Opinion in Cell Biology*, 24(2):262–268.

Spence, R., Gerlach, G., Lawrence, C., and Smith, C. (2008). The behaviour and ecology of the zebrafish, *Danio rerio. Biological Reviews*, 83(1):13–34.

Spitzer, F. (1970). Interaction of Markov processes. *Advances in Mathematics*, 5:66–110.

Spudich, J. L., and Koshland, D. E., Jr. (1976). Non-genetic individuality: Chance in the single cell. *Nature*, 262:467–471.

Stocker, R. (2012). Marine microbes see a sea of gradients. *Science*, 338(6107):628–633.

Stocker, R., Seymour, J. R., Samadani, A., Hunt, D. E., and Polz, M. F. (2008). Rapid chemotactic response enables marine bacteria to exploit ephemeral microscale nutrient patches. *Proceedings of the National Academy of Sciences*, 105(11):4209–4214.

Stokes, D. E. (2011). *Pasteur's quadrant: Basic Science and Technological Innovation*. Brookings Institution Press, Washington, DC.

Strogatz, S. (1994). *Nonlinear Dynamics and Chaos*. Addison Wesley, Reading, MA.

Süel, G. M., Kulkami, R. P., Dworkin, J., Garcia-Ojalvo, J., and Elowitz, M. B. (2007). Tunability and noise dependence in differentiation dynamics. *Science*, 315(5819):1716–1719.

Summers, W. C. (1999). *Félix d'Herelle and the Origins of Molecular Biology*. Yale University Press, New Haven, CT.

Sun, L., Alexander, H. K., Bogos, B., Kiviet, D. J., Ackermann, M., and Bonhoeffer, S. (2018). Effective polyploidy causes phenotypic delay and influences bacterial evolvability. *PLoS Biology*, 16(2):e2004644.

Suttle, C. A. (2005). Viruses in the sea. *Nature*, 437:356–361.

Talmy, D., Beckett, S. J., Zhang, A. B., Taniguchi, D. A., Weitz, J. S., and Follows, M. J. (2019). Contrasting controls on microzooplankton grazing and viral infection of microbial prey. *Frontiers in Marine Science*, 6:182.

Taylor, B. P., Dushoff, J., and Weitz, J. S. (2016). Stochasticity and the limits to confidence when estimating R_0 of Ebola and other emerging infectious diseases. *Journal of Theoretical Biology*, 408:145–154.

Taylor, G. (1951). Analysis of the swimming of microscopic organisms. *Proceedings of the Royal Society A: Mathematical and Physical Sciences*, 209(1099):447–461.

Thattai, M., and Van Oudenaarden, A. (2001). Intrinsic noise in gene regulatory networks. *Proceedings of the National Academy of Sciences*, 98(15):8614–8619.

Thattai, M., and Van Oudenaarden, A. (2004). Stochastic gene expression in fluctuating environments. *Genetics*, 167(1):523–530.

Thomas, L. (1978). *The Lives of a Cell: Notes of a Biology Watcher*. Penguin, New York.

Tilman, A. R., Plotkin, J. B., and Akçay, E. (2020). Evolutionary games with environmental feedbacks. *Nature Communications*, 11(1):1–11.

Ting, L. H., and McKay, J. L. (2007). Neuromechanics of muscle synergies for posture and movement. *Current Opinion in Neurobiology*, 17(6):622–628.

Tiwari, A., Balázsi, G., Gennaro, M. L., and Igoshin, O. A. (2010). The interplay of multiple feedback loops with post-translational kinetics results in bistability of mycobacterial stress response. *Physical Biology*, 7(3):036005.

Tkachenko, A. V., Maslov, S., Wang, T., Elbana, A., Wong, G. N., and Goldenfeld, N. (2021). Stochastic social behavior coupled to COVID-19 dynamics leads to waves, plateaus, and an endemic state. *Elife*, 10:e68341.

Traulsen, A., and Nowak, M. A. (2006). Evolution of cooperation by multilevel selection. *Proceedings of the National Academy of Sciences*, 103(29):10952–10955.

Tritton, D. P. (1988). *Physical Fluid Dynamics*. Oxford University Press, Oxford, UK.

Turner, L., Ryu, W. S., and Berg, H. C. (2000). Real-time imaging of fluorescent flagellar filaments. *Journal of Bacteriology*, 182(10):2793–2801.

Turner, P. E., and Chao, L. (1999). Prisoner's dilemma in an RNA virus. *Nature*, 398:441–443.

Twort, T. W. (1915). An investigation on the nature of ultramicroscopic viruses. *Lancet*, 186:1241–1243.

van der Pol, B. (1920). Theory of the amplitude of free and forced triode vibrations. *Radio Review*, 1:701–710.

van Kampen, N. G. (2001). *Stochastic Processes in Physics and Chemistry*. Elsevier Science, Amsterdam.

Vespignani, A., Tian, H., Dye, C., Lloyd-Smith, J. O., Eggo, R. M., Shrestha, M., Scarpino, S. V., Gutierrez, B., Kraemer, M.U.G., Wu, J., et al. (2020). Modelling COVID-19. *Nature Reviews Physics*, 2(6):279–281.

Vicsek, T., Czirók, A., Ben-Jacob, E., Cohen, I., and Schochet, O. (1995). Novel type of phase transition in a system of self-driven particles. *Physical Review Letters*, 75:1226–1229.

Vicsek, T., and Zafeiris, A. (2012). Collective motion. *Physics Reports*, 517(3-4):71–140.

Volterra, V. (1926). Fluctuations in the abundance of a species considered mathematically. *Nature*, 118:558–60.

von Foerster, H., Mora, P. M., and Amiot, L. W. (1960). Doomsday: Friday, 13 November, AD 2026. *Science*, 132(3436):1291–1295.

Wallinga, J., and Lipsitch, M. (2007). How generation intervals shape the relationship between growth rates and reproductive numbers. *Proceedings of the Royal Society B: Biological Sciences*, 274(1609):599–604.

Webre, D. J., Wolanin, P. M., and Stock, J. B. (2003). Bacterial chemotaxis. *Current Biology*, 13(2):R47–R49.

Wei, W. E., Li, Z., Chiew, C. J., Yong, S. E., Toh, M. P., and Lee, V. J. (2020). Presymptomatic transmission of SARS-CoV-2—Singapore, January 23–March 16, 2020. *Morbidity and Mortality Weekly Report*, 69(14):411.

Wei, Y., Kirby, A., and Levin, B. R. (2011). The population and evolutionary dynamics of *Vibrio cholerae* and its bacteriophage: Conditions for maintaining phage-limited communities. *American Naturalist*, 178:715–725.

Weitz, J. S. (2015). *Quantitative Viral Ecology: Dynamics of Viruses and Their Microbial Hosts*. Princeton University Press, Princeton, NJ.

Weitz, J. S., and Dushoff, J. (2015). Modeling post-death transmission of Ebola: Challenges for inference and opportunities for control. *Scientific Reports*, 5(1):8751.

Weitz, J. S., Eksin, C., Paarporn, K., Brown, S. P., and Ratcliff, W. C. (2016). An oscillating tragedy of the commons in replicator dynamics with game-environment feedback. *Proceedings of the National Academy of Sciences*, 113(47):E7518–E7525.

Weitz, J. S., Park, S. W., Eksin, C., and Dushoff, J. (2020). Awareness-driven behavior changes can shift the shape of epidemics away from peaks and toward plateaus, shoulders, and oscillations. *Proceedings of the National Academy of Sciences*, 117(51):32764–32771.

West, G. B., Brown, J. H., and Enquist, B. J. (1997). A general model for the origin of allometric scaling laws in biology. *Science*, 276(5309):122–126.

West, S. A., Diggle, S. P., Buckling, A., Gardner, A., and Griffin, A. S. (2007). The social lives of microbes. *Annual Review of Ecology, Evolution, and Systematics*, 38:53–77.

White, C. R., Cassey, P., and Blackburn, T. M. (2007). Allometric exponents do not support a universal metabolic allometry. *Ecology*, 88(2):315–323.

Wilhelm, S. W., and Trick, C. G. (1994). Iron-limited growth of cyanobacteria: Multiple siderophore production is a common response. *Limnology and Oceanography*, 39(8):1979–1984.

Wilkinson, G. M., Carpenter, S. R., Cole, J. J., Pace, M. L., Batt, R. D., Buelo, C. D., and Kurtzweil, J. T. (2018). Early warning signals precede cyanobacterial blooms in multiple whole-lake experiments. *Ecological Monographs*, 88(2):188–203.

Wiser, M. J., Ribeck, N., and Lenski, R. E. (2013). Long-term dynamics of adaptation in asexual populations. *Science*, 342(6164):1364–1367.

Wiser, M. J., Ribeck, N., and Lenski, R. E. (2014). Data from: Long-term dynamics of adaptation in asexual populations. Dryad, November 11, 2014.

Woolf, S. H., Chapman, D. A., and Lee, J. H. (2021). COVID-19 as the leading cause of death in the United States. *JAMA*, 325(2):123–124.

World Health Organization ERT. (2014). Ebola virus disease in West Africa—the first 9 months of the epidemic and forward projections. *New England Journal of Medicine*, 371(16):1481–1495.

Xavier, J. B., and Foster, K. R. (2007). Cooperation and conflict in microbial biofilms. *Proceedings of the National Academy of Sciences*, 104(3):876–881.

Yang, G.-Z., Bellingham, J., Dupont, P. E., Fischer, P., Floridi, L., Full, R., Jacobstein, N., Kumar, V., McNutt, M., Merrifield, R., et al. (2018). The grand challenges of science robotics. *Science Robotics*, 3(14):eaar7650.

Yong, E. (2013). Dynasty: Bob Paine fathered an idea—and an academic family—that changed ecology. *Nature*, 493(7432):286–290.

Yoshida, T., Jones, L. E., Ellner, S. P., Fussmann, G. F., and Hairston, N. G. (2003). Rapid evolution drives ecological dynamics in a predator–prey system. *Nature*, 424(6946):303–306.

Zangwill, A. (2013). *Modern Electrodynamics*. Cambridge University Press, Cambridge, UK.

Zehr, J. P., Weitz, J. S., and Joint, I. (2017). How microbes survive in the open ocean. *Science*, 357(6352):646–647.

Zhang, P., Khursigara, C. M., Hartnell, L. M., and Subramaniam, S. (2007). Direct visualization of *Escherichia coli* chemotaxis receptor arrays using cryo-electron microscopy. *Proceedings of the National Academy of Sciences*, 104(10):3777–3781.

Zhang, T., and Goldman, D. I. (2014). The effectiveness of resistive force theory in granular locomotion. *Physics of Fluids*, 26(10):101308.

Zheng, Q. (1999). Progress of a half century in the study of The Luria–Delbrück distribution. *Mathematical Biosciences*, 162(1-2):1–32.

Zimmer, C. (2012). *A Planet of Viruses*. University of Chicago Press, Chicago, IL.

INDEX